Walther Nernst
and the Transition
to Modern Physical Science

Primarily a scientific biography of Walther H. Nernst (1864–1941), one of Germany's most important, productive, and often controversial scientists, this book addresses a specific set of scientific problems that evolved at the intersection of physics, chemistry, and technology during one of the most revolutionary periods of modern physical science. Nernst, who won the 1920 Nobel Prize for Chemistry, was a key figure in the transition to a modern physical science, contributing to the study of solutions, of chemical equilibria, and of the behavior of matter at the extremes of temperature. A director of major research institutes, rector of Berlin University, and inventor of a new electric lamp, Nernst was at once the first "modern" physical chemist, an able scientific organizer, and a savvy entrepreneur. His career exemplified the increasing connection between German technical industry and academic science, between theory and experiment, between concepts and practice.

Diana Kormos Barkan was trained as a chemist in Romania and Israel before receiving a doctoral degree in the history of science at Harvard University in 1990. Her dissertation received the Prix Marc-Auguste Pictet of the Société de Physique et d'Histoire naturelle de Genève in 1992. She has been a member of the Institute for Advanced Study in Princeton and the recipient of NSF, NEH, and Sloan Foundation grants. She has carried out research in many European countries and has been an invited lecturer in Austria, England, Germany, and Israel. Her articles have been published in edited volumes and in *Science in Context* and *Perspectives on Science*.

T0296928

Walther Nernst and the Transition to Modern Physical Science

DIANA KORMOS BARKAN

CAMBRIDGE
UNIVERSITY PRESS

CAMBRIDGE UNIVERSITY PRESS
Cambridge, New York, Melbourne, Madrid, Cape Town, Singapore,
São Paulo, Delhi, Dubai, Tokyo, Mexico City

Cambridge University Press
The Edinburgh Building, Cambridge CB2 8RU, UK

Published in the United States of America by Cambridge University Press, New York

www.cambridge.org
Information on this title: www.cambridge.org/9780521444569

First published 1999

A catalogue record for this publication is available from the British Library

Library of Congress Cataloguing in Publication data
Barkan, Diana Kormos (date)
Walther Nernst and the transition to modern physical science /
Diana Kormos Barkan
p. cm.
Includes bibliographical references and index.
ISBN 0-521-44456-x
1. Nernst, Walther, 1864–1941. 2. Chemistry, Physical and
theoretical – Germany – History. 3. Physics – Germany – History.
4. Chemists – Germany – Biography. 5. Physicists – Germany – Biography.
1. Title.
QD22.N39B37 1998
540'92 – dc21
[b]   98-22028
CIP

ISBN 978-0-521-44456-9 Hardback

Contents

Preface

In his speech during the memorial festivities in honor of the University of Berlin's founder in August 1922, the new rector addressed assembled students, professors, and guests in a bold and unconventional manner: he offered a moving encomium to his murdered friend and longtime associate, the prominent German Jewish industrialist and politician Walther Rathenau, killed two months earlier by reactionary gangs in the Grunewald neighborhood of Berlin. Walther Hermann Nernst, distinguished Professor and Geheimrat, appealed to German academia and intellectuals to expunge violence and dogmatism from public life.

And yet, Nernst had his own record of involvement with violence. He had lost his two sons in the Great War, and had only recently been removed from a list of German war criminals because of his work on chemical warfare. A Prussian by birth and education and, at age fifty-eight, an academic personality of international reputation, Nernst had recently been awarded the Nobel Prize in Chemistry, to the opprobrium of the Allies' press. He had been one of the deposed Emperor's scientific advisors, and yet had become a trusted intellectual of the young Weimar Republic.

In its makeup and language, the speech reflected Nernst's complex and seemingly contradictory personality. A scientist and government employee, he was nonetheless speaking out in public about private friendship with a rather unpopular Jewish politician, drawing on Schiller and Shakespeare for eloquent odes to freedom and indictments of tyranny. He then shifted swiftly to an elaborate, detailed, and technical astrophysical lecture on the birth of new stars and their cosmology. Hence, the lecture comprised two completely different texts that confronted two distinct realms of social discourse: His views on public life were ruled by empathy, political partisanship, sense of civic duty, literature, and philosophy; the other, scientific view, described exploding novas and attempted to balance the energies of the universe and reconcile competing astrophysical hypotheses. Between the two sections, Nernst inserted some reflections on the tragic figure of the early modern astronomer Tycho Brahe, on whose tomb in the Teyn church in Prague is written: *Nec fasces, nec opes, solum artis sceptra perennant.*

vii

"Neither raw force, nor riches, but only the scepter of the spirit lasts forever."

Upon his death almost two decades later in the midst of World War II, the *New York Times* portrayed Nernst as one of the last German scientists "free to think and to say what he pleased, [a] *Gelehrter* . . . honored and not regarded, as now, as an annoyance, from whom nothing socially useful can be expected."

This is primarily a scientific biography of Walther Nernst (1864–1941), one of Germany's most important, productive, and controversial modern scientists. His life and work have until now been treated as a minor chord in the accounts of the triumphant marching song of German theoretical physics and chemistry. I have sought to convey the flavor of daily existence of a German scientist who was immersed and enmeshed in the great scientific, educational, technological, and political debates of his time. His life was one of many rewards and punishments.

The book is designed mainly to address a specific set of scientific problems. When and how they became problems, how they were elucidated experimentally and theoretically, and how this understanding will promote our overall view of developments in modern physical science are the main questions at hand. The most substantive concerns the manner in which events leading to a total reformulation of modern physics may be seen as a complex process, extending synchronically across several disciplines and domains, as well as having their origins in diachronically diverse traditions. Moreover, by looking at Nernst, I expand the circle of scientists concerned with radiation, conductivity, and heat to include a number of chemists, physical chemists, and engineers who have been largely absent from the historiography of early quantum theory.

This approach raises an additional set of questions: In what particular ways are scientists, or groups of scientists, different from one another? That is, can we, or should we, isolate the work of, say, physicists from that of their colleagues simply because we know quantum theory to have been created by a physicist? What exactly was Max Planck's circle of colleagues in the years prior to his enunciation of the quantum hypothesis in 1900? How does the traffic in theories, methods, and instruments across fields affect disciplinary identity? Do even disciplines themselves exist beyond budgets and letterheads? In short, how are individual and group identities formed, and does all this matter at the end of the day?

To get at these questions, I found that I had to make this a book about alternatives and about the benefits of rereading. It tells the stories several times, in several incarnations, and explores ambiguities that ultimately resolve into a new narrative. It is, in many ways, a revisionist book. By focusing my attention on the often plodding reality of everyday scientific practice, I believe that the resulting picture will eventually modify our accounts on a number of issues, especially on the origin of the heat theorem and the relationship among physics, chemistry, and the needs arising from industrial practical application. This is neither a "privileged" account, nor does it seek to describe all aspects of scientific activity in late nineteenth-century physics and chemistry. My aim is to provide a "thicker" description of theory, experiment, and technology by addressing three major areas: the history of physical chemistry in Germany; the formulation of the third law of thermodynamics; and the relationship of physics, chemistry, and industrial concerns as present in the work of Nernst and his contemporaries.

In his later work, Thomas S. Kuhn struggled to understand how exactly we move from an old, established system to a new starting point. It was the study of "displacement" that had remained for him, and still remains for all of us, deeply problematic. In response to his critics, Kuhn still considered it necessary that a story be organized in a spiral way, that some of the points be covered several times. This form of organized reflexiveness, which he found indispensable, was not at the time anything close to the then–already fashionable reflexivity. The only way to explore these displacements is by small steps, not necessarily through microhistory, but by following the reincarnation of questions, methods, and tools in recurring problematic situations. But ultimately, as Kuhn insisted, it matters not where individuals are located but how thought and thought collectives move. By focusing on individuals and their specific re-understanding of the problem on which they are working, we ultimately attempt to explain general movement, that notorious undirected progress.

At the risk of losing the reader after the following introductory explanations, it is necessary at this point to provide a backbone to the book's structure. The various chapters are fairly technical and, although accessible to a nonspecialist, have a recursive pattern that might be opaque. In the tradition of puzzle solving, the book centers around one crucial anomaly: The work, which has been universally brushed aside if not outright derided by those who have known or described Nernst, is in need of "placement" into the context of his scientific career. Between 1894 and 1906, Nernst invented and developed a new electric bulb. This work has been seen as a business proposition, which it was, and as nothing else. The heat theorem, which was formulated at the end of 1905, was connected by biographers to a long-standing chemical theoretical problem: how to predict the feasibility of chemical equilibria. No connection whatsoever was made between the ongoing, intensive work on the lamp and this highly theoretical and hypothetical postulation of a new law of thermodynamics. Furthermore, Nernst immediately began making connections

between his heat theorem and Planck's quantum theory of radiation, and
he organized the First Solvay Congress in Physics in 1911, which was the
first international meeting devoted to the new quantum physics. This
move on his part, too, has been seen as a rather pragmatic, professional
move, which it was as well, but without exploring why Nernst would be
involved in a topic not his own. That is, within the historiography of
quantum theory, the story is as follows: The insufficiency of existing radiation laws prompts by the end of the nineteenth century an examination of black-body radiation, which leads to Max Planck's postulation of
the quantum theory in 1900. This is followed by seminal papers by Albert
Einstein, who develops it into a more comprehensive quantum theory of
light (1905) and matter (1907). In this chronology, Nernst's heat theorem
is "confirmed" by Einstein's 1907 paper; eventually, Nernst's subsequent
experiments at low temperatures are confirmation of Einstein's paper of
1907.

On the other hand, the traditional account of the postulation of the heat
theorem revolves around a different chronology, according to which it
evolved as a result of Nernst's attempt to solve a problem in chemistry:
How do we evaluate chemical affinities and how can we predict on the basis of such knowledge the feasibility of a chemical reaction? Hence, Nernst
devised dissociation experiments of gases at high temperatures and postulated the necessary conditions for understanding the relationship between
chemical work (affinity) and heat. He then moved to low-temperature experiments to prove the heat theorem for the domain of solid substances.

In this scenario, Nernst's work on an electrolytic lamp is an anomaly,
until now an amusing aside. What follows is a scrambling of the pieces
and their reassembling:

The first part, which consists of chapters 1 through 5, sets the stage for
Nernst's education, training, and early researches. It places him in an
eclectic position, of a physicist involved in exploring traditional problems
in physics and electrochemistry. It also introduces a scientific network
of particular people with whose work and personalities Nernst would be
involved throughout his career. Here we meet H. F. Weber, Fr. Kohlrausch,
W. Ostwald, P. Drude, Max Planck, and S. Arrhenius.

The second part, comprising chapters 6 through 9, focuses on what I
have called Nernst's high-temperature electrical work. The major problem to be solved was that of illumination, of obtaining a material which
would emit the brightest light, withstand excessive temperatures, and consume lower amounts of energy than other available lamps. This involved
many subproblems: electrical and electrolytic conduction, the behavior of
solid conductors and nonconductors at high temperatures, the dissociation of gases at high temperatures, the behavior of solids and gases inter-

acting at high temperatures, the problem of measuring extremely elevated temperatures, the design of materials for high temperatures, the design of instruments such as ovens, a calorimeter, photometer, and thermometer for such work. These were all intensely practical problems, each of which will be addressed in turn. The pattern will stay the same: We shall explore problems, the design of experiments, the design of instruments, and problems arising from such instruments.

I will argue that none of the aspects of Nernst's work could be solved – or properly understood – satisfactorily in isolation, and that scientists were enmeshed in a net of puzzle solving. From a theoretical standpoint, all existing tools had to be reevaluated: Matter theory, radiation laws, optics, gas theory, chemical equilibria, conductivity theory were all called to assist and were all found wanting. One central binding theme could be seen to be the problem of specific heats, which runs often invisibly, most often explicitly, throughout the book. Nernst's heat theorem, which is fundamentally about specific heats as well, emerges from these preoccupations, as does a deep sense he acquired of how things really are, that is, how matter behaves under given conditions.

The remaining chapters, 10 through 12, analyze the "canonization" of both quantum theory and the "heat theorem" in the larger scientific community, ultimately resulting in a Nobel Prize award to Nernst. The treatment is not complete. Important aspects of Nernst's late scientific career will be briefly mentioned, since the focus throughout has remained a set of particular problems and their practical resolution.

Acknowledgments

Like the scientists with whose intellectual ancestry the book is repeatedly concerned, I too have elaborated upon the craft and preoccupations of my teachers. Yehuda Elkana opened the field to me and tried to inspire both vision and incisiveness. Erwin Hiebert, whose work on Nernst and on the history of physical chemistry was pathbreaking, may disagree with some of what I have to say, but I learned it all from him. Under no circumstances could I have completed this project without the patient and firm guidance and the genuine friendship of Daniel J. Kevles, to whom I am immensely grateful. Fiona Cowie was a tireless reader, and friend. I owe special gratitude to Martin Klein, for his thoughtful and acute criticism, advice, and support. I also thank Jed Buchwald, Moti Feingold, Peter Galison, Kostas Gavroglu, Mary Jo Nye, Norton Wise, and the anonymous readers at Cambridge University Press for valuable comments on sections of a rather undigested first draft. Alex Holzman, my editor, was especially patient and graceful. Thanks also to Phyllis Berk and Eric Newman at Cambridge. Charles Kormos struggled to smooth some of my prose and to keep me going.

I thank Helga Galvan, Rosy Meiron, Gina Morea, Susan Davis, and the staff of the Millikan Library at the California Institute of Technology. I received generous assistance from the Royal Swedish Academy of Sciences in Stockholm; the Max-Planck Gesellschaft Archives in Berlin; the Archives of the Berlin Academy of Science; the German State Archives in Berlin, Merseburg, and Potsdam; the Leipzig University Library; the Archives of the École de Physique et Chimie, and the Institut de France, Paris; the Museum Boerhaave in Leiden; Nuffield College Library in Oxford; and Lady Francis Simon.

My research has been funded by grants from the American Institute of Physics, the Deutscher Akademischer Austauschdienst, the National Science Foundation, the National Endowment for the Humanities, and the Institute for Advanced Study in Princeton, and by the munificent support of the Division of the Humanities and Social Sciences at Caltech. Special thanks also to Mme. & M. Jean-Michel Pictet.

Ady Barkan has waited much too long for this book to be done. I thank him for his unlimited patience, wit, trust, and love.

1

Nernst, the Historiography of His Science, and Its Context

This is physical chemistry, formerly a colony, now a great, free land.

J. H. van't Hoff

This book is about a set of scientific and technical concerns that emerged toward the end of the nineteenth century at the intersection of physics, chemistry, and technology, and about the role that Walther Nernst (1864–1941) and some of his colleagues and collaborators played in the formulation of novel, even revolutionary, scientific theories and experimental techniques.

Nernst, who won the 1920 Nobel Prize in chemistry, was a key figure in the transition to a modern, quantum theoretical physical science that occurred over forty years or so at the end of the nineteenth and the beginning of the twentieth centuries. He was scientifically and personally engaged with many of the major figures of the period, including Planck, Einstein, van't Hoff, Ostwald, and Arrhenius. His foremost contributions to science include the study of electrolytic solutions, chemical thermodynamics, the theory of chemical equilibria, quantum chemistry, low-temperature phenomena, photochemistry, and his celebrated heat theorem, known also as the Third Law of Thermodynamics. Nernst's career exemplified the increasing connection between German technical industry and German academic science, inasmuch as he himself invented and patented a new type of electric lamp and other devices. Hence, his life and work provide entry into most of the significant and powerful developments in physical science that would profoundly affect the world in the twentieth century.

Nernst's professional career, which spanned the period from Wilhelminian Germany through the First World War and the Weimar Republic to the Third Reich, and from experimental electrical studies through physical chemistry and quantum theory to modern astrophysics, instantiates the connections between theory and experiment, concepts and practice, academia and industry. One of the more prominent representatives of German science, Nernst has been described as the first "modern" physical chemist, as the originator of solid-state physics, and as an able entrepreneur. Yet, little attention has been given to the relationship among the different strands of his scientific life. Therefore, this account examines the interconnections between physics and chemistry, the fluidity of disciplinary allegiances and identities, and the importance of concrete, manageable, and task-directed work in the larger picture of scientific research.

Because of their work on the properties of matter, Nernst and his colleagues in Berlin played an important role in the reception of Einstein's quantum theory of solids and, more generally, in the development of quantum theory and its relevance to chemical properties and reactions. The events surrounding Nernst's organization of the first Solvay conference in 1911 illustrate aspects of the precarious role of quantum theory at the time. The shifts in fundamental conceptions about the nature of matter and radiation, and their correlation to his heat theorem, account, at least in part, for the tortuous path toward Nernst's Nobel Prize.

The centerpiece of Nernst's scientific work was his postulation of the heat theorem, now called the Third Law of Thermodynamics. Trying to solve the problem of predicting chemical equilibria, Nernst postulated at the end of 1905 that at low temperatures, the internal and free energy of the reactants of a chemical transformation become equal in magnitude. This assumption allowed for the elimination of a troublesome integration constant that had previously hampered the ability to predict whether a chemical reaction was feasible. It simplified the problem by allowing the calculation of the equilibrium constant of the reactions from measurements of the specific heats of the reactants at low temperatures.

In its current formulation, the Third Law of Thermodynamics states that it is impossible to reach absolute zero temperature in any finite number of steps, and that the entropy change between states of a system that can be connected by a reversible process vanishes at absolute zero. Accordingly, near the absolute zero of temperature, matter attains a perfectly ordered state. Nernst's observation about the peculiar behavior of matter in the vicinity of absolute zero is considered a milestone in the history of physical chemistry because it furnished an algorithm that enabled researchers to predict the feasibility of chemical reactions from experimentally available thermal data on chemical substances.

After its publication in early 1906, Nernst's heat theorem received several refinements and reformulations. Albert Einstein's paper on the quantum theory of solids, published in 1907, provided proof for Nernst's prediction that specific heats tend to zero at absolute zero temperature. Confirming Nernst's singular intuition that low temperatures in the solid phase are ideal loci of experimentation on peculiar quantum phenomena that do not take place at ordinary temperatures, Nernst's paper and that of Einstein became the basis for the development of an entirely new field of theoretical and experimental studies, later called solid-state physics. A decade later, it became evident that molecular and atomic entropies were critical elements for the understanding of the behavior of matter, leading to quantum statistical treatments of a variety of physical and chemical properties.

The Problems

One important historical puzzle concerns the original formulation and conceptual and experimental sources of the heat theorem. How and why did Nernst arrive at the heat theorem?

An early account, which persists to this day, was presented in the 40th Guthrie Lecture delivered before the Physical Society of England in March 1956 by one of Nernst's best-known students, Sir Francis (Franz) Simon. Having worked on low-temperature experiments in Germany before emigrating to England after the Nazi seizure of power, Simon located the origins of this heat theorem in Nernst's "intense interest in gas reactions, partly because of the relative simplicity of the problem involved . . . and partly because of the economic possibilities" which its solution might bring to the industrial synthesis of ammonia. "Thus there was every incentive for physical chemists to develop the theory of gas reactions systematically." Simon wrote that, "*starting from vague and rather inconclusive beginnings,* [the heat theorem] was at first mainly regarded as a useful rule for calculating chemical equilibria" (emphasis added). This initial work by Nernst was followed by a "period when the quantum theoretical foundations of the theorem were recognized and physicists became interested in it, particularly when quantum statistics made possible a direct calculation of the entropy constants of gases. Later on, further quantum statistical considerations led to a discussion of the general validity of the law when applied to solid phases."

Simon's account is open to question on a number of grounds. The published evidence that Nernst had been substantially interested in predicting chemical equilibria prior to 1906 is at best only indirect. His main line of work between 1895 and 1905 was the improvement of the new electric lamp. An analysis of Nernst's own chronology of events, as well as the record of his own and others' publications, shows that Nernst's research on the heat theorem proceeded simultaneously with other work into electrical conductivity, Maxwellian electromagnetism, heat and light radiation, and the needs of the electro-technological and illumination industries of the day.

According to Simon, after postulating the heat theorem, Nernst left "aside the gas reactions" and "switched to the condensed phases" and the investigation of solids at low temperatures. Nernst then switched back to gas reactions, "for after all these started the whole problem. Moreover nearly all the equilibrium data which were available at that time concerned gas reactions." Later, the obtained specific heat data were applied "to predict a number of chemical equilibria, particularly in galvanic cells and crystallographic transitions."[1] While it is correct that the heat theorem could fundamentally be confirmed only through the study of solids, difficulties arise when one attempts to reconcile Nernst's work until 1905 with this alleged overriding preoccupation with the "calculation of chemical equilibrium constants from thermal measurements."[2] The standard view, as indicated in Simon's account of the origins of Nernst's heat theorem, is misguided. Simon may have been misled by the title of Nernst's first paper on the topic and by the rather rapid publication of two important monographs on the application of the heat theorem to chemical transformations between 1910 and 1920.

The heat theorem's roots lay not only in a "pure" concern with chemical equilibria *per se* but also in the nexus of theoretical as well as practical or, more broadly, technological issues in which Nernst was absorbed at the time. The theorem originated in a thorough blending of electrochemical, electrical, and chemical researches with studies on the constitution of matter that preoccupied the Nernst laboratory, and many others, in the years 1891 to 1909. Accordingly, we shall examine in great detail the precise path of Nernst's researches and the context of the formulation of the heat theorem.

An equally significant historical puzzle concerns Nernst's move, after 1906, to low-temperature investigations. This experimental work has been seen to flow directly from the postulation of the heat theorem. Because it is generally presumed that in physical-chemical research the locus of major interest to chemists at the time was the prediction of equilibria, Nernst's awareness of the importance of low temperatures is viewed as an outcome of his examination of the free energy curves and the internal energy curves of physical and chemical transformations: Nernst assumed, correctly, that in the vicinity of absolute zero these curves coincide asymptotically and become tangent to each other. However, Nernst's perceptive insights and his low-temperature investigations actually emerged

1. F. Simon, "The Third Law of Thermodynamics. An Historical Survey," 40th Guthrie Lecture, *Year Book of the Physical Society* (1956):1–6.
2. Keith J. Laidler, *The World of Physical Chemistry* (Oxford: Oxford University Press, 1993), p. 127, is the latest history of the field to continue this line of explanation.

from a deeper and different understanding of his own previous electrical conductivity researches and their relation to similar work carried out elsewhere. He was "looking at" low temperatures before he "saw" the special feature of the energy curves that helped him formulate the heat theorem. The mind had been prepared in advance.

In sum, a good case can be made for a more textured reading of Nernst's work, both during his apprenticeship in Graz, Würzburg, and Leipzig and later in his laboratories in Göttingen and Berlin. Detailed examination of the work not only of Nernst but also of myriad collaborators and colleagues in various fields and laboratories triggers a departure from previous historical reconstructions. They have read back into its origins the "chemical" benefits that flowed from Nernst's postulation of the heat theorem, and they have taken low-temperature physics as its natural outgrowth. In fact, the low-temperature investigations that followed were not a major shift but rather constituted a program of correlating existing data and eliciting new data for atomic heats, electrical conductivity, and melting points. All this information was useful for both chemical and physical investigations.

In general, histories of the field tend to include a statement on the rather sudden "insight" about the energy balance at low temperatures that occurred to Nernst in late 1905, and describe how this "Eureka" experience – alleged to have taken place during a regular university lecture – consequently led him to an unexpectedly swift transition to the field of low-temperature research. The central point of these accounts, however, has remained the concern about the predictability of chemical equilibria.

I dispute this account. The conventional chronology seems to be a teleological, retrospective narrative, culminating in Simon's own scientific field of expertise, and constructed almost half a century after the postulation of the heat theorem. It is an account of the origins of solid-state, low-temperature work carried out by some of Nernst's former students in the 1930s and afterward at the Clarendon Laboratories in Oxford.

I will show that what ultimately came to be regarded as an important insight, innovation, or solution did not stem from any purposeful attempt to answer precisely the problem eventually solved. Simon and historians following him overlooked some essential features of the heat theorem's etiology. For although Nernst was indeed interested in gas reactions, they were not only those of the industrially desirable ammonia formation, as Simon suggests. And although Nernst did indeed engage briefly in exploring the synthesis of ammonia from elements, the study of the behavior of gases was a major research topic of physicists and chemists concerned with radiation phenomena – with the dissociation of matter at elevated temperatures.

Nernst's ongoing electrolytical and electrochemical research into conduction processes constituted the core of his experimental investigations.
For example, he did not merely "apply" the data gained from the application of the heat theorem to the galvanic cells (batteries), as might be
concluded from cursory readings of his papers and of Simon's account.
Instead, the production of mechanical work by galvanic cells was a long-
standing project, which he had begun during his student years.

Nernst's work from 1886 to 1914, his most productive years between
the ages of 22 and 48, was embedded in the theoretical and experimental
culture of the period, was to a large extent influenced by the particular
laboratory and its environment, and was characterized by the slow, often
piecemeal practice of electrical and thermodynamic studies carried out at
the time. My account, thus, accords with the stress that recent scholars
have placed on the heterogeneity and cultural specificity of both scientific
theory and practice. In particular, scholars have been concerned with explaining the nature of "negotiated" knowledge, how consensus is reached,
and how theories and skills are stabilized, replicated, and multiplied in
the academic, industrial, and larger milieus.[3] Many of these studies address the alternation of consensual intervals in the growth of science, focusing on Kuhnian incommensurability and the disruptions that punctu-
ate placid normal science.[4] Their aim is to understand and to explain how
successful theoretical and experimental candidates survive a presumed
chasm of mutual incomprehensibility; in this effort, studies of general
change alternate with microstudies of power.[5]

It has become clear, however, that our categories of theory, observation,
and experiment have a history of their own, and that the armamentarium
of concepts and criteria that we have so far employed might not be adequate for a description of recent science. In his most recent work, Peter
Galison has insisted that different cultures of theoreticians, instrument
builders, experimentalists, and engineers live in distinct milieus, and that

---

3. On negotiation and scientific controversies, see in particular Bruno Latour and Steve
Woolgar, *Laboratory Life: The Construction of Scientific Facts* (Princeton: Princeton University Press, 1986). On construction and stabilization of the objects of experimental
work, see M. Norton Wise, "Mediating Machines," *Science in Context* 2 (1): 77–114
(1988); David Gooding, Trevor Pinch, and Simon Schaffer, eds., *The Uses of Experiment*,
(Cambridge: Cambridge University Press, 1989). Specifically, for Nernst, Haber, and
chemical industry, see Timothy Lenoir, "Practical Reason and the Construction of Knowledge: The Life World of Haber-Bosch," (MSS).

4. Thomas S. Kuhn, *The Structure of Scientific Revolutions*.

5. For example, by moving from the quantum or relativistic to the larger probabilistic revolution, in Lorenz Krüger et al., eds. *The Probabilistic Revolution*, vol. 1. *Ideas in History* (Cambridge, Mass.: MIT Press, 1987).

the traditions, skills, goals, and vocabularies of these differing segments of the very large community of physicists operate under varying constraints in their mode of argumentation, in their understanding of a problem and of the paths taken toward its solution. In Galison's view, these subcultures coexist but are not diachronically accessible to us across the board. Developments and transformation in one "tradition" do not necessarily reverberate synchronically in all others. Thus, change in "theory" and change in "experiment" do not have simultaneous life histories.[6] One aspect of what follows may serve as further illustration of Galison's important argument for the existence of an inner and an outer life of the experiment: We shall see that in the case of Nernst and his contemporaries, it was still possible to entertain a lively dialogue between the laboratory and the outside, between academia and industry, and that certain instruments circulated, albeit haltingly at times, between the two. Moreover, our description of this circulation of devices between inside and outside will illustrate Galison's argument that theory and experiment cannot be followed along a single, linear story.

The experimental and theoretical network of interactions that figure in Nernst's life may be a test for recent treatments of discursive formations, of local reconfigurations, coexistence, and grouping of statements.[7] However, the particular concepts of disciplinary coherence and regularity do not figure prominently in my account. The search for uniformities and constraints regulating "discipline formation," or research programs, seems to belong to the perennial quest for ordering and systematizing. The accounts of discipline formation and disciplines as a "Ding an sich" are often taken as a given entity in many science studies, rather than as a conclusion.[8]

Recent scholarship has also focused on the question of whether there has ever developed a peculiarly "chemical" as opposed to "physical" view of nature and, as a corollary, whether chemical laboratory practices have ever acquired a life of their own, distinct from other kinds of practices.[9] Many arguments have been advanced in support of a consolidation of disciplines during the twentieth century, and for the consequent stabilization

---

6. Peter Galison, *Image & Logic. A Material Culture of Microphysics* (Chicago: University of Chicago Press, 1997).
7. Timothy Lenoir, "The Discipline of Nature and the Nature of Disciplines," pp. 70–102, in Ellen Messer-Davidow, David Sylvan, and David Schumway, eds., *Knowledges: Historical and Critical Studies in Disciplinarity* (Charlottesville: University of Virginia Press, 1993), p. 5.
8. For a recent discussion, see Mi Gyung Kim, "Labor and Mirage: Writing the History of Chemistry," *Studies in History and Philosophy of Science* 26 (1995): 155–66.
9. Mary Jo Nye, *From Chemical Theory to Theoretical Chemistry,* 1994.

of distinct "cultures" of scientific practitioners.[10] These views draw on studies of professional identity and stress differences between cultures over cross-cultural similarities. Indeed, the enduring quest for national styles of scientific research, styles of reasoning, epistemic styles, and many of their variants derives from similar preoccupations.[11]

Inevitably, however, once a particular tool (theoretical, such as the quantum hypothesis; mathematical, such as the calculus; or instrumental, such as the liquefier) becomes available, it cannot help but permeate scientific practice in many related fields. Therefore, alongside the increasing specialization that characterized all sciences at the turn of the century, one must also take note of the pragmatic appropriations that promoted, rather than inhibited, cross-disciplinary discourse. Physical chemistry, for instance, employed and was modulated by a variety of tools from physics, ranging from electricity and thermodynamics to instrumentation, error calculus, graphs, and other methods. Conversely, various advances in chemistry introduced new goals, practices, and perspectives into the emerging fields of biochemistry and solid-state physics.

The picture that emerges from my study is the ongoing practice of appropriation across disciplinary boundaries, always in tension with the demands of intradisciplinary coherence. Furthermore, disciplines, research programs, and institutions never act in isolation. Scientific activity takes place in a number of intersecting arenas, and each scientist inevitably partakes of multiple spheres of scientific communication.

The work of Nernst, as well as that of Max Planck, Albert Einstein, Heike Kamerlingh Onnes, and many others, was in constant and multidimensional flux. Picking data out of a certain context and applying them in another, or taking theoretical formalisms and expanding them, or looking anew at older data in a fresh light are all processes that negate a simple story of the development of new disciplines. Nor can "new disciplines" straightforwardly be credited with the production of new ideas or new practices. The ability continually to make "the world anew by bringing us back to familiarity with it, by paying attention to what we are doing,"[12] was a pertinent feature of much fin-de-siècle science, as well as art and culture. Thus, innovation occurred not by breaking the shackles or shaking the foundations of an old edifice but by reconfiguring what was meant by force, matter, field, and heat in a variety of specific problems.

10. Peter Galison, *How Experiments End.*
11. For a recent discussion, see Marga Vicedo, "Scientific Styles: Toward Some Common Ground in the History, Philosophy, and Sociology of Science," *Perspectives on Science* 3 (1995): 231–54.
12. Kirk Varnedoe, *A Fine Disregard: What Makes Modern Art Modern* (New York: Harry N. Abrams, 1990), p. 277.

Nernst and Physical Chemistry

The usefulness of the notion of a discipline as an analytic tool for the history and sociology of scientific practice – an issue that has recently attracted the attention of many historians – is exemplified by Nernst and his "school" of physical chemistry. Nernst is considered one of the more successful descendants of the original "triumvirate" of Ostwald, van't Hoff, and Arrhenius and the institutions, journals, and networks they created. But at least in the case of Nernst, the category of physical chemistry is convenient only as a taxonomic shorthand. It often appears that a preoccupation with disciplines is a result of the ethos of specialization and professionalization, which has dominated the histories of the last century, but this focus loses its persuasiveness when placed in the context of the highly "interdisciplinary" practice of the natural and exact sciences within academic as well as industrial settings in both the recent past and the present.[13]

Physical chemistry has become something of an archetype for the new sciences born from the turmoil of a nonphysical, nonmathematical, somewhat premodern chemistry at the opening of the twentieth century. In many ways physical chemistry has afforded scientists and historians a unique opportunity to study modern science in the making. After all, chemistry, physics, and medicine had all been around for centuries. But here was, or so it seemed, a self-conscious effort on the part of the more enlightened, physicalist chemists to rejuvenate a science that was gradually becoming submerged in a surfeit of new chemical elements and substances, with a lack of systematic nomenclature, a profusion of explanatory models, and a wealth of reactions and applications. Did atoms truly exist? Did they react with each other, and if so, how? Did mechanical, electrical, magnetic, or gravitational forces play a role in chemical reactions? What was affinity, and how could it be measured? Was wet chemistry different from dry chemistry? Was inorganic matter different from organic substances in any fundamental way?

Physical chemistry emerged in the mid-1880s as part of modern physical science. By the early decades of this century, physical chemistry was considered to have grown into a lively, well-populated, and well-organized scientific discipline, primarily through the determined efforts of Wilhelm Ostwald, Jacobus Henricus van't Hoff, and Svante Arrhenius.[14] It grew

---

13. Lenoir, 1993, suggests, for instance, that disciplinary programs "adapted locally to the political economy" are more useful categories than monolithic "disciplines."

14. This historiography of physical chemistry has also been criticized because it has purportedly relied too heavily on the self-image of turn-of-the-century Continental physical chemists, who saw themselves as distinctive specialists. R. G. A. Dolby, "Thermochemistry

out of a new understanding of the nature of solutions and the constitution of matter therein. One prominent feature of the historiographical background is the examination of the self-representation and intellectual hagiography of a small number of well-known physical chemists. The theme of disciplinary demarcations is a recurring one, if for no other reason than that it plays a paramount role in the transmission of historical narratives, and has significantly influenced recent histories of physics and chemistry, as well as a number of biographical studies. These problematic self-locations show that later scientific research and controversies – in addition to the subsequent historical analyses of these events – relied on certain retrospective "disciplinary" accounts, which lacked contextual references, and thereby promoted a misleading "single perspective" on the development of low-temperature physics, quantum chemistry, and chemical thermodynamics.

In the mid–nineteenth century, Claude Louis Berthollet called attention to the special role of solutions in chemistry. He had unsuccessfully tried to define chemical affinity by linking it to the masses of reactants in a solution, postulating the existence of forces akin to Newtonian gravitational attraction.[15] In the mid-1880s, Arrhenius and van't Hoff further probed the nature of the processes in liquid solutions. They demonstrated not only that solutions are composed of matter that exists in a state of dissociation but also that different kinds of solute coexist in solutions.

It was assumed, for example, that in the solid state, crystals of table salt are constituted of molecular entities of sodium chloride, in which the atoms of sodium and chlorine are physically bound, or connected. It was assumed that the chemical molecule as a whole continues to exist in solution, as it did in the gaseous phase. But by 1887, in an elaboration of his rather bold and quite speculative dissertation of 1884, Arrhenius developed his ionic dissociation theory of electrolytic solutions: A solution of table salt in water is not composed of molecules of salt distributed in water but of separate ions of sodium and chlorine, in addition to undissociated molecules of sodium chloride. Whereas Arrhenius had argued in his dissertation of 1884 that such ions were present when an electric current was passed through a solution, by 1887 he postulated the existence of "free ions," regardless of the presence of such a current. Furthermore, Ostwald, Arrhenius, van't Hoff, and Nernst showed in the late 1880s that many physical and chemical properties of solutions could be explained

versus Thermodynamics: The Nineteenth Century Controversy," *Hist. Sci.* 22 (1984): 379–80.

15. John W. Servos, *Physical Chemistry from Ostwald to Pauling: The Making of a Science in America* (Princeton: Princeton University Press, 1990), p. 14.

and analyzed both qualitatively and quantitatively with the help of the ionic dissociation theory. These ions – atoms or groups of atoms that carry positive or negative electrical charges – migrate in the solution, and are capable of transporting electricity when a difference of potential is applied to the solution.

These insights were important in explaining electric conductivity, the phenomena of electrolytic deposition of metals and salts, the thermal and chemical effects that occur in solutions, and the various characteristics of multiphase mixtures, as well as some physiological processes, such as the production of electric currents in a gradient of salts that are present in cells and nerves. Most significantly, however, physical chemists came to suspect that the electric conduction that ions in a solution are able to carry out is related to the ability of the same reactants to participate in chemical reactions. Hence, they hoped, the ionic dissociation theory would provide the long-sought answer to the problem of chemical affinity.

Another important foundation of physical chemistry comprised the application of the gas laws to solutions. Extending the gas laws to the new domain made it possible to explain processes that occur in the presence of membranes and their analogues, such as the migration of certain ions through a medium that is impermeable to the other ions in the same solution. Van't Hoff, one of the few scientists of the 1880s to apply thermodynamic considerations to the examination of chemical processes, related the study of chemical equilibrium and kinetics in solution to thermodynamic quantities.

Ostwald, Arrhenius, and van't Hoff became known as the "theoretical" chemists, those concerned with formalizing and mathematizing the study of chemical processes and injecting physical conceptions, methods, and tools into chemistry. They were less interested in the taxonomical chemistry practiced by organic and inorganic chemists. Nor were they concerned with the discovery of new compounds. Despite some resounding successes, however, none of the "founding fathers" of physical chemistry ultimately succeeded in solving the paradigmatic problem of chemical affinities, which several decades later was dealt with in the theory of electronic chemical bonding. Ostwald had spent the first half of his career in Latvia devising methods for the description of molecular properties in the hope of classifying substances in some system indicative of their chemical affinities. Van't Hoff's equilibrium laws, while immensely fruitful in redescribing chemical reactions in thermodynamic terms, were difficult to apply and generalize. And Arrhenius's dissociation laws applied strictly only to weak electrolytic solutions.

A transparent discrepancy between rhetoric and practice thus can be seen to run through the personal visions of a new chemistry. For while the programmatic writings of the group of scientists engaged in physical chemistry present well-formed strategies and retrospective images of a continuous and smooth development, their own research proves more varied and less easily pigeonholed into a "Ding an sich" called "physical chemistry." The narrative endures, however. Ostwald's preoccupations, less with chemical structure and composition and more with reactions, mechanisms, and energetics, remain the leitmotif guiding important recent disciplinary histories.[16]

The story of the birth of modern physical chemistry was "constructed" in a mold that bears the characteristics of what has elsewhere been termed the "invention of tradition." Scientists, like everyone else, were absorbed with the construction of a usable past, with uncovering a genealogy. The invention of traditions came in great part as a response to dramatic social change, which was accompanied by a need to reorder hierarchies, ideologies, and customs. This change affected, to a significant degree, academic and scientific life. It is therefore not by chance that the emergence of specialized departments and chairs in scientific disciplines in Germany came both at a time of rapid industrialization and when student enrollment rose dramatically in all academic and professional schools. The growing appeal of higher education for the increasing number of students rising from the middle classes was motivated by the same drive for tradition that inspired scientists' self-conceptions – only in this case, the immediate goal was the establishment of an "upper-middle-class elite socialized in some suitably acceptable manner."[17] The 1880s and 1890s witnessed the flourishing of physical chemistry, biochemistry, and experimental psychology with their attendant landmarks: the proliferation of chairs, institutes and journals; the influx of foreign students into the European, primarily German, academic graduate seminars and laboratories; the establishment of professional organizations; and more generally, the beginnings of international meetings, congresses, and networks.

Between 1880 and 1914, a desire to define groups of people, objects,

16. Servos, 1990, does indicate that the notion of a discipline is highly specific in time and locale, and often prefers to speak in terms of "mental maps" that guide the scientists within the often fluid boundaries of a discipline. Nye, 1994, traces the changing concept of the "chemical" molecule through the waxing and waning of a particular "chemical feeling," one that differentiated chemists from physicists in the late nineteenth and again in the mid–twentieth century in that they belonged to a distinctive "thought collective." Nye insists that despite a disunified practice, chemists rallied around the concept of affinity, and that physical chemistry developed a culture of its own.

17. Hobsbawm, 1984, examined the mass production of traditions, primarily in Europe, that was most "enthusiastically practiced" in the years 1870 to 1914 by political movements and social groups who invented public ceremonies and the mass production of public monuments.

and natural phenomena in terms of a common language was embraced by the expanding number of scientists preoccupied with nomenclature, classification, the unification and standardization of measures and weights, and the delineation of the standard scientific paper. Just as small nations relied on the compilation and standardization of their languages and insisted on ethnicity-based education, poetry, and dictionaries, so scientists and other professionals engaged in vigorous attempts toward homogenization.

Identity making, or what is usually in a highly laudatory tenor called professionalization, played a prominent role in the landscape of the scientist's life during the second half of the nineteenth century. Professionalization certainly involved the creation of journals, societies, and organized gatherings, of honors, prizes, and hierarchies. By setting intellectual and practical standards, it also involved the exercise of control over access to the profession, over the leadership and mores of the guild. But this ideal of professionalism – the scientist's aspiration to the status of the traditional professions of theology, law, and medicine – was often also an expression of a genuine cosmopolitanism. Technology fulfilled a new desire for communication, for play on the world scene, and for expanding the inner circle beyond national and local boundaries. Between 1840 and 1900, some six hundred international meetings, conferences, and congresses took place mainly in Europe, but also in the United States and South America. Facilitated by the railway and steamship, these transnational meetings were initially organized around universal expositions that celebrated the height of the British Empire or the centenaries of the American and French revolutions. National scientific meetings and associations were formed in the early 1800s. In the 1840s and 1850s, charitable institutions, pacifists, and some isolated professional groups were primarily the ones who convened. Only in 1860 did an international group of scientists meet for the first time at the Karlsruhe Congress of Chemistry; only in 1899 did the French Physics Society recommend that an international congress of physics be organized in Paris in 1900. The physicists hoped to present a comprehensive view of the definitive state of scientific knowledge at the turn of the century, a summary of the ideas and hypotheses by which one explains the constitution of nature and the laws that govern it.

Recently, historians, too, seem beset by classificatory impulses similar to those exhibited by the scientists themselves. But does a recognition that a larger science – of chemistry or physics – contains within itself separate, more circumscribed sets of participants, practices, and rules justify the claim of greater cohesion based on these subdivisions? Probably not. It is taxonomy – the same identificationist urge that led to the conscious creation of official languages and shared holidays for old and new nations –

that privileges in-group similarity at the expense of intergroup difference. A proclivity to seek and assert the epistemological – rather than practical – unity of science is another, ever-present theme in the historiography of chemistry and physics. Resisting the familiar binary controversies of philosophers and scientists alike, William Thomson (Lord Kelvin), for instance, insisted in 1885 that "there is no philosophical division whatever between chemistry and physics." In his view, the two sciences are separated only by the "different sets of apparatus" employed for the investigation "of different properties," and he urged that any distinction between them "must be merely a distinction of detail and of division of labour,"[18] a claim repeated almost verbatim a decade later by Nernst at the inauguration of his Göttingen institute for electrochemistry in 1896.

This drive to find in science some paradigmatic practice is anchored in a conception of the theoretical preoccupations of an idealized, homogeneous community of scientists. How is it possible to balance this account of coherence with the picture of local knowledge, local agendas, local skills that contemporary students of science paint? Does the move toward the creation of international specialized societies, standards, journals, meetings, and so on imply the imposition of uniformity? Or, more plausibly, does an effort toward better communication become a necessity when variability, splintering, and separation threaten isolation, lack of recognition, and immersion into a sea of anonymity?

Contemporary accounts of the status of physical chemistry at the turn of the century give the impression that for a significant period of time, physical chemistry was not perceived as a separate body of chemical knowledge. A wide range of definitions was put forth, with varying emphases on methods (physical, mathematical) or content (electrochemistry, chemical phenomena and properties, physical transformations) and occasional claims for a comprehensive and independent science. Physical chemists were drawn, over many decades, from diverse backgrounds. (Arrhenius, for example, had hoped to become a lecturer in physics at Uppsala University and designed his dissertation accordingly.) The experimental and theoretical apparatus they employed varied significantly. The research goals and organizational structures that developed were similar or identical to those of other ongoing projects. The appeal of physical chemistry around the turn of the century consisted of providing a "progressive" scientific agenda and a growing nexus of institutional support that issued from successfully promoting – individually and collectively – precisely this

18. William Thomson, *Popular Lectures and Addresses*, Vol. 2 (London: 1891–4), p. 484. Quoted in Crosbie Smith and M. Norton Wise, *Energy and Empire: A Biographical Study of Lord Kelvin* (Cambridge: Cambridge University Press, 1989), p. 335.

self-identified progressiveness. But the enhanced status and funding were the result of concerted efforts by individual scientists who were favorably located, not of any perception of an absolute or intrinsic value to just one field. Physical chemistry defies categorization; it also stands as a counterpoint to the sometimes facile delimitation of theoretical versus experimental traditions. As is true of chemists in general, physical chemists are difficult to classify into theoreticians and experimentalists.

Chairs for theoretical chemistry are a relatively novel phenomenon, dating from the post–World War II years, and largely reflect the heightened complexity of integrating quantum mechanics with chemical theory. Even theoretical physics became established as a separate academic discipline only toward the end of the nineteenth century. The first independent chair of physical chemistry at a German university was created for Heinrich Kopp in the 1860s, followed a decade later by the inauguration of the first instructional laboratory at a German university by Gustav Wiedemann in Leipzig in 1871. Kopp had performed individual research only in this field, whereas Wiedemann expanded his activities to teaching experimental physical measurement methods for the investigation of chemical phenomena. In 1869 he had been appointed successor to O. L. Erdmann, an organic chemist. The majority of the faculty searched for a traditional organic chemist to fill the position, while an eventually successful minority opted for a specialist. Physical chemistry, it was hoped, would provide a strong basis for the development of a theoretical chemistry. When in 1887 Wiedemann took over the chair for physics, the proposed list of candidates for his position as director of the Second Chemical Institute was already composed primarily of physical chemists: H. Landolt, L. Meyer, Ostwald, van't Hoff, and H. Brühl. As a result, Ostwald moved to Leipzig. There a section for physical chemistry was approved in 1894, completed in 1897, and officially inaugurated in 1898.[19] By then, however, Ostwald's was the second institute nominally devoted exclusively to physical chemistry; the first, led by Nernst at the University of Göttingen, had opened in 1896.[20]

Departments and chairs for inorganic chemistry in Germany, for instance, grew from the efforts of "physical chemists" under the aegis of their professional association (the Bunsengesellschaft, or German Electrochemical

19. At its inauguration in January 1898, Ostwald held a lecture entitled "Das Problem der Zeit," using the occasion for further expounding his views on energetics, on irreversibility, and other controversial issues that had taken main stage at the 1895 Naturforscherversammlung in Lübeck. W. Ostwald, *Abhandlungen und Vorträge* (Leipzig: 1904), pp. 241–57.
20. The expenses for the new institute were: land 63,000, new construction and renovation 42,000, internal construction 60,000, coming to a total of 165,000 Reichsmarks.

Society) in the late 1890s.[21] As late as 1899, the distinguished scientist W. Hittorf pleaded for the reintroduction and recognition of the once powerful quantitative analytic methods. He advocated the study of minerals and the decomposition and classification into chemical elements at a time when there existed only two professorships in inorganic chemistry in Germany. All these subjects had been eclipsed in the second half of the nineteenth century by the preoccupations with the constitution of organic matter, the debates surrounding the role of the so-called "Lebenskraft" or vital force, and the synthesis of new organic substances never before observed in nature. Hittorf deplored the ever-growing specialization of organic chemists and the abandonment of inorganic chemistry, the result of an exponential growth in the number of organic substances and methods that overwhelmed the profession. "There is no one among us, except Viktor Meyer," Hittorf told the assembled Bunsengesellschaft in 1899, "who still practices inorganic chemistry while being an organic chemist." Organic chemists were "forced to restrict themselves" to their field and ignore developments elsewhere. Hittorf recounted how he had tried, in vain, to persuade one of his fellow organic chemists seriously to consider how the theory of the migration of ions in a solution seemed to contradict existing conceptions of affinity. But communication with his friend was "impossible." On the contrary, his friend dreamed of "completely" reforming inorganic chemistry by applying insights gained from organic chemistry.

It was in such an atmosphere of apparently mutual incomprehension that the executive board of the Electrochemical Society went so far as to urge that parliament approve the "fostering of education in scientific inorganic chemistry" by creating new chairs and laboratories.[22] The sessions of this important meeting were held in Nernst's institute in Göttingen, attended by several of his most devoted supporters. Wilhelm Ostwald could not refrain from voicing a candid and outspoken manifesto: that the new chairs in inorganic chemistry be populated with physical chemists, since he personally saw the two subjects as quite interchangeable.

The concept of an element, the elucidation of the relation between the structure of matter and its properties, the interpretation of combustion phenomena, and the establishing of a chemical nomenclature are all significant disciplinary chemical problems. These concepts evolved in parallel, often disjointedly, but were mostly interconnected. For the historian,

21. Prof. Witte, Technische Hochschule Darmstadt, Physikalisch-Chemische Abteilung. The Deutsche Bunsengesellschaft Archives were lost during World War II. Emma Nernst, Walther's wife, also provided information for other biographical essays published after the war. AAW Ostwald Papers and Historische Abteilung, Personalia.

22. "VI. Hauptversammlung der deutschen Elektrochemischen Gesellschaft, am 25. bis 27. Mai 1899 in Göttingen." *Zeitschr. f. Elektrochem,* Nr. 2 (13 July 1899): 27–33.

the analysis of these subjects is at times exclusionary since the choice of a problem as *the* relevant one relegates equally pervasive, often competing motifs to the sidelines. Affinity, for instance, is a vast topic, and *how* to approach affinity has remained a legitimate, and to my mind unsolved, methodological problem. To see a concern with affinity or, alternatively, with equilibria as definitive of the work of physical chemists is to misrepresent their more immediate interests. It is important to remember that Ostwald and Arrhenius were both immensely interested in the problem of affinity, especially at the time when they became acquainted with each other's work in the early 1880s. While still in Riga, Ostwald was investigating the properties of acids, hoping to build a table of relative chemical activities. Upon reading Arrhenius's work, he deemed it as one of the most significant contributions to the study of chemical affinity.[23] While it is true that Nernst continued the thermodynamic line of thought initiated by van't Hoff, in which the balance of energies in a chemical reaction may give insights into affinity, he, as well as other physical chemists, was never explicitly concerned with either the theory or the experimental or quantitative definition of affinity.

The difficulties encountered in producing both rational reconstructions and rich contextualizations of science have recently been compounded by the growth of cultural relativism in the social sciences. This proliferation of "truths" has created tensions and pressures in the history of the physical sciences where previously an extreme progressivism, and to a certain extent positivism and empiricism, had reigned. In contradistinction to this contemporary trend, some late-nineteenth-century physical chemists, much like other scientists, were hoping to gain insight into their own work by applying the "historical" method to their scientific work. The traditional physical sciences – physics and chemistry – had supposedly reached a satisfactory stage by the 1870s and 1880s, after which no further fundamental theoretical advances seemed possible or even necessary. All that remained was the provision of precision measurements and detailed descriptions.[24] Thus, it was essential for scientists to articulate an adequate

23. Crawford, 1996, p. 51.
24. David Cahan writes that "The progress made by [Helmholtz, Kundt, Warburg, Magnus, Kohlrausch] . . . encouraged the belief, not uncommon by the 1880s, that the fundamental principles of physics had already been discovered and that, hence, their major task lay in the refinement of physical laws and constants to the highest degree of precision possible." David Cahan, "The institutional revolution in German physics, 1865–1914," *HSPS* 15 (1985): 38. This attitude contributed to the establishment not only of the Physikalische Technische Reichsanstalt, Germany's foremost institution for precision measurements, but also to the expansion of experimental physics institutes and added funds for updated and sophisticated instrumentation.

justification for the elaboration of a new theory, or subfield, or especially a "new" science, which would address foundational issues. In the mid–nineteenth century, positivists had insisted that the social sciences emulate the methods and practices of the physical sciences. By the last third of the century, unease with such reductionism led to a call from several directions for a unique methodology of the social sciences. Among others, neo-Kantians framed the distinction between *idiographic* and *nomothetic* sciences, that is, between sciences aiming at historical and social specificity, as opposed to those dedicated to the formulation of universal natural laws. While these antitheses have been the subject of extensive scholarship, the enlistment of historical models and insights by scientists for the development of a physical scientific discipline has remained generally unexamined.

The historicist program embraced by the positivists, in particular by Georg Helm, Ernst Mach, and Ostwald, engendered debates concerning the interpretation of the Second Law of Thermodynamics that were heavily dependent upon earlier reductionist attempts – such as Helmholtz's – to trace the bases of thermodynamics to mechanical principles.[25] In the same spirit, physical chemists at the turn of the century, much like Lavoisier and others before them, sought to justify the construction of a space for their new discipline along historical lines, emulating recent precedents set by Mach, as well as by Ludwig Boltzmann, in his effort to bolster atomism. Borrowing from past historical episodes helped physical chemists vindicate their labors. Several texts written between 1890 and 1915 by influential physical chemists, such as the "founders" and their disciples, exemplify the use of history in their quest to recover or create a scientific tradition. Their historical accounts vary according to their timing and to the authors' backgrounds and agendas. Each writer claimed different scientific ancestors. Yet, their texts generally reflect an increasing reliance on physics as the legitimate ontological precursor to physical chemistry, minimizing the role of chemistry while simultaneously seeking entry into mainstream, established chemistry.

Texts, such as those written by Ostwald, Pierre Duhem, van't Hoff, and Nernst, help to demonstrate how "métier made" professional, scientific mentalities are created by the scientists in the process of distancing themselves from a perceived center to a new periphery, how the "exile from Eden" syndrome, while inevitable, is transformed by the "emigrés" into a new life, one that maintains links with the past but has to come to terms with ruptures.[26]

The new discipline of physical chemistry was furthered by the need for

25. Wise, 1983.
26. Clifford Geertz, *Local Knowledge* (New York: Basic Books, 1983). See especially "The Way We Think Now: Toward an Ethnography of Modern Thought," pp. 147–67.

a more systematic theoretical approach, for standardized measurement methods, and for a comprehensive study of physical properties, such as specific heats, specific volumes, refractivity indices, the power of rotation of plane polarized light, the various spectroscopic properties, selective absorption of light, dielectric constants, and surface energies.[27] These trends incorporated the continuation of the electrochemical tradition, which itself had been greatly enhanced and aided by the technological needs of an expanding electrical and electrotechnical industry. Moreover, there was an equally dramatic growth of organic chemistry (known substances, new syntheses, biologically active compounds) and industry (dye industry, pharmacology, and extracting industries, etc.) where a state of near chaos existed in the nomenclature and classification of organic compounds. Eventually, these subfields were profoundly affected by some aspects of the general, or theoretical, or physical chemical approach, even though many original theoretical postulates of the founders were discarded or transformed.

Physical chemistry arose from a diversity of practices and theoretical constructs, as well as technological aims, that belie strict disciplinary demarcationism. Essentially, disciplines are pragmatic devices aimed at nurturing institutions and carving out identities. Preoccupations with identity come in many shapes; whether in integrative, dissipative, or separatist incarnations they surely depend on time and place and on who is doing the work of identification. An individual may associate with or disassociate from traditions, nations, professions, and religions, without necessarily completing a successful construction of a novel identity. The tension of belonging and not-belonging, of remaining part of and yet fleeing from the past, of turning outward and toward the future, relates to the symbolic space occupied by scientists. To deny that such struggles impinge upon the daily labors at the desk or the bench would be foolish. To claim that they decisively determine the emerging scientific views of nature would be equally imprudent. In fact, the literary productions and the institutional wrangling impelled by the need to forge a disciplinary and professional "identity" must be balanced against the record of the actual doing of science.

As far as Nernst himself is concerned, I believe that we shall find in what follows that he navigated during his lifetime a highly personalized course, and that his work addressed major scientific problems rather than disciplinary agendas. In 1886, during Nernst's study years in Graz, Boltzmann

27. Erwin N. Hiebert, "Developments in Physical Chemistry at the Turn of the Century," in *Science, Technology and Society in the Time of Alfred Nobel,* ed. C. G. Bernhard, E. Crawford, and P. Sörbom (Oxford: Pergamon, 1982), p. 100.

delivered his inaugural lecture on the Second Law of Thermodynamics upon admission to the Vienna Academy of Sciences. He there avowed that, since all attempts to save the universe from *heat death* had failed, he too would refrain from such an undertaking. Many years later, in his last major publication in 1921, Nernst, then 57, wrote:

> [Boltzmann's] passage [on heat death,][28] which I read as a student, made the greatest impression on me, and since then my sight has been set on the question whether a solution couldn't be found somewhere. [ob nicht irgendwo ein Ausweg sich zeigte.]

We may thus understand how and why Nernst's rather convoluted trajectory from electrochemistry to astrophysics was anchored in an enduring interest in the "bigger" questions of science, in a larger perspective beyond the specific scientific investigations performed during his thirty-year career.

28. Ludwig Boltzmann, *Populäre Schriften* (Leipzig: Amb. Barth, 1905), p. 25.

2

Nothing Is More Practical
Than Theory: Beginnings

Burdened by contradictory reminiscences of contemporaries, scattered statements in Nernst's own scientific writings, and a meager correspondence, the portrait of Hermann Walther Nernst, one of Germany's foremost representatives of Wilhelminian science, has remained fragmented and incomplete. Unfortunately, like other prominent scientists of his generation, Nernst commented sparingly on the sources of his scientific work. He wrote even less about his nonscientific views and personal life. Moreover, he destroyed his papers before the end of the Second World War.[1] A more complex image of Nernst and his work can be reconstructed through scientific publications and the scattered correspondence and memoranda found among the papers of other scientists and in the archives of various scientific and governmental institutions.[2]

1. Nernst apparently destroyed his personal papers and correspondence shortly before his death in November 1941, as attested by two letters from his wife, Emma Nernst, in April–May 1942, to Wilhelm Ostwald's daughter and biographer, Grete Ostwald. AAW, Ostwald Nachlass.

2. Kurt Mendelssohn, who as a student briefly encountered Nernst in Berlin and became a second-generation member of the Nernst school of low-temperature physics at the Clarendon Laboratories in Oxford, has provided much useful information on Nernst's family, highlights of his career, and a general description of the academic and scientific environment during Nernst's lifetime. Primarily Mendelssohn's personal reminiscences, the book lacks references to original sources. It relies on published articles and necrologies, and on a brief memoir provided to the author by Nernst's wife, Emma Nernst, and his daughter, Edith v. Zanthier. Kurt Mendelssohn, *The World of Walther Nernst: The Rise and Fall of German Science, 1864–1941* (London: Macmillan, 1973). Hendrik Casimir commented on Mendelssohn's account: "I also attended some sessions of the main colloquium. Kurt Mendelssohn has described this institution in enthusiastic terms as the place where the most prominent physicists of the day pronounced judgment on the most recent developments. It did not strike me that way at all; compared to Ehrenfest's colloquium, discussions were both formal and perfunctory. But it was an experience to listen to Walter Nernst." Hendrik Casimir, *Haphazard Reality: Half a Century of Science* (New York: Harper & Row, 1983), pp. 131–2. A second, nontechnical biography, Hans-Georg Bartel, *Walther Nernst,* Biographien hervorragender Naturwissenschaftler, Techniker und Mediziner, vol. 90 (Leipzig: Teubner, 1989), explores new sources from the Humboldt University Archives in Berlin.

Nernst has been justly recognized in textbooks, homilies, and histories as one of the major figures in the development of physical chemistry. These sources recount his eclectic background and training, and his exposure to the techniques and research programs of several established physics laboratories. They also discuss his teaching duties and the diverse academic environments that shaped his work. Nernst's penchant for theoretical physics, balanced by his extraordinary experimental skills and intuition, defy his classification as either an experimentalist or a theoretician. His career also defies categorization because of his wide-ranging preoccupations and collaborations with numerous disciples whose research fields ranged from electricity, electrochemistry, and chemical equilibria to low-temperature experimental studies and astrophysics.

How did Nernst's work in the years before the enunciation of the heat theorem at the end of 1905 fit into any "disciplinary" pattern or program? Certainly his contact with traditional chemical problems was minimal. He did indeed explore some topics of interest to physical chemists and electrochemists.[3] Nevertheless, he belonged to a group that undoubtedly considered itself, and was perceived by others, as being active on the frontiers of physical science more generally. Between 1886 and 1939, Nernst published more than two hundred articles and a number of books.[4] His most productive years fall into two separate decades: Between the ages of 22 and 33 (1886–97,) he published sixty-two articles, two books, an encyclopedia chapter, and two memoirs on the Electrochemical Institute in Göttingen. Between 1904 and 1914, Nernst authored ninety articles, as well as his *Silliman Lectures*. The intervening years between 1898 and the end of 1903 were taken up mainly with research on the electrolytic lamp, during which time he produced some twenty-seven publications. The years following the First World War were less intense, and after 1920 he published mainly essays, of which a third were devoted to popular expositions and memoirs on colleagues such as Clausius, Gibbs, Arrhenius, Ostwald, Tammann, Bodenstein, and Urbain.

While at Leipzig between 1886 and 1890, Nernst worked primarily in electrochemistry. In 1889 and 1890, he devoted five articles to molecular weights, and penned his first piece on low-temperature measurements. Between 1892 and 1902, most of his articles dealt with electrolytic dissociation, electrolytic conductivity in solids, and chemical equilibria. These were

---

3. Erwin Hiebert has written that "Until he was about forty years of age, Nernst's efforts were directed predominantly toward the refinement of methods to explore principles already current among chemists. After moving to Berlin [in 1905], however, he became totally involved, theoretically, in the exploration of new ideas in thermodynamics." E. N. Hiebert, "Walther Nernst," *Dictionary of Scientific Biography*, p. 436.
4. For a complete bibliography, see in particular Lindemann and Simon, 1942.

followed between 1902 and 1907 by several papers on high-temperature work. In 1906, he wrote on the heat theorem, electrochemistry, the electrical conductivity of liquids, and equilibria in nitrogen oxide that were relevant to the synthesis of ammonia. That year, Nernst also delivered the Silliman Lectures at Yale University.[5] The following period was devoted primarily to writings on thermodynamics, with the start of publications on specific heats in 1910. Low-temperature experimentation and the application of the heat theorem to solids continued through 1914. During the war, Nernst volunteered as an expert in explosives, and later became infamous for his work on gas warfare. He spent substantial periods on both the Western and Eastern Fronts. With the changed political circumstances after the war, Nernst's own scientific research virtually stopped, although a large number of students continued important work under his direction.[6]

During 1921, the year in which he became rector of the Berlin University, Nernst published only a survey article on Helmholtz's electrochemical work.[7] During this year, however, he first ventured seriously outside his domain and wrote a book-length exposition of his views regarding cosmology and the heat-death of the universe in *Das Weltgebäude im Lichte der neueren Forschung* (*The Structure of the Universe in Light of Recent Research*). In 1923–4 he coauthored three additional articles on photochemical processes, but in the following years he published only sporadically on a variety of subjects: electrochemistry, thermodynamics

5. In 1907, only two articles appear, one being a brief history of general and physical chemistry during the past forty years – "Die Entwicklung der allgemeinen und physikalischen Chemie in den letzten 40 Jahren," *Ber. Chem. Ges.* 40: 4617, also published in the *Smithsonian Institute Report* 9 (1908): 245.

6. In 1916, see: "Über einen Versuch, von quantentheoretischen Betrachtungen zur Annahme stetiger Energieänderungen zurückzukehren," *Verh. Phys. Ges.* (1916): 18, 83, and also "Krieg und deutsche Industrie," *Intern. Monatsschrift f. Wissenschaft, Kunst und Technik,* 10 (1916): 119. There are no publications for 1917, except an article on ballistics in a popular newspaper, *Leipziger Illustrierte Zeitung* ("Innere und äussere Ballistik der Minenwerfer," 22 November 1917). In 1918, only one article, "Zur Anwendung des Einstein'schen Aquivalentgesetzes. I.," *Z. Elektroch.* 24: 335, and an article "Quantentheorie und neuer Wärmesatz," *Naturwissenschaften* 6, 206, and publishes *Die theoretischen und experimentellen Grundlagen des neuen Wärmesatzes,* Halle, 1918. In 1919, publishes article on the theory of gas degeneration, "Einige Folgerungen aus der sogenannten Entartungstheorie der Gase," *Berl. Ber.* (1919): 118, and three papers coauthored with his students: with Th. Wulf, "Über eine Modifikation der Planck'schen Strahlungsformel auf experimenteller Grundlage," *Verh. der Phys. Ges.* 21: 294; with Walther Noddack, "Zur Kenntnis der photochemischen Reaktionen," *Physik. Z.* 21: 602; and with K. Moers, "Zur Konstitution der Hydride," *Z. Elektroch.* 26: 323.

7. "Die elektrochemischen Arbeiten von Helmholtz," *Die Naturwissenschaften.* 9: 699 (1921).

of dilute solutions, and thermodynamics of chemical reactions, as well as a dozen articles on astrophysics.

In his sensitive biography of Einstein, Philipp Frank compared the personalities of Nernst and Planck, the two scientists who exerted a decisive influence on Einstein's appointment to the Berlin Academy of Sciences in 1913. To Frank, Planck personified the upper-class scientist who "accepted the philosophy of his social class; he believed in the mission of the Kaiser to make the world happy with his conception of German culture, and in the right of his class to provide the leaders for Germany and to exclude people of other origins from such functions." On the other hand, Planck was "an ardent adherent of Kantian philosophy in the diluted form in which it had become the common religion of the German academic and government circles. . . . He also believed in the international mission of science. . . . But since his immediate emotional reaction was to respond in terms of the philosophy of the Prussian bureaucracy, an appeal to his reason was necessary to make him recognize the rights of aliens. As he was conscientious and an idealist, such an appeal was usually successful."

Quite in contrast, in Frank's view, Nernst:

> although a great scientist and scholar, exhibited the mentality characteristic of a member of the merchant class. He had no national or class prejudice and was imbued with a type of liberalism that is often peculiar to businessmen. He was short, active, witty, and quick of apprehension. He occasionally utilized his craftiness in professional life, and his students jokingly referred to him as the "Kommerzienrat," a title conferred in Germany on successful businessmen. There was a story about him that he was the only physicist who had ever signed a contract with an industrial firm in which the advantage was not on the side of the firm.[8]

In sum, Frank's portraits perfectly contrast Nernst and Planck, two very dissimilar personalities.

The anecdotes Frank recounts were dear to the young physics students who met Nernst in the late 1920s. For Hendrik Casimir, a student in Berlin when Nernst was no longer active in intensive research, the aging scientist struck him "as a ridiculous figure, and that impression was strengthened by the many stories that circulated about him and by his being a 'Bonze' [an academic German mandarin]." Almost half a century later, however, Casimir was impressed on rereading Nernst's work: "True, there

8. Philipp Frank, *Einstein. His Life and Times* (New York: A. Knopf, 1947), p. 107. (Translated from a German manuscript by George Rosen. Edited and revised by Shuichi Kusaka.)

were some irritating mannerisms and his mathematics was shaky, but his work shows throughout a remarkably clear and often prophetic vision." When delivering a paper in 1964 on the occasion of Nernst's centenary anniversary, Casimir was pleased to have "an opportunity to atone in public" for his youthful error of judgment.[9]

In the late 1920s and early 1930s, at the zenith of his professional life, Nernst was the director of the Physics Institute at the Berlin University. At the time, many students who were to become eminent representatives of the new quantum physics saw Nernst as one of the grand old men of the field. It was in his institute, formerly directed by W. Rubens, that the famed physics colloquia, inaugurated by Gustav Magnus – who, born of a Jewish family, had been baptized into Protestantism by Nernst's Lutheran grandfather – took place every Friday afternoon. Regardless of the presentations, "the main performance" was provided by the audience. The front rows were occupied by Einstein, Max von Laue, Planck, Erwin Schrödinger, Gustav Hertz, and Nernst, joined by Otto Hahn, Lise Meitner, and many younger luminaries. The seminars attracted students from around the world, eager to witness "modern physics in the making."[10]

If Frank's portrait of Nernst distills him almost to the point of caricature, can it be reconciled with his position in German physical science? If a Kantian like Planck, even slightly outdated, is the prototypical German academic mandarin, where is Nernst to be placed in the context of the traditional Berlin scholarly establishment? Was he an outsider, his rise to eminence somehow contradicting traditional German values and customs? Frank's sketch of Nernst is true only to the extent that it excludes all those features that he underscores in his portrait of Planck. Nernst was as much a part of the Prussian establishment as Planck, Fritz Haber, and his other Berlin colleagues. He knew the Kaiser and believed in the international mission of science, and subscribed to the empire's political goals. He was a well-known figure in chemical industrial and commercial circles, an influential organizer, and a representative of German science abroad. Although many of his contemporaries have been praised for their scientific achievements and often criticized for their political and social views, Nernst was among the few, as Frank's analysis reveals, to have gained points in collective memory chiefly on account of his unconventional manners and his equally unequivocal openness and lack of prejudice. As we shall see in later chapters, Nernst's many Jewish friends, students, and collaborators were sometimes used as foils against him in professional wrangling.

Rudolf Peierls recounted, less flatteringly, his exposure to Planck and Nernst when he enrolled at the Berlin University in 1925. Planck's lectures

9. Casimir, p. 132.   10. Mendelssohn, pp. 2–3.

on theoretical physics, arranged in a three-year cycle, were "the worst he ever attended." Planck "would read verbatim from one of his books," and although Peierls knew that he was "a very famous man," he had "no idea what he was famous for." In later years, Nernst, too, became a rather dull lecturer. The students knew Nernst as the inventor of the "Nernst lamp," "a not-very-successful alternative to Edison's carbon filament bulb" according to Peierls, who himself remembered Nernst as "a great physicist of rather small stature and an even smaller sense of humility." But Nernst's arrogance was always tempered by his wry sense of humor. He would preface discussions of his lamp by saying: "Gentlemen, when Bunsen had to mention the Bunsen burner in his lectures, he always referred to the 'non-luminous burner.' In the same spirit I shall now talk about the 'electrolytic lamp.'"[11]

Nernst's theories, research methods, accuracy, and behavior in public and private debates have sometimes been criticized. Nonetheless, his openness, his bourgeois inclinations, and his modernity have scored well with a posterity that has faulted German academic leaders for having conservative or lax attitudes in the face of political and social changes taking place during the Weimar Republic and the ensuing debacle of the 1930s.

Youth

Nernst was born to a middle-class family on 25 June 1864 in the small town of Briesen in Western Prussia. His grandfather was a pastor, and his father a country judge. Nernst had two older sisters and a younger brother. A third sister had died during the cholera epidemic of the 1860s. Like his equally famous contemporary, Fritz Haber, Nernst lost his mother at a young age. His father died when Nernst was only twenty-four. Enrolled in the local humanistic gymnasium of Graudenz, a small town on the river Vistula (or Weichsel), now in Poland, he received a conventional education. Berlin, the closest major German city, could be reached in half a day's trip by changing trains once onto the major European railroad that linked Paris to Moscow. But Nernst spent most of his vacations on the estate of his grandparents, some six miles away from his hometown, where he acquired a lifelong passion for hunting boar and rabbits.

During Nernst's high school years in the 1880s, Germany experienced a profound economic crisis, the reverberations of which echoed well into the 1890s. The leveling off of industrial investment, the near stagnation

11. Rudolf Peierls, *Bird of Passage: Recollections of a Physicist* (Princeton, N.J.: Princeton University Press, 1985), pp. 19–20.

of railroad expansion, the steep decline of prices, and a dramatic rise in unemployment led to a pervasive hopelessness among the growing masses of salaried workers. Even people in liberal circles began wondering whether the dreaded socialist reforms advocated by the left would not be an acceptable solution.[12] Among the repercussions resulting from Bismarck's fears of socialism and materialism was the prohibition in 1879 of the teaching of Darwin's and Haeckel's evolutionary theories. When Bismarck decided to abandon the *Kulturkampf* (cultural struggle) against Roman Catholicism and turned the Socialists into his declared enemy that same year, he reached a reconciliation with the newly elected pope, fired the reformist minister of education, and appointed instead, however briefly, the archconservative Robert von Puttkamer. As a result, biology disappeared from the curricula of Prussian high schools.

The consequences of Bismarckian politics were also evident in other scientific fields. In a highly virulent attack, the antistructuralist organic chemist Hermann Kolbe accused August Kekulé, the venerable discoverer of the structure of benzene, of delivering an "almost illiterate" rectoral address, the "obvious product of a former Realschüler" whose chemistry was "colored by crude Haeckelism."[13] Kolbe opposed the modernization of the German high school system, which had allowed the coexistence of the traditional humanistic Gymnasium and the newer Realschulen – schools that emphasized science, mathematics, and modern languages at the expense of Latin, Greek, and Hebrew.

These tendencies notwithstanding, technical education became an imperative for the expansion of the industrial, unified German state. In 1872 the minister of education, Karl Schneider, instituted new "General Regulations Concerning the Volksschule [elementary school] and Teacher Training in Preparatory Schools and Seminars," which prescribed different minimum standards for urban and rural schools. As with the higher educational trends in the United States and Britain, the emphasis was on modernity. History and German language and literature "were to replace religion as the core of the curriculum."[14] Oberrealschulen – vocational-oriented alternatives to the Realgymnasia and the humanistic Gymnasia – were established in response to the growing unemployment within universities and offered the hope that the number of students pursuing academic careers would be reduced. They were one part of the major restructuring of the educational system in 1882. By then, the number of students enrolled

12. Hans-Ulrich Wehler, *Das Deutsche Kaiserreich 1871–1918* (Göttingen: 1973), p. 46.
13. Rocke, 1993, p. 331.
14. Karl A. Schleunes, *Schooling and Society: The Politics of Education in Prussia and Bavaria, 1750–1900* (Oxford/New York/Munich: Berg, 1993), p. 177.

in the Realgymnasien had increased rapidly to 26,000 in eighty-five schools, whereas the number of students at the classical Gymnasia rose dramatically to 75,000 enrolled in 251 schools, twice as many institutions compared to a generation earlier.[15] The various reforms, followed by the stricter admission policies into the upper classes of high school in 1890, gave promise of more scientists and fewer academicians to be trained by the educational system. The Gymnasium taught more science, the Realgymnasium less Latin, and the Oberrealschulen received as many as fourteen additional weekly hours of physics over the course of four years.[16]

In 1890, when Nernst was at the beginning of his academic career in Göttingen, virulent opposition to the admission of the new Realschüler into universities led to an education conference, whose participants were stunned by the unwonted intervention of the recently enthroned Kaiser Wilhelm II. Speaking to the educators assembled, the Kaiser enjoined them to remember that the schools would best serve healthy national and social goals by inculcating "love for the Fatherland and the fear of God." Wilhelm explicitly stated his aim of restructuring the schools in order to "counteract the spread of socialist and communist ideas." From then on, German language and history became the curricular core. Although he strongly championed the abolition of the Realgymnasium and favored the separation of education into the classical at the Gymnasium and the scientific at the Oberrealschulen, the tripartite system survived. The teaching of physics suffered little, but the requirements for the final science examinations were diluted.

On average, Nernst would have studied a little over an hour of physics and about half an hour each of chemistry, mineralogy, and biology per week during his high school years in Prussia.

What would he have learned? The Prussian higher schools that Nernst attended were the only ones that consistently included mathematics and natural history as part of their unitary school programs. But even so, chemistry played practically no role even in Prussian Gymnasia, although it was taught more intensively in the Realgymnasia.[17] Outside Prussia the Gymnasia had nonexistent or minimal science courses.

15. Margaret Kraul, *Das deutsche Gymnasium 1780–1980*, Suhrkamp (1984), p. 86.
16. Friedrich Paulsen, *Geschichte des gelehrten Unterrichts auf den deutschen Schulen und Universitäten vom Ausgang des Mittelalters bis zur Gegenwart mit besonderer Rücksicht auf den klassichen Unterricht. Leipzig 1885*, 3rd ed., vol. 2. (Berlin/Leipzig: 1921), p. 585. Quoted in Carsten Schuldt, *Die Enwicklung des Physikunterrichts der höheren Schulen Preussens in der Zeit zwischen dem Suvernschen Unterrichtsgesetz und den Richterschen Reformen unter besonderer Berücksichtigung der Einführung von Atomistik und Hypothesen*, Ph.D. diss. (Hamburg: Hamburg University, 1979), p. 87.
17. Gert Schubring, *Bibliographie der Schulprogramme in Mathematik und Naturwissenschaften 1800–1875* (Bad Salzdetfurth: Verlag Barbara Franzbecker, 1986).

Jacob Heussi's 1879 textbook for physics, which Nernst most probably used at age fifteen, was meant to promote "formal education . . . not the training of physicists." Heussi placed scientific and humanistic education on an equal footing, especially since physics was not enjoying "at the time the best of favor." The distinguished Leipzig chemist Hermann Kolbe also advocated chemistry as a liberal art, criticizing those who considered it a "crude and empirical craft," and he foretold that "all will be expected to study chemistry" in order to become "truly educated."[18] Heussi decried the lack of understanding and foresight that deprived the young generation of the greatest recent progress in science.[19]

Heussi introduced the atomic hypothesis in the first section of his textbook, relying on knowledge derived from the chemistry of the known sixty-five elements, constituted of molecules, themselves made of atoms. He also discussed the theory of heat as molecular and atomic motion. He presented both Faraday's laws and electrolysis, as well as Bunsen's and Kirchhoff's spectroscopy. In view of Nernst's later interest in this problem, it is suggestive that even in these elementary texts, Heussi expressed doubts about the validity of the law of Dulong and Petit, which states that all solid elements have the same heat capacity per mol – that is, that equal amounts of energy will raise the temperature of one mol of any substance by an equal number of degrees.

This problem became the crux of Nernst's and Einstein's work in the early years of the twentieth century, since this property of substances is an indication of the behavior of atoms and their components – a property related to the constitution of matter and the nature of the kinetic motion of particles.

Heussi provided instructions for physics experiments, but acknowledged a regrettable lack of properly trained teachers and of apparatus. Gymnasium teachers were accepted into university seminars only reluctantly and mainly for financial reasons. They usually received a "general education" (allgemeine Bildung). Their final examinations included only German literature, religion, philosophy, and pedagogy. Science-oriented educators began to publish a small number of specialized journals, beginning in 1869 with the *Zeitschrift für den mathematischen und naturwissenschaftlichen Unterricht*, edited by Immanuel Carl Volkmar Hoffmann. Even so, physics education at Germany's prestigious Gymnasia was limited to occasional, droning lectures, which would be interrupted only, according to a contemporary, when "a student had to be wrenched in time from

18. Rocke, 1993, p. 282.
19. Jacob Heussi, *Lehrbuch der Physik, für Gymnasien, Realschulen und andere höhere Bildungsanstalten*, 5th ed. (Leipzig, 1879), p. vii. Quoted in Schuldt, pp. 93 ff.

Morpheus's arms. . . . A swift walk through the physical cabinet, which occurred at most once a semester, was sufficient illustration" of theoretical concepts.[20]

In short, Nernst and his generation of scientists did not come from a scientifically exigent secondary school system. Their school years by no means provided them with the wealth of knowledge and skills with which their colleagues in the humanities were equipped. To choose a career in science in the 1880s was a novel proposition even in Germany, the country that by 1900 was to dominate the international technological, industrial, and scientific landscapes. But the opportunities that a scientific education was going to afford were clearly anticipated by the very debates regarding school reform and the fierce social and political battles of the Gründerjahre – the founding years of the empire.

Nernst entered German universities at the height of turmoil in higher education. Uncertain career prospects, the "overcrowding" of academic professions, and the fear of creating an "academic proletariat" were met by the equally disturbing change in the social composition of the student population. In 1881, Bismarck decried the rising numbers of students, fearing a mushrooming of "dissatisfied contenders" who would make demands on the state based solely on their educational background. Yet by 1890, toward the end of the first Great Depression, the situation in science eased significantly.[21] Nernst thus entered the university in 1883 at a time when career prospects were grim, and the waiting period for a government position – such as a teaching appointment – often extended to eight years.

He received his high school diploma at the end of 1882 and soon thereafter, at age nineteen, began the traditional Wanderjahre of the German student. He intended to study "primarily physics, then chemistry and mathematics," traveling successively to Zürich, Berlin, Graz, and Würzburg. It is interesting, yet not surprising, that Nernst first traveled to Zürich, where the Polytechnic Institute at that time, and for many decades thereafter, had one of the best laboratories for experimental physics in Europe, training engineers with better-than-average career prospects. Fritz Haber, Walther Rathenau, and Albert Einstein were among those who also studied there.

In the two existing versions of his early vita, Nernst singled out the main figures who influenced his academic training. He wrote that he was in-

---

20. Hans Keferstein, "Physik an höheren Schulen," *Encyklopedisches Handbuch der Pädagogik*, 2nd ed., ed. Wilhelm Rein (Langensalza: 1907), p. 836. Quoted by Schuldt, p. 103.
21. Jarausch, pp. 71ff.

debted for his "scientific education primarily to the teachers of physics: H. F. Weber, Boltzmann, Fr. Kohlrausch; of chemistry: Landolt, Wislicensus; of mathematics: H. Streintz, Prym."[22] One year later, in his application for a position in Göttingen, the order of priorities and personalities changed slightly, with a greater emphasis on chemistry: "My teachers were," he wrote, "primarily: Landolt, Wislicenus, Merz (Chemistry); Hettner, Prym, A. Meyer (Mathematics); H. and Fr. Streintz, Klemencis, Krazer (mathematical Physics); Boltzmann, v. Ettingshausen, Fr. Kohlrausch (experimental Physics)."[23]

The first evidence of Nernst's independent work is to be found during his stay in Graz during the academic year 1885–6, where Ludwig Boltzmann was lecturing at the time. Boltzmann's influence looms larger than anyone else's in Nernst's scientific biography. He had been equally important during those years in providing inspiration to the young and beautiful Viennese student, Lise Meitner, who was to become Nernst's colleague in Berlin two decades later. And through the years, Boltzmann and his successor, F. Exner, trained an impressive number of extremely bright and enthusiastic young physicists in Vienna and Graz. In 1899, Professor Nernst, by then quite well known, greeted the young student Emil Abel by saying: "You come from Vienna, the city of Boltzmann! Always remember his wise saying: 'Nothing is more practical than theory.' You can see that in the ophthalmoscope of Helmholtz, the gas light of Auer, Rontgen's rays and my lamp."[24]

In the mid-1880s, Boltzmann had already been the director of the Physical Institute at the University of Graz, a "model institution in the physical-technical sense" for more than a decade. Albert von Ettingshausen was Boltzmann's experimental and administrative assistant, assuming responsibility for the laboratory exercises that accompanied Boltzmann's experimental physics lectures. Heinrich Streintz had recently been promoted to Ordinarius, a full professorship for mathematical physics.[25] Yet, contrary to many later accounts, the students who studied with Boltzmann were, in their great majority, future middle school teachers, not particularly interested in advanced physics education. Boltzmann therefore limited his teaching over the years to only six courses: in experimental physics (five

22. UAL-PA 773, Bl.2, Nernst Vita, Leipzig, March 1889.
23. Göttingen University Archives, Personalakten 1732, Nernst, 12 April 1890, photocopy from Karlsruhe University, Bunsen Gesellschaft Archive.
24. E. Abel, *Osterr. Chem.-Ztg.* 55 (1954): 151, quoted in Engelbert Broda, *Ludwig Boltzmann* (Woodbridge, Conn.: Ox Bow Press, 1983), p. 105.
25. Christa Jungnickel and Russell McCormmach, *Intellectual Mastery of Nature: Theoretical Physics from Ohm to Einstein*, vol. 2 (Chicago/London: University of Chicago Press, 1986), pp. 65–6.

hours per week each term), physical laboratory (four to five hours), mechanical heat theory, gas theory, wave theory, and electromagnetic light theory. But he offered the latter specialized courses only on very few occasions. Despite the fact that Boltzmann perceived the general lectures as an imposition, he nonetheless seems to have enjoyed them and to have paid particularly close attention to the experimental part of each lecture. He always made sure that the most interesting and engaging demonstrations were set up for his students, and that everyone in the large auditorium had a complete and unhampered view of the demonstration table. Moreover, Boltzmann never assumed prior knowledge, and always explained new concepts and experiments in the simplest, most intuitive fashion. He loved to emphasize the beauty and the characteristic style of each mathematician and physicist, likening their personal imprint to the readily recognizable first notes of a Mozart or Schubert piece.

Nernst remembered his years in Graz with great fondness. It was there that he first encountered a well-organized institute, in which "teachers and researchers collaborate exceptionally well." Boltzmann's figure was "luminous," and his mere presence made up for the lack of specialized lectures.[26] Among the other advanced students was Svante Arrhenius, with whose research Nernst became acquainted during the academic year 1886–7, when both were working in the laboratory of the Physical Institute at the Würzburg University. In March 1887 the Würzburg institute closed for several months, and the two men visited Graz for a term. They were eager to "study as much as possible of the modern physical mode of thought," and Nernst recommended "Boltzmann's impressive institute."[27] The summer of 1887 may well have been the most productive in Boltzmann's career. Intimations of a coming emotional crisis had already been felt two years earlier, but in 1887 he published the largest number of articles and vigorously directed his institute. It was also the year of Arrhenius's most important work.

Nernst gave a brief and clear-sighted description of the chain of events that led to Arrhenius's formulation of the dissociation theory. Earlier that year, Arrhenius had presented at the institute's colloquium the main ideas of his doctoral dissertation completed in Stockholm, *Sur la conductibilité des Électrolytes*. But at the time, Nernst recalled, "this project hardly influenced the development of science, even though it reveals the author's strong research ability and undeniably transcends in some important points the knowledge of the chemical behavior of electrolytes at the

26. W. Nernst, "Albert von Ettingshausen. Eine Erinnerung und meine Grazer Studienzeit," in *Elektrotechnik und Maschinenbau* 48(13): 279–81 (1930).
27. Arrhenius to Felix Ehrenhaft, quoted in Boltzmann, *Leben und Briefe*, p. I 73–4.

time."[28] The audience, including Nernst and Kohlrausch, did not understand anything in Arrhenius's lecture beside the fact that "he used as a guiding principle Clausius' hypothesis of electrolytic dissociation." As is revealed in what is now a well-known document, Arrhenius's letter of 30 March 1887 to van't Hoff, it was only under the strong impulse of van't Hoff's work that Arrhenius's ideas were consolidated.[29] Nernst remembered that one day Arrhenius told him he had received an extremely important letter from van't Hoff. Although not acquainted with Arrhenius's work on dissociation, Nernst recalled the ardent importance that Arrhenius accorded to the exchange of letters. He retired for a whole day in order to draft a reply to van't Hoff. Arrhenius wrote that he was planning to visit van't Hoff in the near future. He was extremely grateful for, and flattered by, the attention that his dissertation had elicited. But he acknowledged that he still had difficulty in understanding the relationship between his "electro-kinetic conception" and van't Hoff's "thermodynamic view of the [dissociation] processes," although he had "no doubts that such an intimate connection exists."[30]

Even more important for Nernst's development than his brief exposure to the incipient ionic dissociation theory was Arrhenius's "lively enthusiasm" for Ostwald, which led Nernst to contemplate a "pilgrimage" to Riga, a journey that he had wanted to make, particularly after having met Ostwald when he was briefly in Graz. However, his plans were foiled by Ostwald's appointment to the University of Leipzig in the fall of 1887.

Ostwald came to a vibrant scientific center, a wealthy university with a cosmopolitan student body, which could "offer perhaps the largest and most modern scientific and medical facilities" in the 1880s.[31] In the late 1860s, Hermann Kolbe had built there the most dynamic chemistry institute in Germany, where as many as 170 Praktikanten, or advanced students, studied each semester. As a consequence, enrollments in chemistry had skyrocketed throughout Germany. Prominent organic chemists had "created decentralized institutes where much basic chemical instruction –

28. W. Nernst, "Zum 50. Geburtstage der elektrolytischen Dissoziationstheorie von Arrhenius," *Z. f. Elektrochem.* 43 (1937): 146–8. For Arrhenius, see in particular the recent exhaustive biography by Elisabeth Crawford, *Arrhenius: From Ionic Theory to the Greenhouse Effect* (Canton, Mass.: Science History Publications/USA, 1996).

29. Ibid., p. 146. J. H. van't Hoff, "Lois de l'equilibre chimique dans l'état dilue, gazeux ou dissous," *K. Svenska Vetenskaps-Akademiens Handlingar* 21, no. 17 (1886); "Une proprieté generale de la matierè diluée," ibid.; "Conditions electriques de l'équilibre chimique," ibid.

30. Arrhenius to van't Hoff, Würzburg, 30 March 1887, Arch. 200f, Museum Boerhaave.

31. Alan J. Rocke. *The Quiet Revolution: Hermann Kolbe and the Science of Organic Chemistry* (Berkeley/Los Angeles/London: University of California Press, 1993), p. 269.

and virtually all of it in the burgeoning field of organic chemistry – was tendered by Extraordinarien and Privatdozenten" (associate professors and assistants who subsisted primarily from students' fees, rather than steady governmental paychecks).[32] Nernst was called to be an assistant in Leipzig, a position for which he felt "as well qualified as if Rubens had appointed me one of his collaborators in his painting studio." Nernst's entry into physical chemistry was somewhat fortuitous. It provided him with an academic position at a time when he was lacking any systematic training either in chemistry or in physical chemistry.

Despite the historical significance accorded to Arrhenius's and van't Hoff's work during the spring of 1887, when they jointly formulated the ionic dissociation theory, Nernst had no memory of any serious conversation on this subject having taken place even as late as the summer of 1887, when Arrhenius and Nernst were again working together in Boltzmann's Physical Institute in Graz. Perhaps Arrhenius "avoided" the subject because no definitive scientific publication on the dissociation theory had yet been published. Only after joining Ostwald and after reading van't Hoff's article on the subject did Nernst become an "Ionist" – a term used with "slight irony by August Kundt when referring to the 'theoreticians of electrolytic dissociation.'"[33]

Before reaching Ostwald's laboratory at the Second Chemical Institute in Leipzig, Nernst had already published a number of important papers, primarily on electrical conductivity, drawing on his research with Kohlrausch in Würzburg and later with A. von Ettingshausen, his adviser in Graz.

Nernst passed his oral doctoral examination on 10 May 1887. Kohlrausch examined him on the "main subject in physics," while Prym and Emil Fischer examined him on the secondary subjects of mathematics and chemistry, respectively. He received his doctorate on 5 August 1887 and shortly thereafter returned to Graz to complete his work on the "effects of magnetism on phenomena of electricity and heat," which he had begun with von Ettingshausen in 1885–6. His training in electromagnetic experimental physics made him an exceptional candidate for electrochemical work. According to Nernst, the strong opposition to the electrolytic dissociation theory of Arrhenius and van't Hoff was due in part to the chemists' simple ignorance "of purely electrical problems" – precisely those for which he had been well prepared during the years from 1885 to 1887.[34]

---

32. Rocke, 1993, p. 281.
33. Nernst, "Zum 50 . . ." (1937): 147. See van't Hoff, Z. *physik. Chem.* 1 (1887): 501.
34. AAW, Historische Abteilung, Akten der Preussischen Akademie der Wissenschaften 1812–1945, Personalia/Mitglieder, Sign. II-III.85, pp. 60, 147.

At the age of twenty-two, Nernst was an unassuming, diminutive, almost homely-looking young man, his fair hair neatly parted down the middle, his round face adorned by a then customary, still-budding mustache. Arrhenius and Nernst, both rather corpulent at an early age, became close friends during the months they spent together. Their Würzburg group around Kohlrausch included Italian, English, and German students.[35]

Kohlrausch, whose interests lay primarily in precision measurement of electromagnetic, electrochemical, and elastic phenomena, had a "talent, if not an obsession, for counting." He had taught briefly at the Zürich Polytechnic Institute in 1870–1 and had moved to Würzburg in 1875, where he stayed for thirteen years. He established a new physics institute and gained a "reputation as a good supervisor of advanced physics students." Nernst must have been attracted by Kohlrausch's pedagogical reputation and inventiveness in building numerous devices for the measurement of electrical, magnetic, and thermal phenomena, and less daunted than Arrhenius by the apparent drudgery of exact measurement.[36] Among all his colleagues, Arrhenius seemed to find Nernst's work on the heat conduction of air the best among that of the recent arrivals to the Institute.

One of Kohlrausch's important research interests was the measurement of electrolytic conductivities in dilute solutions, a topic that was to become Nernst's first area of expertise. During the 1870s, Kohlrausch and his collaborators had conclusively shown that electrical conductivity was influenced by changes in the temperature and the concentration of solutions. After moving from Darmstadt to Würzburg, Kohlrausch built a new Physics Institute, with such success that by the mid-1880s when Nernst arrived, twenty students per semester were working in the well-equipped, insulated, and disturbance-free laboratory. Kohlrausch had by then completed the laborious task of stabilizing the instrumentation needed for the measurement of the mobility of ions in solution. "From 1880 on, every student participating in an elementary physics laboratory course could learn from Kohlrausch's *Leitfaden* the simple and inexpensive procedure for using a Wheatstone bridge with a telephone hookup for measuring electrolytic conductivity, a measuring problem that barely two decades earlier had stumped Europe's best experimental physicists and chemists."[37] Kohlrausch amassed significant data before eventually publishing the

35. See a group portrait of Kohlrausch's students in Ernst H. Riesenfeld, *Svante Arrhenius,* in *Grosse Männer. Studien zur Biologie des Genies,* vol. 11, ed. W. Ostwald (Leipzig: Akademische Verlagsgesellschaft, 1931), after p. 24.
36. Cahan, *An Institute for an Empire,* pp. 128–9. Crawford 1996, 66.
37. David Cahan, "Kohlrausch and Electrolytic Conductivity," *Osiris* 5 (1989): 182.

conclusive data on electrolytic dissociation in extremely dilute solutions, data which Arrhenius had not been able to obtain, either from Kohlrausch himself or from any other source while working on his dissertation published in 1884.[38] And although Kohlrausch's gentle criticism invalidated some substantial findings of his own regarding the equivalence of the behavior of acids and bases in extremely dilute solutions, Arrhenius nonetheless seems to have resisted adopting Kohlrausch's laboratory methods. Whereas Kohlrausch had, as mentioned earlier, obtained reliable data with his telephone method, where one could literally "hear" the electric current passing through solution, Arrhenius continued to use his "method of zeroes," which used the galvanometer as a device for measuring the current generated by the dissociation process.[39] The problem of dilute solutions was to plague scientists, but in particular Arrhenius, for at least another decade.

The first important research results of Nernst's apprenticeship years were published in collaboration with his advisor von Ettingshausen during 1886 and 1887. Albert von Ettingshausen (1850–1932) had been appointed Extraordinarius in Physics upon the second appointment of Boltzmann in Graz. In the summer of 1876, Boltzmann, then 32 years old, became professor of physics and director of the Physikalisches Institut in Graz. Ettingshausen, who had been educated in Graz and who, according to letters of recommendation, was the only physicist fully acquainted with the arrangements in the university laboratory, thus became, at age 27, Boltzmann's adjunct. He was appointed because Boltzmann's reputation, his busy schedule, and his failing eyesight convinced the Ministry of Education that a competent and energetic young coworker was needed in order to maintain the national prominence of the Graz institute. While Boltzmann held the main lectures and supervised the laboratory, Ettingshausen was actually in charge of physical experimental exercises and research, as well as special experimental seminars. But neither Boltzmann nor Ettingshausen undertook any major expansions or modernizations of the laboratory. In fact, when Boltzmann left Graz, his successor complained to the ministry that "the projection apparatus and the battery installation were defective, that only one usable balance was available in the whole institute, but without a usable set of weights, that no decent thermometer or barometer were available, etc."[40]

Under Boltzmann's supervision, Nernst and Ettingshausen had examined the phenomenon of the rotation of the electric field lines under the

---

38. Friedrich Kohlrausch, "Über das Leitungsvermögen einiger Elektrolyte in äusserst verdünnten wässrigen Lösungen," *Ann. der Phys. und Chem.* 26 (1885): 161–226.
39. Crawford, 1996, p. 55.   40. Boltzmann, *Leben und Briefe*, p. I 58.

influence of magnetic forces, known as the "Hall effect."[41] The effect occurs when a rectangular metal plate traversed by an electric current is placed perpendicularly to the field lines of an electromotive force. Two points formerly on an equipotential line will show a potential gradient after the excitation of the field. The maximum effect is achieved when the two points are at the edges of the plate, so that the effect is proportional to the distance between the points. The direction of rotation and the intensity of the effect varies substantially with the metals used. The Hall effect, named after the American physicist Edwin H. Hall (1855–1938), was the subject of intensive investigations at the time. Boltzmann had hoped, as early as 1880, to be able to determine through the measurement of the Hall-constant the speed of moving charges.[42]

In 1886 Ettingshausen and Nernst published the results of their inquiry into some new Hall effects: They demonstrated the effect of the induction of a transverse difference of electric potential when a magnetic field is applied to a heat conductor, a phenomenon now called the "Nernst effect," and the production of a transverse difference of temperature due to action of the magnetic field on an electrical conductor – the "Ettingshausen effect."[43] In a concluding paper in 1887 they attempted to generalize these results in accordance with recent theoretical calculations by Boltzmann. They first examined the variability of the effect with the thickness of the metal plates, then with their shape, and explored in particular the dependency of the induced current on the magnetic field for some twenty-one different substances.

The two explanations generally offered at the time related the Hall effects either to the influence of the magnet on the electric current or to its influence directly on the conductor. Adopting the latter hypothesis, a number of scientists, such as John Hopkinson and A. Righi, suggested that the resistivity varies along different directions of the metal plate – or the resistance coefficients vary – according to the direction of the magnetic field. In this model, the magnetic field produces changes in the metal plate in

41. W. Nernst and A. v. Ettingshausen, "Über das Auftreten elektromotorischer Kräfte in Metallplatten, welche von einem Wärmestrome durchflossen werden und sich im magnetischen Felde befinden," *Wied. Ann.* 29 (1886): 343; W. Nernst, "Über die elektromotorischen Kräfte, welche durch den Magnetismus in von einem Wärmestrome durchflossenen Metallplatten geweckt wird," *Wied. Ann.* 31 (1887): 760; A. v. Ettingshausen and W. Nernst, "Über das Hall'sche Phänomen," *Ber. Wien. Akad.* (1887): 560–610; W. Nernst and A. v. Ettingshausen, "Über das thermische und galvanische Verhalten einiger Bismuth-Zinn-Legierungen im magnetischen Felde," *Wied. Ann.* 33 (1888): 474. Nernst expressed his devotion to his teacher by later dedicating his textbook, *Theoretische Chemie* (1893), to von Ettingshausen, a patient and humorous mentor.
42. Boltzmann, *Leben und Briefe*, p. II 90, fn. 439.
43. A. v. Ettingshausen and W. Nernst, *Wien. Ber.* 94 (1886): 560.

such a way that the conductor "takes on a peculiar structure" that leads to the rotation of the electric current lines. Another possibility under the same hypothesis, propounded by English scientist Shelford Bidwell, was that the Hall effect was produced by the unequal heating of individual parts of the conductor, due to deformations caused by heating (or Peltier) effects.

Nernst and von Ettingshausen found that their experiments confirmed the first hypothesis, because the Hall effect does not depend on the shape of the metal or the placement of the primary electrodes. In an addendum to their experimental paper, they presented a calculation of the electromotive forces (EMFs) that arise when the primary and the Hall electrodes are exchanged on a circular metal plate, rather than a rectangular, cross-shaped sample. This calculation, which Boltzmann performed for them after they submitted their experimental data to him, confirmed that the value of the EMF will not change due to the inversion of the primary and the Hall electrodes.[44]

This set of investigations was much more important than a mere verification of a previously discovered effect. Work on the Hall effect carried fundamental implications for the reformulation of Maxwellian field theory. It ultimately contributed to the emergence of a true microphysics, of a particulate electrodynamics. Hall and his mentor Henry Rowland, however, thought that, in accordance with Maxwellian theory, he had discovered a new force "no different in kind than any other electromotive force," in effect a new electric field that was generated by the interaction of the current with the magnetic force applied to the conductor (or, Rowland held, even to a dielectric).[45]

Ettingshausen and Nernst executed an impressive number of measurements with rectangular plates made of aluminum, antimony, lead, cadmium, cobalt, iron, gold, copper, magnesium, sodium, silver, nickel, palladium, platinum, steel, zinc, carbon, and tellurium, and they focused in particular on experimenting with a large number of bismuth plates, the ones which Hall had originally examined. Following Boltzmann's calculations of 1886, Ettingshausen and Nernst looked at three kinds of circular plates, and while their data gave ample indications that the effects recorded had something to do with the position of the electrodes on a circular sample, they dismissed these variations as being due to "inhomogeneities of the plate, small deviations in their thickness etc." On cutting transversally through the plates, they found that the slicing could be per-

44. A. v. Ettingshausen and W. Nernst, 1887, pp. 603–6.
45. Jed Z. Buchwald, *From Maxwell to Microphysics* (Chicago/London: University of Chicago Press, 1985), p. 89.

formed easily in some places, while it encountered resistance in other sections of the plates. The inspection of these slices under the magnifying glass also reinforced their view that differences in the structure of the conductor were "quite clearly distinguishable." Had they pursued the matter further, they might have arrived at completely different interpretations of the Hall effects. Even Boltzmann seemed somewhat disappointed, intimating that Ettingshausen was not yet capable of independent scientific work.[46]

They did, however, comment on the fact that the Hall effect was more pronounced when the Hall electrodes were placed on the periphery of the disk, rather than when they were in the interior. When the electrodes were placed on lines that were not perpendicular to the current, the Hall effect (the rotation of the field lines) was of "uneven magnitude," a phenomenon they thought "would be explained by changes in resistance." They also undertook a novel experiment. A number of rectangular bismuth plates of identical dimensions except for varying thickness were soldered together. The primary current was applied at the shorter ends of the entire resulting plate, whereas Hall electrodes were applied transversely on each individual section. They found that the Hall effect did not correlate well with the thickness of the individual pieces of bismuth, and although they noted that the effect was more pronounced for the inner two sections, they simply attributed the variability to the differences in the primary current within the various sections.

Ettingshausen and Nernst did not probe the anomalies they had observed. They were satisfied to confirm that the reversal of electrodes did not change the magnitude or the sign of the Hall effect, thereby invalidating Bidwell's theory. Boltzmann, too, did not pursue the matter any further. But he had drawn attention to the fact that his calculations of the Hall effect, as well as experimental data, pointed to inconsistencies. Boltzmann insisted that he considered the "material theory of electricity only as a picture which facilitates intuition, [a picture] whose consistent elaboration is useful at least as long as no other equally lucid theory of electricity is available."[47]

We can thus date quite precisely the time of Nernst's collaboration with

46. Letter of Boltzmann to A. Toepler, 14 February 1888, in Boltzmann, *Leben und Briefe*, pp. I 83 and II 120.
47. Ludwig Boltzmann, "Zur Theorie des von Hall entdeckten elektromagnetischen Phänomens," *Wien. Ber.* 94 (1886): 644–69. Also in Boltzmann, *Wissenschaftliche Abhandlungen*, vol. 3 (Leipzig: Barth, 1909), p. 190. His first note on the subject, in which he calculates the Hall effect for a circular plate, is "Notiz über das Hallsche Phänomen," *Wien. Anz.* 23 (8 April 1886): 77–80, in Boltzmann *Wissenschaftliche Abhandlungen*, vol. 3, pp. 182–6.

Boltzmann, since it is evident that the period of his most intense work on the extremely complicated Hall experiments falls into the spring and summer of 1886. On this occasion, Nernst probably studied the pertinent literature in great detail, and must have become well acquainted with Boltzmann's careful analyses and rigorous mathematical approach. Nernst's experimental paper on the Hall effect pays exquisite attention to the difficulties of obtaining a variety of metal samples and preparing them in a number of shapes and sizes. It also displays a fine awareness of possible factors influencing the instrumental setup, as well as much inventiveness in employing the most appropriate measuring devices. Ettingshausen and Nernst had, to all intents and purposes, examined all possibilities for the registering of the Hall effect: They had varied the placement of electrodes, the intensity of the current, the magnitude of the magnetic field, and the properties of samples. Their discovery of new thermoelectric phenomena related to the Hall effect established an early name recognition for Nernst and prepared him well for research on electrical conductivity, which was to dominate both his impending solution research in Leipzig and his later studies on the electrolytic lamp.

With this set of electrical papers, which provided the basis for future electrochemical work, Nernst proceeded to Würzburg in 1886. He there returned to his earlier research on electric conduction in liquids, and rapidly completed his formal university education under the watchful eye of Friedrich Kohlrausch.

3

The Early Researches

When Ostwald arrived as the new director of the Physical-Chemical Laboratory of Leipzig University in the fall of 1887, the chemical laboratory, which was divided into three sections, lacked an assistant for physical-chemical experimentation. In the other sections for pharmaceutical and analytical chemistry, Ostwald retained his predecessor's assistants. Reminded of Nernst, whom he had briefly met in Graz, and "of whose talent and knowledge Arrhenius had given a very good opinion," Ostwald recalled that he had invited him to Leipzig to fill the empty slot in physical chemistry.[1] Nernst, however, recounts that he came to Ostwald as Assistant for Physics and was only later introduced by his "boss into the domain of physical chemistry."[2] This discrepancy in their recollections is significant only to the extent that it indicates Ostwald's purposes at the time: He acquired an assistant in physics but evidently remembered him as a future physical chemist. It also fits Nernst's confession that he was as ill-prepared for physical chemistry as for painting portraits, and it illustrates the rather fluid state of affairs at a time when Ostwald was just beginning a challenging new professorial appointment. Since Nernst had no prior exposure to chemical instruments, he took advantage of the fact that only two students needed instruction, and gradually familiarized himself with Ostwald's new techniques and research.

Nernst's somewhat picaresque journey toward a position as a physical chemist is, properly speaking, neither surprising nor unusual for a person situated as he was. At age twenty-three, Nernst was an experimental physicist with minimal chemistry training. Since the German educational system in general did not yet produce particularly focused postgraduates, he embarked upon an academic career initially very similar to that enjoyed by the generation of his physicist teachers: He sought employment on the basis of specific research or instrumentation skills and general ability, rather than specialized education in a narrow field. Einstein also

1. Ostwald, *Lebenslinien,* vol. 2, p. 36.
2. Nernst Vita Göttingen, 14 April 1890, quoted above.

sought an assistantship with Ostwald more than a decade after Nernst, although his training in experimental physics and chemistry was less advanced, compared to his already substantial theoretical work. Arrhenius, who had not been able to obtain the highest distinction for his doctoral dissertation in 1884, feared that the only career open to him would be a teaching post in the Gymnasium. Einstein had at first to content himself with precisely such meager subsistence as temporary teaching assignments.

At the time, the options available to fresh graduates were the completion of a teaching certificate for employment in secondary schools and high schools or a technical-managerial position in industry, which was similar to an apprenticeship in banking or any other profession. Employment as a researcher in industrial chemical laboratories was not yet a fully established career choice in the mid-1880s. The first such laboratory had been formally instituted at the Badische Anilin- und Soda Fabrik (BASF) under the direction of Heinrich Caro in 1877, the same year that Germany enacted its first patent law. But even so, previous experience in academic research was a crucial requirement for employment in the expanding laboratories of both physics and chemistry, as well as related industries. Until the turn of the century, chemical dye-stuffs companies, for instance, preferred to recruit university professors, lecturers, or assistants for their research laboratories, rather than mere recent graduates.[3]

More specifically, the deep transformations, which also took place in the political, social, and cultural spheres of Wilhelminian Germany, were characterized by rapid specialization and splintering. In order to link his own identity to the support of science, Wilhelm II created a persona set apart from big business and the high bourgeoisie. Prodded by his powerful education minister, Friedrich Althoff, the Kaiser was largely responsible for the expansion of science outside the traditional venues of industry and university. His deep personal involvement in the creation of the Kaiser Wilhelm Institutes for research in the exact and natural sciences after the turn of the century destabilized the power relationships of the traditional triumvirate of state, industry, and academia, the three pillars of the scientific establishment.[4] While Wilhelm's motive may have been "the greater good," Emperor Friedrich III's widow Victoria is said to have remarked of her willful Willy: "Don't for a moment imagine that my son does anything from any motive but vanity."[5]

3. Ernst Homburg, "The Emergence of Research Laboratories in the Dyestufts Industry, 1870–1900," *BJHS* 25 (1992): 91–111.
4. Schiera, p. 273.
5. Quoted by Rocke, 1993, p. 363, from Koppel S. Pinson, *Modern Germany: Its History and Civilization,* 2nd ed. (New York: Macmillan, 1966), p. 278.

In Leipzig, Ostwald proposed to Nernst that he investigate the heats of formation of mercury salts, a project designed to verify the thermodynamic relationship, established several years earlier by Helmholtz, between the heats of formation and the electromotive forces measurable in galvanic cells. But the experimental data obtained through calorimetric methods by Julius Thomsen, the foremost representative of thermochemistry, did not correlate well with Helmholtz's theoretical predictions. The deviations were particularly marked in the case of mercury. Ostwald assigned Nernst a redetermination of Thomsen's data. Nernst's experiments, his first "chemical" work, confirmed Helmholtz's theory. His results, together with a rerun and verification by Thomsen himself, who agreed with Nernst's conclusions, were published in the second issue of the *Zeitschrift für physikalische Chemie.*[6] This was the first of Nernst's contributions to Ostwald's journal, in which he was published regularly during the following few years of happy collaboration and friendship with Ostwald. But after the stormy events of the 1895 Lübeck conference at which he sided with Boltzmann rather than Ostwald in the debate over the role of energetics in physics, Nernst abruptly disappeared from the journal's list of authors.

Many have argued that Ostwald's physical chemistry was mainly oriented toward a "general method of measuring affinities by introducing new physical methods." Ostwald's search for a variety of chemical and physical investigative tools, methods, and mathematical algorithms and his use of "different theories and models"[7] certainly describes well the practice of his assistant Nernst. The question therefore arises whether Nernst fits into the same mold of an affinity-motivated agenda. Nernst certainly came under his tutelage precisely at a time when Ostwald, in 1885, articulated an expansive project for his fellow chemists: "to erect in a relatively short time a comparative theory of affinity" through electrical conductivity studies well correlated with other methods and preexisting data.[8] But Ostwald's and Arrhenius's projects were not entirely successful, and Nernst, in many ways, contributed to the dismantling of his peers' efforts. The dependence between electrical conductivity and affinity, or between dilution and conductivity, proved to be tenuous and limited to specific cases. He could not have remained an enthusiastic "wet" electrochemist and physical chemist because affinities were not easily forthcoming. By

6. Ostwald, ibid., pp. 36–7. In fact, Ostwald had known of the apparent mistakes in Thomsen's work, and had assigned Nernst a quite trivial experiment, one which Ostwald had previously assigned to a Riga student before arriving in Leipzig. W. Nernst, "Bildungswärme der Quecksilberverbindungen," *Z. f. phys. Chem.* 2 (1888): 23.

7. Most recently among them Mi Gyung Kim in her excellent treatment of nineteenth-century chemistry. Kim, pp. 268ff.

8. Kim, p. 273.

1900, "the entire concept of affinity, for a century and a half regarded as a crucial problem in chemistry, seemed to have lost its magic; . . . it was no longer considered interesting to search for measures or models of chemical affinity."[9] After dealing Thomsen a forceful blow in 1887, Nernst remained bounded in his research by his physical-electrical-thermodynamical matrix and rarely spoke of affinity, then or in the future.

In the late 1880s and 1890s, Ostwald began what is perhaps the first comprehensive effort at justifying the historical and disciplinary integrity and independence of physical chemistry. His impressive history of *Electrochemistry*, published in 1896, was to be emulated in many texts by van't Hoff, by Nernst to a certain extent, and by their students.[10] Ostwald belongs to the group of leading representatives of German positivism engaged in producing empirically derived laws of broad applicability. In his myriad publications, Ostwald mapped out a grand strategy to formulate, if possible, natural laws for the development of scientific thought. Such universal systems were popular endeavors among nineteenth century thinkers, the more noted among them being Auguste Comte, Herbert Spencer, and Karl Marx. For Ostwald, the history of science parallels its logical development and provides, if not the quickest, at least "the most successful and attractive" mode of scientific education. As was characteristic of his wide- and deep-ranging interests, Ostwald aimed to render more than a coherent history of a specific discipline; in *Electrochemistry* he also wished to answer the broader question of whether "it be at all possible to establish general laws for history." He used the example of recent work in physical chemistry to illustrate how science develops, while simultaneously hoping to extract regularities for a theory of history. It was a dialectical experiment in this direction, and, by his own verdict, a successful one.[11]

Ostwald compared the growth of science to the process of crystallization of substances from a liquid solution, an *Auskristallisation*. Yet, only rarely will this product be "chemically" pure. It will contain numerous impurities and will continue to do so for many decades, due to the per-

---

9. Helge Kragh, "Julius Thomsen and Classical Thermochemistry," *BJHS* 17 (1984): 255–72.

10. W. Ostwald, *Elektrochemie. Ihre Geschichte und Lehre* (Leipzig: Verlag von Veit & Co., 1896), 1150 pp. Ostwald's book set standards for future similar histories of science, such as Partington's *History of Chemistry*. See G. S. Morrison, "Wilhelm Ostwald's 1896 History of Electrochemistry: Failure or Neglected Paragon," in *Selected Topics in the History of Electrochemistry*, ed. G. Dubpernell and J. H. Westbrook, Proceedings of the Electrochemical Society, vol. 78 (Princeton, N.J.: 1978), pp. 213–25.

11. Ostwald, *Elektrochemie*, pp. v–vii.

sistence of old, sometimes even mistaken, conceptions. Conscious, active interventions will bring about a vigorous reformulation and an unequivocal theoretical breakthrough.[12]

Ostwald had for a long time entertained hopes for a completely hypothesis-free chemistry, as he had postulated earlier in his *Überwindung des Wissenschaftlichen Materialismus* [*Transcending Scientific Materialism*] of 1895, and was to elaborate later in the *Prinzipien der Chemie* [*Principles of Chemistry*] in 1907. For Ostwald, the fundamental subject of scientific inquiry was the exploration of human sense perceptions, namely our knowledge of phenomena. Such knowledge he considered inherently relative, and the search for essences – for the true causes of phenomena – futile and misplaced. In line with Ernst Mach, Henri Poincaré, and other contemporary positivists, Ostwald deemed metaphysical preoccupations fruitless, since there was no way to correlate knowledge thus derived with observation. In contradistinction to early Comtian positivism, Ostwald and Mach postulated in an empiricist spirit a science "completely free of any arbitrary hypothesis," related to empiricism, which would allow for the existence of real objects, knowledge of which could be gained though perception.[13] His antiatomistic stance and his later preoccupation with the development of energetics as an overarching conceptual schema for all sciences, physical and otherwise, made Ostwald into a revered yet contested scientific and cultural figure of the German-speaking world, an apostle and a heretic in Europe, England, and America.

Despite the monism, or holistic world view for which he is well known, what is rarely spelled out is Ostwald's tolerance toward a multiplicity of explanatory frameworks: He emphatically denied, for example, that energetics would become the only or sufficient vehicle for an understanding of nature:

> Even though the energetic world view has immense advantages over the mechanistic or materialistic one, a few points that are not covered by the laws of energetics indicate that other principles exist that transcend energetics. *Energetics will remain side by side with these new laws . . . and will constitute a special case of more general conditions,* of whose future shape we hardly have any premonition at the present time.[14] [Emphasis added.]

12. Ostwald, ibid., p. 2.
13. W. Ostwald, *Überwindung des wissenschaftlichen Materialismus,* lecture held at the third general session of the Association of German Scientists and Physicians in Lübeck, 20 September 1895 (Leipzig: Verlag von Veit & Co., 1895), p. 25. See Weyer, p. 119.
14. Ostwald, *Überwindung,* p. 35.

Ostwald's history of electrochemistry constituted the first major attempt to create space for the new discipline of physical chemistry through historical exposition. Entitled *Electrochemistry: Its History and Theory,* this massive tome is a serious and still unsurpassed attempt at a total disciplinary history. Ostwald's main line of argument was that an understanding of the etiology of scientific problems constitutes the most effective means toward a sound scientific education. Moreover, insight into the vagaries of history, into the inevitable hostility encountered by new theories, justified primarily by the new theory's crass contradiction with established views, was invariably and paradoxically followed by the claim that the new ideas are in fact old, and that forerunners exist to most any innovation. Ostwald's critical attitude toward such "knee-jerk" hostility to new ideas was by no means uncommon. A similar critique was later leveled by Whitehead, who remarked that "Everything of importance has been said before by somebody who did not discover it" and that "there is a considerable difference between a vague awareness of a particular problem and systematic research into it."[15]

Ostwald had begun work on his *Electrochemistry* in 1893, but by the time of its publication in 1896, no one could avoid associating its rhetoric with the shattering experience that had been the Lübeck meeting of 1895, at which his new science and principle of energetics had been vehemently opposed by the foremost physicists of the time.[16] But it is to Ostwald's merit that the extremely favorable and detailed analysis of Nernst's work given in his history of *Electrochemistry* seems not to have been altered after Lübeck, where Nernst sided with Boltzmann and Planck in rejecting Ostwald and Helm.

The self-definition of German-speaking physical chemists at the time rested heavily upon the acceptance of the Ionist program. Its advocates aimed to persuade physicists and chemists that material entities called free ions exist in solutions, and that electrical and chemical phenomena can be accounted for only if such ions are assumed and proved to exist. Hence, a major part of Ostwald's book was devoted to the work of Alessandro Volta and the ensuing rivalry between those supporting a physical as against a chemical interpretation of Volta's theory of the galvanic cell – the so-called voltaic pile, or electrical battery. Volta thought that the production of an electric current between two or more metal plates immersed in a solution is solely the result of the differences in potential between the

15. Quoted by Peter Burke, "Introduction," pp. 1–21, in *The Social History of Language,* ed. Peter Burke and Roy Porter (Cambridge: Cambridge University Press, 1987), p. 17.
16. For discussions see Hiebert, "The Energetics Controversy"; Root-Bernstein, *The Ionists;* and Körber, vol. 1, pp. 79–83, 119–20; vol. 2, pp. 138–40, 252, 352.

two metals, without any conductive role of the solvent. His viewpoint was further explored by a number of physicists, such as Ohm; whereas another group, including Davy and Faraday, focused on the source of the electric current and found it in the chemical processes taking place in solution.[17]

While decrying the pernicious effect of Volta's contact theory on the subsequent development of the field, Ostwald explicitly denied that Volta's "inability" to perceive the correct role of the solvent should be criticized. Volta had interpreted "the phenomena such as they presented themselves to him in the context of his time . . . [and they] corresponded completely to the regular pace of scientific development." Ostwald reminded his readers that there had once been no reason to doubt the geocentric system solely on the basis of "appearance," and only the scientific examination of the problem, "the need to understand this phenomenon in the context of other related phenomena," produced a reversal in our conceptions. Similarly, only the necessity to *explain the processes* taking place between metals and solutions later produced misgivings about Volta's experiments. Volta was, therefore, not to be faulted, since he persisted in his contact theory primarily because of his prior commitment to a physicalist, as opposed to a vitalist, interpretation of electrical phenomena. Ostwald emphasized, however, that posterity has to be blamed for persevering in accepting Volta's explanations despite "sufficient and decisive" evidence to the contrary; the fault lies in vehemently defending, rather than thoroughly reexamining, an inadequate theoretical interpretation. The longevity of the dispute that lasted until the latter part of the century was, in Ostwald's opinion, due to the lack of a rigorous theoretical basis for chemistry. To Ostwald, this gap had only lately been filled by "an all-round satisfactory chemical theory of the Voltaic pile" provided precisely by the new physicochemical theories of the 1880s.[18]

Ostwald placed the debates over the validity of the ionic dissociation theory within a broad historical perspective: Its origins reached into a

---

17. In 1797, Volta put forth the theory that the electromotive forces generated in the battery are produced by the potential difference originated by the contact of dry metals only. In Volta's theory, the solution in which the battery electrodes were immersed (usually water) played only the role of an electric conductor. Opposition to Volta's contact theory and the proposal that chemical reactions are responsible for the electric current arose soon thereafter, and the split over the correct interpretation has persisted into the twentieth century. Thus, "on the side of Volta were Davy, Pfaff, Peclet, Marianini, Buff, Fechner, Zamboni, Matteuci, Kohlrausch, Pellat and [Lord] Kelvin," while in opposition stood "Fabbroni, Ritter, Wollaston, Parrot, Oersted, Ritchie, Pouillet, Schoenbein, Becquerel, De La Rive, Faraday, Nernst, Ostwald, and Lodge." Sidney Ross, "The Story of the Volta Potential," in *Selected Topics in the History of Electrochemistry,* ed. George Dubpernell and J. H. Westbrook (Princeton, N.J.: The Electrochemical Society, 1978), p. 260.

18. Ostwald, *Elektrochemie,* pp. 65–6.

significant past, although they could only with difficulty be associated with present conceptions. The history of electrochemistry contained two parallel chains of events; each pitted against the other opposing scientific worldviews – the physicalist against the chemical. And although the great themes of scientific thought were continuous and rarely reconcilable, only with the advent of the new physical chemistry could the old enmities be finally resolved. Ostwald explicitly claimed that physical chemistry was a natural descendant and the culmination of the maturation of electrochemistry.

This view of history was a common one at the time, but was anathema to Ostwald's nemesis, Ludwig Boltzmann. In an obituary of the famed mathematician and physicist Josef Stefan, Boltzmann wrote, more than half a century before Thomas S. Kuhn, that "the idea of continuous development [in theoretical physics] . . . is mistaken. The development of theoretical physics has always taken place in jumps. Often a theory underwent development for decades, indeed for more than a century, so that it offered a rather lucid picture of a certain class of phenomena. Then new phenomena became known that stood in conflict with this theory, and attempts were made in vain to reconcile it with these phenomena." Electricity was Boltzmann's paramount example of abrupt and discontinuous change. Boltzmann, the historical relativist, argued forcefully against the older historical explanatory models, which simply stated that the "older views" were false, a model which "implies that the new view had to be absolutely correct." Boltzmann stood on the side of contemporary views that a new theory is "a better view," reserving, however, the possibility that "the new theory can in turn be displaced by a still more suitable one." Boltzmann subscribed to an evolutionary historical model. Boltzmann's account, similar to Ostwald's schema, included the theme of struggle. He noted the frequent vilification of innovators by the old guard, a process not limited, according to him, to theoretical physics, but one which "seems to recur in the evolutionary history of all branches of human activity."[19]

In the final chapter on "The Theory of Electrolytic Dissociation," Ostwald did not distinguish between electrochemistry and the "new physical-chemical theory." He used the terms interchangeably. He concluded with the remarkable statement that, by virtue of the ionic theory of Arrhenius and van't Hoff, the future holds in store a "revolution [*Umwälzung*] comparable to that brought about by the steam engine."[20] The seeds of self-conscious historical writing, later picked up by students of the Ostwald

19. Ludwig Boltzmann, *Josef Stefan* (1895) and *On the Development of the Methods of Physics* (1899), quoted in Broda, pp. 108–9.
20. Ostwald, *Elektrochemie*, p. 1148.

school, were sown when Ostwald compared the work of a handful of scientists that "emerged from the rooms of the Leipzig physical-chemical institute" to the "events connected with the elaboration of the antiphlogistic theory by Lavoisier and his collaborators. . . . This research found solutions for old puzzles," he writes, "and annexed region after region to the empire of scientific electrochemistry."[21] Thus, almost a decade after the foundation of the *Zeitschrift für physikalische Chemie*, Ostwald was able to close the circle: In both cases, "[a] few men, at first opposed with scorn and anger, eventually made their way through the world and gave science a new face."[22] In 1887 he had written that "physical chemistry is not just a branch but it is the flower of the tree." To many readers of Ostwald, physical chemistry was many things at the same time: the rational theoretical foundation of chemistry, a branch of chemistry, a direct descendant of electrochemistry, the crowning achievement in the history of chemistry.

The atmosphere in the Leipzig laboratory was much influenced by new developments in Arrhenius's and van't Hoff's work. Many doubted at that time the ionic state of dissolved substances in solutions. It was difficult to visualize and accept that such substances would behave like a gas in solution, and that they would exert strong pressures. As Ostwald recalled it, Pfeffer's experimental proof of such pressure was not yet known to the Leipzig physical chemists in the late 1880s, although Pfeffer taught at the same university and eventually built an osmotic cell for the Ostwald group at their request. Nernst and Ostwald discussed the phenomenon of diffusion, the mechanics of osmotic pressure, and the behavior of dissolved molecules and ions at the surface of solutions. Nernst was apparently puzzled by the question of why these molecules, which exercise such a strong pressure on the inside of the solution surface, did not escape from the solution into the atmosphere above it. The physicist Josef Stefan had shown that the solute molecules do not escape the solution at the surface with high speeds, due to the very strong reversed attraction forces that are immediately formed when the freed molecules leave the solution.

Nernst argued that the addition of a pure solvent layer would neutralize these strong surface forces, and the molecules would "rush into the

21. W. Ostwald, *Electrochemistry: History and Theory*, trans. N. P. Date (Washington: Smithsonian Institution and the National Science Foundation, Amerind Publishing Co., 1980), vol. 2, p. 1123.
22. Ibid., pp. 1147–8. Ostwald's name could therefore be added to the list of scientists "who said explicitly that [their] contribution was revolutionary or revolution-making or part of a revolution" drawn up by I. B. Cohen, and which contains sixteen scientists from Robert Symmer to Benoit Mandelbrot. I. Bernard Cohen, *Revolution in Science* (Cambridge, Mass.: The Belknap Press, 1985), p. 46.

solvent layer" like the powerful escape of a gas into a vacuum. His problem was: Why did the diffusion process for dissolved molecules take days or even weeks, whereas for gases the process is swift? Ostwald maintained that the liquid – as opposed to a vacuum – presents a much higher friction resistance that significantly slows down the process. He recalled: "I left it at that; in Nernst's mind, however, the process formed into a visualizable picture, which he pursued analytically until he arrived at his discovery of the electromotive effect of ions. After a few months he presented his views in a paper, and their further development soon led to his theory of the galvanic cell." But Nernst had been a cautious and critical student; it took time, and persistence, and he did not embrace the analogy with gases "without putting up some internal resistance."[23]

He seems to have had enough misgivings about the physical interpretation of the phenomena of diffusion to visit Helmholtz, seeking advice, in January 1888. Helmholtz had assumed the presidency of the newly founded Physikalisch-Technische Reichsanstalt (PTR hereafter) in Berlin only three months earlier, and was reigning supreme, at age 68, as the "Imperial Chancellor of German Science." An excellent administrator, Helmholtz had brought young workers to Berlin and "made them feel that they were at the forefront of their field," encouraging a *Forschergeist,* a spirit of inquiry, and continuing his teaching duties at the university.[24] Nernst showed Helmholtz his treatment of electromotive forces in solution and explained "that the electrolytic 'solution pressure' [*Lösungstension*] is for me mostly a formal quantity, and that a possible physical interpretation is not in the making." Although we do not know precisely what Helmholtz had to say, Nernst returned encouraged and overwhelmed by the meeting with the illustrious scientist. The 24-year-old Nernst assured Helmholtz that their conversation would remain unforgettable in his memory.[25]

Among the young apprentices in Ostwald's budding laboratory in 1887, there was reluctance to engage in physical research. Ostwald acknowledged that very few German students came to study with him. The ideas proposed by the "ionists" were exotic and, to a great extent, damaging from a professional point of view. He recalled that some foreign students, among them the Scot James Walker and the American A. A. Noyes, had come to study during that early period, but German students were

23. Ostwald, *Lebenslinien,* vol. pp. 38–9.
24. Cahan, *An Institute for an Empire,* p. 71.
25. Nernst to Helmholtz, 15 January 1888, Nachlass Helmholtz Signatur 326, AAW. The meeting took place on 13 January 1888, since Nernst refers to "unsere vorgestrige Unterredung," – "our conversation of the day before yesterday."

mostly those for whom other laboratory spaces were unavailable, and who spent only a brief time in the physical-chemical section.[26]

Nernst's paper on galvanic cells, published in 1889, one year after its embryonic formulation, represented a significant transition from investigation of the electromagnetic properties of solids to solutions and solution phenomena. For the first time it involved "wet" chemical research. Before that, Nernst spent the summer in Heidelberg to recover from an acute, and unexplained, sore throat, and returned afterward for a second year as Ostwald's assistant. Nernst, whose ailments may have been connected to his father's death, was forbidden to enjoy his most beloved drink, beer, and any sensitizing foods, a diet which greatly displeased him. Nernst was apparently well known among his friends for his alcoholic excesses, as well as his fondness for romantic adventures. When these were restricted for health reasons, he apparently pursued his scientific research with renewed vigor, an enthusiasm for which Arrhenius praised him: "The little Nernst . . . flourishes in his scientific progress, in which he uncovers with wonderful audacity one secret after another."[27]

Arrhenius enjoyed reporting Nernst's excesses. One month before Nernst's departure for Heidelberg, after he had spent the night banging on Arrhenius's door, Arrhenius fabricated a story according to which during that night Nernst narrowly escaped a police arrest. Nernst was immensely upset. It was only a month later that Arrhenius and his friends disclosed the hoax to Nernst, who had ambivalent feelings of relief and anger.[28]

During September 1889, Arrhenius spent a week as Nernst's guest in Heidelberg, where physical chemistry celebrated a "real triumph," in that the young men elicited the attention and favorable appreciation of senior colleagues. Yet at the time, Nernst was apparently quite dissatisfied with Ostwald, who had omitted to mention Nernst's work on diffusion in his recently published *Grundriss der Allgemeinen Chemie*.[29] But he continued work, and during 1888 and 1889, Nernst published the results of a joint investigation with Morris Loeb on the kinetics of solution processes.[30] In the first of these papers, Nernst exhibited what was to become a characteristic style: the use of analogies from related domains in

26. Ostwald, *Lebenslinien*, vol. 2, pp. 50–1.
27. Arrhenius to Tamman, Leipzig, 31 January 1889, 2 pages, Arrhenius Papers, KVA.
28. Arrhenius to Tammann, 16 April 1889, 2 pages, Arrhenius Papers, KVA.
29. Arrhenius to Tammann, 8 October 1889, 2 pages, Arrhenius Papers, KVA.
30. W. Nernst, "Zur Kinetik der in Lösung befindlichen Körper. I. Theorie der Diffusion," *Z. physik. Chem.* 2 (1888): 613–37; "Zur Kinetik der in Lösung befindlichen Körper. II. Überführungszahlen und Leitvermögen einiger Silbersalze," *Z. physik. Chem.* 2 (1888): 948–63; reprinted as "Rates of transference and the conducting power of certain silver salts," *Am. Chem. J.* 11 (1888): 106.

order to clarify processes for which he had not yet formed firm and intuitive notions.

Cross-domain analogies were not necessarily a novelty. The voltaic cell had served as a mnemonic device in the late 1840s, when DuBois Reymond compared the action of a galvanometer immersed in an electrolytic solution to the activity of muscle current in physiology.[31] The long theoretical paper of 1888, however, was not devoid of concrete descriptions of the electrolytic dissociation processes, despite Nernst's nervous caution during his meeting with Helmholtz half a year earlier; it started from van't Hoff's recent research on osmotic pressure, in which he showed the similarity between the behavior of gases and that of the molecules of a liquid solution exercising an osmotic pressure. Nernst's main purpose was to develop a theory of the diffusion processes in solution and to evaluate numerically precisely those forces of resistance of which he had talked to Ostwald during 1887. These forces would then explain the relatively slow rates of diffusion in liquid solutions, as opposed to the almost instantaneous expansion of gases.[32]

Nernst concluded that a system of five different forces was acting on the ions of a solution: electrostatic forces, forces due to osmotic pressure, gravitational and centrifugal forces, para- and diamagnetic forces, and electromagnetic forces. In the conclusion to this remarkable first paper of 1888, he predicted that if the calculations and experimental results were to prove the existence of such a system of forces, then an electrolyte in a magnetic field, traversed by a current, would "have to exhibit phenomena . . . which are strikingly analogous to the Hall effect, and to the effect discovered by von Ettingshausen and by me."[33]

It is significant to note here that Nernst seems to have been thoroughly immersed in the discussions on the various forms of electrical phenomena of the day, both by virtue of his previous work on the Hall effect, and by his invoking a broad spectrum of electromagnetic phenomena to account for electromotive forces in solutions.

Nernst's 1889 paper was published in response to a challenge. August Kundt, the Berlin physicist, had de facto forbidden discussions of the ionic dissociation hypothesis among his students, even threatening to fail them in their examinations if any mention of ions were made. Ostwald and Nernst undertook to visit Kundt, who challenged them to cite a particular example, an experimental proof, of the existence of excess positive ions

31. Timothy Lenoir, "Models and Instruments in the Development of Electrophysiology, 1840–1912," *HSPS* 17: 1 (1986): 1–54.
32. Nernst, *Z. f. physik. Chem.* 2 (1888): 614.
33. Ibid., pp. 636–7.

in an acidic solution. Ostwald and Nernst apparently devised a simple experiment that same night on their return train ride to Leipzig and were able to wire the successful results to Kundt the very next morning. Kundt, however, remained skeptical.[34] In their paper, Ostwald and Nernst stated that it was their "duty to remove any possible reservation by experimental evidence, since doubts [on the existence of free ions in solutions] have been expressed by an authoritative" personality.[35] This kind of experimental creativity, which he always seemed to exhibit when challenging theoretical problems arose, was characteristic of Nernst's work. Yet he never became a pure empiricist. Instead, he kept theory and theoretical speculations in the foreground of his activities, and most of his experimental and instrumental innovations arose from theoretical elaborations of practical problems, where practice meant solving the questions at hand.

Nernst's next project was the problem of electromotive forces, on which he wrote a long paper published in May 1889 that convincingly exhibited his research talents. Nernst proved that free ions in solution will generate electrostatic charges, thus producing a galvanic current. An electric current in a solution will be created only if ions are in movement, and the direction and speed of such movement must be accounted for by postulating the existence of certain forces. Since such a state of movement leads to the accumulation of free electricity, Nernst reasoned that additional forces would have to be postulated to account for the relative displacement of the ions according to the electrostatic laws. "These additional forces," he wrote, "are the ones which, produced through adequate arrangement of electrostatic charges, lead to the production of a potential gradient within the solution." Nernst developed a simple formula, which necessitated only the knowledge of the ion mobility in a voltaic chain for the calculation of this potential, and demonstrated how various physical phenomena, such as the electromotive forces, the transport numbers, and the gas laws, are connected. These were "magnitudes between which one has never before suspected a connection."[36]

Whereas Ostwald and Arrhenius were primarily concerned with electrolytic phenomena, Nernst's research on the electromotive forces of galvanic cells was located within a distinct, primarily physical, tradition. The operation of galvanic cells relied on the existence of such newly understood

34. Ostwald, *Lebenslinien*, vol. 2, pp. 121–2.
35. Wilhelm Ostwald and Walther Nernst, "Über freie Ionen," *Z. f. physik. Chem.* 3 (1889): 120. They proved that hydrogen is liberated at a mercury electrode immersed in a sulfuric acid solution without introducing a second electrode of the electric circuit into the solution.
36. W. Nernst, "Die elektromotorische Wirksamkeit der Ionen," *Z. f. physik. Chem.* 4 (1889): 145.

forces, and it was to Nernst's credit that he not only acknowledged but also emphasized the deep connections between the pursuits of the ionists and those of previous electrical researchers. He indicated that the hypothesis established by him about the electromotive force of ions is related to the magnitude of the Hall phenomenon in electrolytes or of the electrostatic charges appearing in a rotating electrolyte, subjects which he had examined in his doctoral research. The problems which he faced were indeed of a different nature from those with which many of his other contemporaries were concerned: How to account for electromagnetic waves and electric currents in a vacuum or a wire were the topics of utmost interest to physicists at the time.

Nernst was concerned with fundamental questions; he stressed that no one had yet seriously investigated the underlying causes of the "mechanics of the formation of electric currents in a galvanic cell, and if they have done so, it was only in the most general outlines or by accepting totally unproven assumptions." Helmholtz had succeeded in calculating for a few cases the direction and magnitude of these forces from experimental data, first in 1847 and then again three decades later. For currents produced by galvanic elements immersed in solutions, Helmholtz used the Second Law of Thermodynamics. He derived the electromotive force of induction from the electrodynamic interaction between conductors and magnets, one of the brilliant results of the principle of the conservation of energy (that is, the First Law of Thermodynamics). Helmholtz had shown that in the thermodynamical treatment of energy transformations, a distinction should be made between the changes of total energy U and those of the "free energy" A. The latter could be transformed with the help of adequate apparatus into other energy forms, such as electrical energy. His treatment was rigorously thermodynamical. It was a great achievement, but applied only to few cases. And, as Nernst pointed out, "as is inherent in such a successful treatment of physical problems, the mechanics of the investigated phenomenon were not analyzed at all. Even though the intimate relationship between electromotive forces and vapor tensions of the solution at the [electrodes] . . . was securely proved, one still lacks an explanation for the manner in which the current comes into being in such a cell or where the seat of the electromotive forces should be sought."[37] Thus, while Nernst was obviously well acquainted with Helmholtz's work, he recognized and responded to its deficiencies. In later years, Arrhenius attacked Nernst for an apparent lack of originality and in an article of 1901, for reliance on Helmholtz in his work on EMF, an incident of many consequences, to be discussed further below.

37. W. Nernst, ibid., pp. 129–30.

Essentially, Nernst applied the gas laws and van't Hoff's newly developed theories of osmotic pressure to liquids, searching for the fundamental explanation of electrical phenomena in solutions. He focused not only on obtaining valid means for calculating the effect of his newly postulated forces but also on providing a more tangible physical picture of their action. Making an analogy with the mechanism of diffusion through semipermeable membranes, where osmotic pressure drives the processes of transfer through the membrane, Nernst reasoned that a "solvation pressure" (*Lösungstension*) will arise at the contact surface between metals and the metallic salt solution. This pressure differential will transport metal ions in both directions, and will cause a positive or negative accumulation of charges on the metal plate.

He continued this mode of reasoning in his next paper, "On the reciprocal influence of the solubility of salts," published in 1889 while he was spending a semester at the University of Heidelberg. In it he argued, with the aid of thought experiments, that the maximal amounts of work obtainable by dissolving one gram molecule of any given substance in a saturated solution at a given temperature are independent of the nature of the solute and the solvent. These work quantities are equal to those obtainable in the case in which the solvent is completely absent, and the solute simply transforms into a saturated gas. This reasoning allowed Nernst to refute the existence of attraction forces between solute and solvent for the case in which the solute has the "same molecular constitution" in various solvents and in the gaseous state. This condition was fulfilled in most cases, he argued, since Raoult had shown that molecular weights calculated both from vapor density and from freezing point measurements do coincide in general. The heats of dissolution are indeed small quantities. Nernst claimed that the gas laws apply at the level of individual molecules in solution, when one gram molecule of a relatively nonsoluble substance is transported into a saturated solution. His argument is somewhat obscured by his vague definition of "molecular" level, a term which he apparently employed casually, referring solely to the undissociated or nonhydrated state, or the lack of any polymerization. But he did not take molecular structure, as understood today, into account.[38]

The search for a *physical interpretation* – in the sense of tangible material processes, particles, and forces – of the phenomena of electrical transport, available energy and work, and the nature of molecular processes in solutions is the most salient feature of Nernst's endeavors up to this point. Had his results been a mere confirmation of Helmholtz's

---

38. W. Nernst, "Über gegenseitige Beinflussung der Löslichkeit von Salzen," *Z. f. physik. Chem.* 4 (1889): 374.

calculation of the electromotive forces by yet another theoretical route, they would not have aroused the interest that his work seemed to cause not only in the ionists' circle but also among physicists and even Helmholtz himself. Nernst's paper is one convincing example of the involvement of physicists with the newly opened field of what was then a modern physical chemistry, quite distinct from thermochemistry, but closely related to electrical research and thoroughly infused with thermodynamic conceptions.

These early papers constituted a solid foundation for a successful career. All biographical essays, newspaper items, honorary degrees, and prizes single out Nernst's work on the electromotive force carried out at age 25 as one of his most memorable achievements, second only to the – often controversial – heat theorem that he enunciated in late 1905.

In the summer of 1889, after his apprenticeship with Ostwald, Nernst spent a semester as an assistant in the chemical institute of the University of Heidelberg under the directorship of Professor Brühl. The laboratory had become famous through Robert Bunsen, who in the 1850s had developed a number of important novel instruments devoted to what was then already known as physical-chemical analytic work. He had succeeded in producing rare metals through electrolytic separation methods and had constructed the first rudimentary spectroscope – the Bunsen burner – in collaboration with Gustav Kirchhoff. Nernst was immensely pleased to become acquainted with Bunsen's apparatus and laboratory arrangements, which he hoped to duplicate with Ostwald's approval on his return to Leipzig.[39] He was yet to know how closely his researches would eventually bring him to Bunsen's sphere of interests.

Moreover, Nernst was fortunate to meet "old" Bunsen himself, whom he found by then "tired of chemistry after many years of activity." In Heidelberg, Nernst also met A. F. Horstmann, H. Kopp, and V. von Richter, all of whom provided a receptive audience for the "Ionist" ideas Nernst propounded.[40] Hermann Kopp also belonged to an older generation of chemists active in some form of physical-chemical research, primarily devising methods for the physical and chemical characterization of substances.[41] During those spring months, Nernst prepared for the "Habilitation" – the teaching licensure for prospective university professors – at the University of Leipzig. In May 1889, he was already being proposed for a teaching position in Göttingen. He thanked Ostwald in a letter for

39. It is unclear whether Nernst specifically referred to the spectroscopic apparatus, but Ostwald indeed later performed numerous spectroscopic investigations, "photographing ions" as he called it.
40. Nernst to Ostwald, Heidelberg, 24 April 1889; Nernst to Ostwald, Heidelberg, 25 April 1889. Ostwald Nachlass, AAW, Berlin, Signatur 2129.
41. Servos, p. 11.

this "good news." He thought, however, that he would have preferred Arrhenius or Gustav Tammann to be in line for the job, rather than that his own future be changed so unexpectedly. With uncharacteristic benevolence Nernst exclaimed: "*Unter Kameraden ist es ganz egal, wer die Braut kriegt!*" ("Among friends it is totally unimportant who gets the bride").[42]

Nernst submitted his paper on the electromotive force of ions as his *Habilitationsschrift*, but, as if belying the apparent brightness of his professional prospects, reservations were voiced by Carl Neumann and Gustav Wiedemann. By 1 June 1889, the faculty agreed unanimously to admit Nernst to the teaching corps. At his colloquium on 1 July Nernst dealt successfully with his examinations by Wiedemann in physics and by Ostwald in "experimental and physical chemistry." The third part, given by Neumann in analytical mechanics and the "foundations of the mathematical theory of electricity," did not pass smoothly despite his excellent performance in the oral examination. Nernst was asked to "present the mathematical part of his dissertation (pages 13 to 16) to Professor Neumann for another revision before publication." For the final part of the Habilitation process, the *Probevorlesung*, or test lecture, scheduled for 23 October 1889, Nernst proposed three subjects: "On the recent developments of the dissociation theory," "On molecular weights," and "Applications of the mechanical heat theory to electrical phenomena." The committee chose the second topic, an indication that a classical-chemical subject, less controversial at least in its title and with no partisan references to either physical-chemical theories or thermodynamics, was still preferable.[43]

Nernst's early papers on electrical conductivity in solutions came at a time of major changes in the conceptions of German and British scientists on the nature of electromagnetism, optics, and heat. At the same time that Nernst published an extension of Helmholtz's work on the EMF in galvanic cells during 1888 and 1889, Helmholtz's brilliant student Heinrich Hertz produced for the first time electromagnetic waves in the laboratory, a discovery that brought momentous reformulations to Maxwellian electromagnetic theory. Nernst's conception of the transport of electricity in solution was, without his always saying so, a combination of Maxwellian notions of accumulation of charges on a conductor with earlier notions of electrostatic transport, coupled with an "atomic," Weberian impulse to seek for the fundamental unit of electric transport.

---

42. Nernst to Ostwald, Heidelberg, 20 May 1889.
43. Nernst to the Dean of the Philosophical Faculty, 1 July 1889, with signatures of the Dean, Profs. Maurenbrecher, Wiedemann, Neumann, and Ostwald. UAL-PA 773, Bl.6–12.

4

The Göttingen Years

At Easter 1890, the theoretical mathematical physicist Eduard Riecke, then well known, asked Nernst to be his assistant in Göttingen. Riecke (1845–1915), who had been directing the experimental physics section since 1881, was soon to publish an important paper on electric conduction. He was at the time teaching experimental physics four hours a week, as well as directing advanced laboratory exercises for as many as forty-eight hours a week. The *Privatdozent* Paul Drude, who had also studied in Leipzig and who was to publish his path breaking theoretical work on electronic conduction theory in 1900, was teaching electricity and electromagnetism for two hours a week. In 1891, Göttingen ranked fourth in the total budget of German physics institutes, after Tübingen, Strassbourg, and Königsberg. It was one among only nine of the thirty German universities and technological institutes with a separate building for physics instruction, and it expended the third-largest budget per attending student.

Riecke apparently promised Nernst a separate research department within the Physical Institute of the Göttingen University – a promise that was fulfilled in 1896, sooner than many had expected. During the spring semester, Nernst taught a colloquium on physical-mathematical methods in chemistry entitled "On the application of mathematics to chemical problems." Attended by fourteen students, it was largely a repetition of a similar course he had given the previous semester while still in Leipzig. His main effort during this first semester at Göttingen was invested in a seminar on an "Introduction to the recent theory of solutions," attended by only eight auditors. Among them, however, were two full professors, Eduard Riecke and Elias Müller, and his physicist colleague, Paul Drude. This was, in Nernst's words, "*ein Apostelkolleg*" – "an apostolic seminar." As a messenger of the new creed, he carefully tailored the course in order to minimize opposition and elicit enthusiasm from the distinguished audience. He intended to include a wealth of experimental material that could be safely subsumed under the course title, and was confident that his dedication to the new theories advanced by physical chemists would result in a "conversion" of his audience.[1]

Nernst also performed osmotic experiments with membranes whose efficiency increased with the lowering of the solution pressure. The results of these experiments led him to try a thermodynamical treatment of the solubility experiments, an endeavor he expected to be "no easy task." He also carried out some experiments of molecular-weight determinations in ether for a student of Göttingen's chemistry professor, Otto Wallach, and was proudly hopeful that these would contribute as well to Wallach's "conversion" (*"Bekehrung"*) to the ionic theory. During that year, Nernst was asked to write a chapter for a new textbook, Dammer's *Handwörterbuch der anorganischen Chemie,* an enterprise similar to the celebrated organic chemical compendium *Beilstein.*[2]

During the following year (1891), Nernst published on diffusion, osmotic pressure, and molecular weight determinations, and established a sufficiently serious reputation to be proposed in March 1891 for the position of associate professor (*Extraordinarius*) of physical chemistry at Giessen. Moreover, he had been proposed to fill a possible new professorship of physical chemistry at Munich. Ludwig Boltzmann was one of Nernst's most influential sponsors. During negotiations regarding his own possible move to Munich, Boltzmann wrote repeatedly to Friedrich Althoff, the influential undersecretary in the Prussian ministry of education, about Nernst's outstanding scientific abilities, as well as the importance of the new field of physical chemistry. He described Nernst as "by far the most original and creative younger researcher in the domain of mathematical-physical chemistry." Althoff, however, was already well informed. He was convinced of Nernst's qualities and promise, gathering information about him through his "lecture-spies."[3] Because of the momentum gathered by further interest expressed by the Berlin University as well, Nernst was retained in September on the faculty at Göttingen.[4]

While associate professor in the Physics Institute at Göttingen, he established in 1891 a small section for physical-chemical research. In 1894, a plan for an independent institute was drafted by the senate and faculty in conjunction with the Prussian government.

In 1893, Ostwald had founded the Deutsche Elektrochemische Gesellschaft (German Electrochemical Society). It became the first organization

1. Nernst to Ostwald, Göttingen, 17 May 1890, Ostwald Papers, AAW Berlin.
2. Nernst to Ostwald, Göttingen, 6 June 1890, Ostwald Papers, AAW Berlin.
3. Boltzmann, *Leben und Briefe*, Boltzmann to Althoff, 15 August 1891, p. II 163.
4. Nernst to Ostwald, Göttingen, 1 March 1891 and 15 September 1891, Ostwald Papers, AAW Berlin.

of physical chemists in the world, constituted largely of German-speaking scientists, and van't Hoff became its first president. The society aimed at more than the discussion of academic scientific progress in electrochemistry: It had as a chief goal the collaboration of scientists and representatives of industry. Within a decade, the society had six hundred members, representing all the important chemical industries of Germany and other countries. In 1900, Ostwald, together with van't Hoff, Nernst, and Georg Bredig, engineered a change in the name of the Deutsche Elektrochemische Gesellschaft to Deutsche Bunsengesellschaft für Angewandte und Physikalische Chemie (German Bunsen Society for Applied and Physical Chemistry,) primarily in order to avoid conflicts over publications for which both the *Zeitschrift für Elektrochemie* and the *Zeitschrift für physikalische Chemie* were vying.[5] They hoped to attract papers in physical chemistry that would otherwise have been submitted to Ostwald's journal.

During the Göttingen years, Nernst's work was devoted mostly to the theory of solutions and the kinetics of electrolytic processes.[6] Teaching and research in Nernst's institute began in the spring of 1895. One of this institute's special features was that it provided no instruction in introductory physics or chemistry. Since only advanced students were accepted for doctoral degrees, there was a heavy emphasis on independent research in physical chemistry, which made Nernst's institute truly one of a kind.

Outside Germany, van't Hoff's program in Amsterdam remained through the years mainly a purely chemical one. Svante Arrhenius performed independent research at the Stockholm Hogskola and, after 1905, in the Nobel Institute for Physical Chemistry at the Swedish Academy of Sciences in Stockholm, with rather limited resources. Yet an "Arrhenius" school or laboratory never developed. On many occasions Arrhenius expressed his gratitude to Ostwald for sending him students, gestures which appar-

---

5. See J. H. van't Hoff–W. Ostwald correspondence for 1900, in H-G. Körber, ed., *Aus dem wissenschaftlichen Briefwechsel Wilhelm Ostwalds* (Berlin: Akademie Verlag, 1969), vol. 2, pp. 294–7. By 1913, the Bunsengesellschaft had a membership of twenty European physical chemistry institutes and laboratories and another ten electrochemical research institutions.

6. See in particular E. N. Hiebert, "Nernst and Electrochemistry," in *Selected Topics in the History of Electrochemistry,* ed. G. Dubpernell and J. H. Westerbrook (Princeton, N.J.: The Electrochemical Society, 1978), pp. 180–200, and Kurt Schwabe, "Die Bedeutung von Walter Nernst für die Elektrochemie," a lecture on the occasion of Nernst's 100th birthday, Humboldt University, Berlin, and Chemische Gesellschaft der DDR, in collaboration with the German Academy of Sciences. Other biographical material: Paul Günther, "Zu Walther Nernsts 75. Geburtstag," *Z. f. Elektrochem.* 45 (1939): 433–5; M. Bodenstein, "Gedächtnisrede auf Walther Nernst von Hrn. Bodenstein," *Jahrbuch d. Preuss. Akad. d. Wiss.* 1942 (1943): 140–2; Richard Lepsius, "Zur hundersten Wiederkehr des Geburtstages von Walther Nernst," *Chemiker Ztg.* 88 (1964): 603–6.

ently enhanced his standing with the university officials in Sweden. In 1892, shortly after being appointed associate professor, Arrhenius thanked Ostwald for sending him Dr. J. Shields: "I am very much obliged to you [for sending him over] since the presence of a foreigner brings much fresh air into the laboratory and in practical matters he will help me very much." Arrhenius had only one assistant at the time, and in December, he wrote again to Ostwald that he would be happy to receive more visiting students for the summer session (*Ferien-cursus*), "since there is nothing here [at the Hogskola] that impresses as much as foreigners."[7]

In the early 1890s, the appointments of Nernst, E. Beckmann, and G. Tammann to academic positions in physical chemistry, together with Arrhenius's slow rise in Sweden and van't Hoff's growing prestige in Europe, enhanced the visibility of the new physical chemists. The ever increasing number and volume of articles published in the *Zeitschrift für physikalische Chemie* led to an increasing dissemination of physical-chemical views among both physicists and chemists. It would, however, be erroneous to conclude that by the end of the century, physical chemistry had become a homogeneous body of knowledge or a well-defined community of practitioners. Among themselves, these pioneers differed quite substantially in their aims and methodologies, and in the primacy which they accorded to certain fundamental theoretical views. Ostwald's rejection of atomism, his chemical descriptionism, and his peculiar brand of energetics, which viewed all natural as well as many social phenomena as capable of being subsumed under a "science of energy," confused both his contemporaries and historians. These differences led to a split between the chemically inclined Ionists – such as Arrhenius, van't Hoff, and Ostwald – and the physicalists – Planck, O. Lodge, and Nernst.

In March 1890, Arrhenius traveled to Graz and worked throughout the summer in Boltzmann's "great Physical Institute" on gas conduction and related experiments "on salt vapors in the Bunsen flame," essentially spectroscopic examinations of vapor composition. As had been suggested by van't Hoff's analogy between the gaseous state and that of dilute solutions, Arrhenius expected gases also to be electrolytically dissociated, and hoped that his work with Boltzmann would provide decisive proof in this direction. After returning to Uppsala in August, Arrhenius busily prepared a paper for the *Annalen der Physik und Chemie*, Germany's premier scientific journal, and planned to examine diffusion processes in Hammersten's institute in order to "obtain new insights on the kinetic nature of osmotic pressure."

Writing to van't Hoff in late October, Arrhenius reported his impressions

7. See S. Arrhenius–W. Ostwald correspondence, in Körber, vol. 2, passim and pp. 107, 115.

gathered during the year. He thought that German scientists had come to "generally agree that the new conceptions on the nature of solutions have led to the most beautiful results." The analogy between gases and solutions, as well as the analogy between gases and electrolytic dissociation, had been accepted. German scientists, however, did not as yet "want to make the final step and admit the kinetic nature of osmotic pressure." It was the "great authority" of Helmholtz behind whom German scientists could "hide." Despite not yet having published anything on this controversial subject, Helmholtz had nonetheless "let a few words slip on behalf of chemical attraction." Because Helmholtz's standing was so high as well as influential in creating a climate favorable to the reception of physical-chemical ideas, these "few words" led, in Arrhenius's view, to inaction on the part of German scientists: "They say: Helmholtz thinks the thing over, that means it will soon all become clear and we don't need to worry about the theoretical elaboration, which has probably been already beautifully worked out."

Arrhenius was vexed that Planck had recently published theoretical articles on this subject (to be further analyzed below), and that Ostwald seemed to be "very conciliatory in this respect" by allowing a physicist to dominate the discourse in physical chemistry. Arrhenius's letter to van't Hoff provides a fascinating entree into any attempt to clarify the process by which physical-chemical views and allegiances shifted ground around the turn of the century. With much verve and humor, Arrhenius outlined what amounts to a European strategy of conquest:

> *Now it is only you, Dr. Nernst and me who are kineticists.* What a pity that there are no bets in scientific trotting races. Because the Germans will wager as much as they can on Helmholtz and we can thus make a nice profit. In Germany they have resigned themselves to a . . . fate in which the newer ideas will soon come to power, after they have put up some frictional resistance in the beginning. In Austria it is naturally almost the same as in Germany. The interest is however not so great. The Italians . . . will undoubtedly also go with the Germans. . . . The Russians are in general very sympathetic towards the new direction, it corresponds so much to the Russian imagination. Only the great Mendeleev strongly holds back and the majority dare not reply to the great master and patriot. . . . These are in a state of hypnotism in which they have to obey the hypnotist without resistance. And the analogy between gases and solutions . . . lies deep in their sensibility. In France Herr Bouty[8] has made so many discover-

8. Edmond Bouty, in the early 1880s professor of physics at the Lycée St. Louis in Paris, did research on electromotive force measurements in dilute solutions, and published conductivity data for very dilute solutions of salts, acids, and bases in 1884, the year of

ies on the conduction of gases and their importance for chemical problems that others are darkened by them. The first symptoms of a change in this regard have only recently come to light, but the new views do not yet have a foundation of their own. . . . [In Sweden] they are gradually preparing themselves for the arrival of these new ideas. . . .[9]

When Nernst visited Holland in 1890, van't Hoff judged him "an exceptional man."[10] But shortly after moving to Berlin in early April 1896 to become a permanent member of the Prussian Academy of Sciences, van't Hoff expressed his misgivings about the disputes in Lübeck. He subsequently reported to Ostwald:

> When I meet Nernst in person I get a very pleasant impression; but when I hear this and that about his actions, I wish he were more conscious of how much he owes you, or that he would express this consciousness better; any representative of our discipline must feel this in regard to you, but Nernst in particular, since he profited not only from your writings and organizational activities but also from personal acquaintance with you. I must try to impress this upon him again.[11]

Nernst was evidently less respectful than van't Hoff had hoped, or Ostwald expected. Their explicit differences regarding the nature and role of physical chemistry as a profession and scientific discipline surfaced later that year. When Nernst's physical-chemical institute opened in Göttingen, his inaugural speech made clear that he was disengaging himself from Ostwald's views: Physical chemistry was a necessary and welcome bridge between physics and chemistry, not a "unificationist" theory but a method, an approach to scientific inquiry that would profit from both disciplines and train students in the best of both traditions.

If historians try to map the allegiances during the last decades of the century, they must analyze them in terms of very specific problems and the personal agendas, backgrounds, and scientific expertise and knowledge of each individual. Grouping men together is a more subtle and difficult task than appearances suggest. The legacies of Mach and Boltzmann were transmitted in fragments to the next generation. Planck, Ostwald, and Einstein were each influenced by Mach. Mach, however, disagreed

Arrhenius's inaugural dissertation. Bouty also worked on reversible heat effects in voltaic cells. See Partington, vol. 4, pp. 670, 676, 696.

9. Arrhenius to van't Hoff, 28 October 1890, Boerhaave Mus. Archief 200f.
10. Van't Hoff to Ostwald, Amsterdam, 17 October 1890, in Körber, vol. 2, pp. 227–8.
11. Van't Hoff to Ostwald, Berlin, around 8 May 1896, in Körber, vol. 2, pp. 255–6.

with Ostwald's energeticism. Until the mid-1890s, Planck's thermodynamical treatment of thermochemical and chemical problems was applauded by Ostwald, but Planck became increasingly critical of Ostwald, from whom he dissociated himself. Arrhenius therefore saw himself, together with van't Hoff and Nernst, as the last "kineticists," who did not succumb to the pure thermodynamicism and early antiatomism of Planck. Arrhenius and van't Hoff have even been hailed as the paramount example of unifiers of thermodynamics and the kinetic theory of gases. In the long run, however, each of these scientists was committed to a tool appropriate for the particular problem at hand, while on a larger scale they cannot be grouped into opposing camps.[12] Thus, toward the end of the 1890s, Planck altered his attitude toward Boltzmann's molecular kinetic views, a change which deeply influenced his research program on the black-body problem.[13]

Meanwhile, Nernst faced his own difficulties in reconciling thermodynamics with the molecular kinetic hypothesis in the first half of the 1890s. Thermodynamics deals with macroscopic systems and with macroscopic magnitudes, such as energy, temperature, and work, without explaining anything about the behavior of individual participants in the given systems. Atoms, ions, and molecules have no individually defined properties, only statistical significance insofar as a large number of particles participate in the physical or chemical transformations under investigation. The kinetic theory, however, seeks to account for specific properties of such particles, for their speed, energy, and size, and for their participation, for instance, in creating an electrical current in a battery or a transport of substances through a membrane.

As mentioned earlier, Nernst devoted several papers to a further exploration of the sixth chapter of his *Habilitation,* namely the "process of

12. Hiebert has provided a detailed analysis of Planck's conceptions of thermodynamics and his debates with Boltzmann and Ostwald, of the "Machian anti-atomism and anti-kinetic views with which he had flirted for twenty years," and has brought a wealth of original materials pertaining to Lübeck, 1895. Erwin H. Hiebert, *The Conception of Thermodynamics in the Scientific Thought of Mach and Planck,* Ernst-Mach-Institut, Bericht Nr.5, Freiburg i.Br., 1968, pp. 51–65, 70.

13. From T. S. Kuhn's remarks one might even surmise that the change was brought about by Planck's more serious acquaintance, in time, with Boltzmann's work. Kuhn writes: "Though he did not follow its technical development closely, Planck knew at least the main lines of gas theory. . . . Together with other evidence . . . [the] significant confrontation with Boltzmann [on the occasion of Planck's edition of Kirchhoff's *Lectures on the Theory of Heat*] suggests that, until the last years of the century, Planck's knowledge of Boltzmann's own gas-theory papers was spotty. . . ." Thomas S. Kuhn, *Black-Body Theory and the Quantum Discontinuity, 1894–1912* (Oxford/New York: Oxford University Press, 1978), p. 21.

dissolution of solids," related to his experimental determinations of molecular weights performed for Wallach. In these he expanded the *analogy* of solution processes with the gas laws and relied on and strengthened van't Hoff's theory of solutions. Van't Hoff had been the first to articulate this "fundamental analogy" of gaseous pressure produced by "the impacts of the gas molecules on the containing walls" and the "impacts of the dissolved molecules on the semipermeable membrane" of a solution.[14] In his two papers on the thermodynamics of chemical processes of 1882, Helmholtz had already touched, albeit in passing, on the possibility of such an analogy.

In two theoretical papers, which utilized primarily experimental data previously obtained by other authors, Nernst proposed a "solubility formula," based on the theorem that: "the maximum work obtainable during the dissolution of a substance in any given solvent, producing a saturated solution at the same temperature, is independent of the nature of the substance and the solvent, and has the same magnitude as if a solvent were not present at all and the substance would simply transform itself into its saturated vapor." He then expanded on the significance of this proposition: Previously, he reasoned, it had been assumed that during the dissolution process, attractive forces act at the contact surface between the solute and the solution, as for example in the case of sugar and water. If such specific forces were to exist, then the amount of work necessary to bring the sugar into a saturated solution would vary according to the nature of the solute and that of the solution. However, since this was generally not the case, "such forces either do not exist or they play only a secondary role. The extraordinary simplification experienced by the solution theory's excluding such specific attraction forces, and the much greater simplification still ahead, constitutes in my view the theory's greatest successes."[15]

Nernst emphatically and consistently eschewed speculation about the nature of the forces acting in the various solutions to which he applied the conclusions and elaborations developed from his theory of electromotive forces. He insisted that physical interpretations of the electromotive force and the galvanic currents produced by the dissolution of ions were necessary complements to the mathematical thermodynamical treatments of

14. Quoted by J. R. Partington, *A History of Chemistry* (London: Macmillan & Co., 1964), vol. 4, p. 654.

15. W. Nernst, "Über gegenseitige Beeinflussung der Löslichkeit von Salzen," *Z. f. physik. Chem.* 4 (1889): 374. Van't Hoff reiterated this position in his *Vorlesungen über theoretische und physikalische Chemie* (Brunswick: 1903), p. 27, where he writes that "it is thus clear that the cause of osmotic pressure is to be sought in kinetic reasons and not in attractions."

Helmholtz and Planck. In his 1889 paper on electromotive forces, he had conceded that the nature of the forces that lead to the transport of electricity between metal and electrolyte, occurring through the transport of material particles, was "a question which can to a certain extent temporarily be left undecided." For Nernst it was much more important to calculate the work produced in the process. He believed that "one is allowed to operate with the concept of 'electrolytic solvation pressure,'" even though "its physical interpretation is still unclear, in the same way in which one introduces osmotic pressure as an almost undisputed fact, without closely examining the question whether such a concept originates from attractive forces between the solvent and the solute, or in the mutual collisions between the molecules of the solute."

It was clear to Nernst that, although he was providing a slightly more acceptable, concrete, physical interpretation of the phenomena related to measurable macroscopic quantities, fundamental questions remained unanswered and made speculation about the nature of the forces he postulated sterile. He therefore insisted that "as long as the questions regarding the production of the osmotic effect, the manner in which the ion binds electricity, and finally the nature of electricity" were far from being answered, he did "not even consider it appropriate to discuss the nature of the forces which propel the ion from the metal into the solution, and we have to be satisfied with the test whether the formal relationships which result from the introduction of the electrolytic solvation pressure coincide with the facts."[16] Thus, despite his endorsement of the molecular hypothesis and the ionic dissociation theory, like Planck and other thermodynamicists, Nernst refrained from introducing or speculating about material entities still unknown or insufficiently explored. His contentment with *formal relationships* constitutes another characteristic of Nernst's work in general, one which we will encounter again in connection with the postulation of the heat theorem and its correlation to quantum theory.

The expanded solubility research was presented first in abstract to the Göttingen Science Society in February, and published in extenso in the *Zeitschrift für physikalische Chemie* in June 1890. Using the theoretical formalism developed in 1889 for solution theory, Nernst applied the solution pressure principle to the elaboration of a new method of molecular weight determinations. He used the example of a substance that dissolves in two different, almost immiscible, solvents: He envisioned, on the one hand, a system composed of a solution of a substance N in solvent A

16. Nernst, "Die elektromotorische Wirksamkeit der Ionen," *Z. f. phys. Chem.* 4 (1889): 153.

in contact with solvent *B,* and on the other, a contact between pure solvent *A* and solvent *B.* By applying van't Hoff's formula for osmotic pressure, and by postulating an imaginary reversible isothermal cycle, Nernst calculated the relative solubilities of *A* and *B,* the influence of the solute *N,* and stated the following theorem:

> The relative solubility lowering which a solvent experiences with regard to a second liquid by addition of a foreign substance is equal to the ratio of the number of dissolved molecules of the foreign substance to the number of molecules of the solvent:
>
> $$(a - a')/a' = n/100$$

where *a* and *a'* are the solubilities of the pure solvent and of the solution in the liquid *B,* and *n* is the number of molecules of the foreign substance (solute) per 100 molecules of solvent *A.*[17] For the case in which the second liquid *B* is replaced with a vacuum or a gas, the vapor pressure of the solvent can be substituted in the above formula for the solubility, thus transforming Nernst's theorem into the van't Hoff–Planck formula. Nernst retraces his steps when he concludes:

> The analogy between the processes of solvation and evaporation is again clearly exhibited in this case, similar to the manner in which I have previously [in the above discussed article of 1889] arrived at my solubility formula. It is obvious that one should look for similar relationships for the case of "electrolytic solvation pressure."[18]

This distribution formula, well known today, was developed during 1890 and 1891, and was thereafter called the "Nernst partition coefficient." It relied on previous researches by Marcellin Berthelot and Emile C. Jungfleisch, who stated that "a solute distributes itself between two immiscible or partially miscible solvents in such a way that the ratio of its concentrations in each at a particular temperature is constant: $c_1/c_2 = k$."[19] Nernst, however, showed that the molecular weight of the dissolved substance might be different in the two phases – due, for example, to ionization or association. Therefore, his partition coefficient, which took into account the molecular weight of the solute in the different phases, represented a more general case from which the Berthelot-Jungfleisch formula could be derived.[20] In his earlier paper, Nernst had applied the analogy

17. W. Nernst, "Über ein neues Prinzip der Molekulargewichtsbestimmung," *Z. f. physik. Chem.* 6 (1890): 22–3.
18. Ibid., p. 23.
19. Partington, vol. 4, p. 637. Nernst quotes Berthelot and Jungfleisch, *Ann. Chim. phys.* 26 (4): 396 (1872) and Berthelot, ibid., p. 408.
20. See also Hiebert, "Walther Nernst," *DSB,* p. 434.

with evaporation to the electrolytic solvation pressure concept, without, however, employing the mathematical formalism developed in the paper of 1890. It is remarkable that two years after introducing this novel concept, Nernst was still often, though not consistently, placing the term *Lösungstension* in quotation marks. This terminological indecision mirrors his ambivalence about the lack of a satisfying physical interpretation of the solvation phenomena, and was probably intended as a cautionary measure against what had almost certainly been strong reservations expressed by Helmholtz in 1888.

What were the methodological premises and purposes of this set of investigations? As we have seen, these were derived from prior work on the Hall effect and its application to electrolytes. Nernst's commitment to the atomic theory was in 1890 a qualified one. The introduction to his paper on molecular weight determination contains an explicit statement of position: He equated van't Hoff's formulas relating the pressure, volume, and temperature of a dissolved substance in solution ($pv = 0.00819\ T$ for one g-molecule of the substance) with the similarly "established experimental fact" expressed by the formula that gives the amount of work necessary to compress a body in solution from the volume v_1 to v_2. This quantity, given by $p_0\ln(v_1/v_2)$ included a proportionality factor p_0, which "is the same as if the substance in the gaseous state were to be compressed from volume v_1 to v_2."

Nernst considered these laws, much like those elaborated by van't Hoff and Planck in 1887, to belong to thermodynamics and to be based on the "secure foundation of direct experience."[21] Secure foundations notwithstanding, he pointed out that seeking conclusions regarding the molecular dimensions of the substance in solution from the above "established" correlations is a "different situation. . . . In this case, the hypothetical enters naturally into play, since everything that has to do with assumptions regarding the discrete distribution of matter in space has a hypothetical character."

He acknowledged that the "modern theory of solutions allows now for a clear and unambiguous distinction between the direct – or as good as direct (thermodynamically utilized) – result of experiment and its hypothetical extension." He was also aware that this distinction had until then not always been fully taken into account. In consequence, he insisted that, while the extension of Avogadro's hypothesis to solutions – that is,

---

21. Nernst referred here to van't Hoff, *Z. f. physik. Chem.* 1 (1887): 488, and M. Planck, "Über das Prinzip der Vermehrung der Entropie, III.," *Wied. Ann.* 32 (1887): 462–503, also in M. Planck, *Physikalische Abhandlungen und Vorträge* (Braunschweig: Friedr. Vieweg & Sohn, 1958), vol. 1, pp. 232–73.

the assumption that equal volumes of isotonic solutions contain the same number of molecules – may be well supported by the experimental data, it still remained "hypothetical." For, he argued:

> first, the transfer of a hypothesis to another domain, totally foreign, remains at least as hypothetical as in the original case; secondly, one would have to accept the possibility that even in the case of the correctness of the molecular weights of gas molecules as given by Avogadro, a gas might be absorbed according to the Henry law in a given solvent, and still pass into another molecular state. It lies in the nature of things that one cannot arrive at conceptions about molecular weights through thermodynamical considerations; *the molecular hypothesis and the laws of the mechanical heat theory have been leading until now a completely independent existence.*[22]

Nernst emphasized that experimental results concerning solutions had led, of necessity, to the elaboration of a molecular hypothesis sustained in particular by the electrolytic dissociation theory. The experimental results of this hypothesis were extremely fruitful, and therefore the molecular views were becoming "almost a fact," as van't Hoff put it. In related work, Nernst also proposed a new method for the measurement of osmotic pressures, which would provide a new avenue for molecular weight determinations. "The experiment must rely here too on some simple thermodynamic considerations," he continues, "which relate to cycles which can be performed with the help of semipermeable membranes." Although he agreed that the thermodynamic proof was provided "with greater mathematical elegance and generality" by Planck in his elaboration of the entropy principle, Nernst insisted that "the examination of a visualizable [*anschaulich*] cycle should provide more protection against errors in calculation or logic."[23] Nernst here insists on seeking intuitive, workable thermodynamic cycles in addition to theoretical calculations.

Nernst thus required that thermodynamic experiments be conducted in order to validate correlations between the molecular theory and thermodynamics. His seminal paper on molecular weights, where he elaborated an alternative method for measuring indirectly the absolute magnitude of osmotic pressure with the help of the solubility of a solvent in a second solvent, served precisely this goal.

The principles involved in constructing the osmotic apparatus were extensions of methods developed much earlier for demonstrating and measuring the macroscopic magnitude of osmotic pressure. Applied in the

22. Nernst, "Über ein neues Prinzip," p. 17.
23. Ibid., p. 19.

construction of the first semipermeable membranes by Max Traube in the late 1860s, these membranes, which allowed for quantitative measurements, had been used by Georg Tammann, with whom Nernst published a number of articles during his Göttingen years. He used a second apparatus containing a semipermeable animal membrane, saturated with water, to separate an ether solution from a solution of benzene in ether. The expected phenomenon of diffusion of the ether from the dilute benzene solution into the concentrated benzene solution (from the pure ether into the ether containing benzene) was made visible by adding a trace amount of iodine into the ether. With this apparatus, Nernst hoped that isotonic solutions – that is, solutions with equal diffusion pressures – could be investigated, since they too relied on the principle of diffusion of a solute between immiscible solutions (benzene cannot penetrate the membrane since it is insoluble in water, whereas ether will diffuse through the membrane).

On the basis of this experiment, Nernst expressed his first interest in physiology. He later contributed a theory of nerve excitation and a polemical intervention in the debate over the applicability of physical-chemical concepts to the newly developing area of immunochemistry. He drew an analogy between his osmotic cell and a plant cell: The benzene solution represents the cell "liquid," the membrane is analogous to the cell wall, the water layer corresponds to the protoplasm, and the ether mimics the solution surrounding the cell. Nernst also proposed a concrete mode of investigating diffusion processes in plant cells by examining to what extent "pure water or an aqueous solution of known molecular content is soluble in the protoplasm, and the diffusion constant of water in the protoplasm."[24] And although they were explored to some extent by fellow scientists, Nernst's physiological speculations came at a time when the structure of the plant protoplasm was not yet known, and their fundamental significance and correctness came to light only during the following decades.

Nernst further verified his solubility law during November 1890, and restated it as follows:

> The relative lowering of the solubility (solvation pressure) experienced by a solvent (e.g. ether) with regard to a second (e.g. water) as a result of the addition of a foreign substance is equal to the ratio of

24. Nernst, "Ein osmotischer Versuch," *Z. f. physik. Chem.* 6 (1890): 37–40. Nernst referred in a footnote to the "beautiful experiment of [Hugo] de Vries," *Z. f. phys. Chem.* 2 (1888): 415, and 3 (1889): 103. De Vries had described in 1888 a method for determining the concentrations of isotonic (isosmotic) solutions, related to plant cells.

the number of dissolved molecules of the substance to the number of the solvent molecules.[25]

Nernst expected that his findings on the properties of solutions should also be valid "without any reservations, to the change of solubility of crystals by isomorphous additions."[26] By the end of 1891, at a time when he was extending the consequences of solubility research into the domain of mixed crystals, Nernst could write to Ostwald that he had "lately been more of a physicist."[27] That year he published two papers concerning further applications of his solubility formula, and consistently urged the desirability of simplicity and visualizability of processes even when these conflicted with a purely thermodynamic treatment where such interests were secondary. In the conclusion of a paper on the distribution coefficient of a substance between two solvents, for example, Nernst weighed the advantages of molecular-kinetic views. Although this chosen version of the distribution law relied on molecular conceptions, "it can equally be independent of these. All consequences of this law would still remain valid even if one were not to accept the validity of Avogadro's law for solutions." He argued that a substance may or may not have the same molecular dimension in two phases and "is linked to no other conception" than that the given substance is under partial pressures in the two phases that correspond to its concentration. Or more generally, the free energy of a given quantity of substance, when in two different phases, will change proportionally to the change in concentration.

Nernst enjoyed being erudite and humorous. He insisted that "the methodological offense which I perpetrate against the fundamental principle of science, never to operate with a greater expenditure of hypotheses than are absolutely necessary for the given task, should be excused," because by employing molecular notions, "the presentation gains unusually in visualizability [*Anschaulichkeit*] and the expression is much more concise."[28]

25. Nernst, "Über eine neue Anwendung des Gefrierapparates zur Molekulargewichtsbestimmung," *Z. f. physik. Chem.* 6 (1890): 573.

26. Ibid., p. 577.

27. Nernst to Ostwald, Göttingen, 27 November 1891, Ostwald Papers AAW, and Nernst, "Über die Löslichkeit von Mischkristallen," *Z. f. physik. Chem.* 9 (1892):137–42.

28. In the introduction, he referred to the successful thermodynamic derivation of the distribution law by Riecke, whose theoretical work Nernst was going to strengthen experimentally and expand. Nernst, "Verteilung eines Stoffes zwischen zwei Lösungsmitteln und zwischen Lösungsmitteln und Dampfraum," *Z. f. physik. Chem.* 8 (1891): 111, 139. Related papers appeared during 1891 and 1892: Nernst, "Über das Henry'sche Prinzip," *Nachr. Ges. Wiss. Göttingen* (1891): 202–12; "Über die mit der Vermischung konzentrierter Lösungen verbundene Änderung der freien Energie," *Nachr. Ges. Wiss. Göttingen* (1892): 428–38.

The years 1891–2 were eventful ones for Nernst. He met his future wife, Emma Lohmeyer, to whom he became engaged in early 1892. In addition, the Permanent Secretary of the Prussian Ministry of Education, Friedrich Althoff, began to express an interest in Nernst's future career. Althoff importuned Hermann von Helmholtz, at the time vacationing in Italy, with an inquiry regarding Nernst's qualifications for an associate professorship both in Giessen and in Göttingen.[29] Ostwald congratulated Nernst for being a top candidate and expressed the hope that the happy event would compensate for Nernst's move – apparently in response to Nernst's mild criticism about the small-town atmosphere of Göttingen.[30] It might well be that Nernst felt removed from the heart of physical-chemical research in Leipzig or that he had hoped for a different academic position. But his years in Göttingen were extremely productive and enhanced his professional reputation.

During the next decade, when Nernst was in his thirties, a period considered the most fruitful in a scientist's career, he continued to work mainly on solution theory,[31] osmotic pressure,[32] and electrolytic dissociation and electrochemistry.[33] The index to Partington's "History of Physical Chemistry" shows that Nernst's articles were scattered among all fields and subcategories: thermochemistry and thermodynamics, solutions, electrochemistry, and photochemistry. However, Partington's chapter of most interest to chemists, "Affinity," carries no mention of any contribution by Nernst. This strengthens the impression that Nernst was a physicist applying physical and chemical methods and theories to the investigation of phenomena that had been the joint domain of physics and chemistry for several decades. These were subjects originally approached by physicists,

29. Althoff to Helmholtz, 15 August 1891, Helmholtz Nachlass, Sign 12., AAW.
30. Ostwald to Nernst, Leipzig, 6 March 1892, Ostwald Papers, AAW.
31. Nernst, "Über die Beteiligung eines Lösungsmittels an chemischen Reaktionen," *Z. f. phys. Chem.* 11 (1893): 345–59; Nernst and R. Abegg, "Über den Gefrierpunkt verdünnter Lösungen," *Z. f. phys. Chem.* 15 (1894): 681–93.
32. Nernst, "Osmotischer Druck in Gemischen zweier Lösungsmittel," *Z. f. phys. Chem.* 11 (1893): 1–6; Nernst and E. Bose, "Ein experimenteller Beitrag zur osmotischen Theorie," *Z. Elektrochem.* 5 (1898): 233–5.
33. Nernst, "Elektrostriktion durch freie Ionen," *Z. f. phys. Chem.* 15 (1894): 79–85; "Über Flüssigkeitsketten," *Z. Elektrochem.* 1 (1894): 153–5; "Über die Auflösung von Metallen in galvanischen Elementen," *Z. Elektrochem.* 1 (1894): 243–6; "Über das chemische Gleichgewicht, elektromotorische Wirksamkeit und elektrolytische Abscheidung von Metallgemischen," *Z. f. phys. Chem.* 22 (1897): 539–42; "Die elektrolytische Zersetzung wässriger Lösungen," *Ber. dtsch. Chem. Ges.* 30 (2): 1547–63 (1897); Nernst and E. H. Riesenfeld, "Über elektrolytische Erscheinungen an der Grenzfläche zweier Lösungsmittel," *Nachr. Ges. Wiss. Göttingen* (1901): 54–61, and *Ann. Phys.* 8 (1902): 600–8.

but after Volta's work in particular, they became of interest to a number of chemists. Davy and Faraday, like Nernst, advanced these subjects and – again like Nernst – were also difficult to pigeonhole along disciplinary lines. The interweaving of these developments suggests that disciplinary demarcations are not always helpful historical categories.

In Nernst's own textbook of theoretical chemistry where he indexed all his theoretical and experimental papers, we find that he refers most frequently to work in the following fields (according to the number of entries):

I. Atom and Molecule: solutions, osmotic pressure; kinetic theory, specific heat of gases; molecular weight; physical properties and molecular structure; electrolytic diffusion; electrolytic theory
II. Electrochemistry
III. Transformations of energy (Theory of Affinity II): thermochemistry, incomplete equilibria; electrochemistry
IV. Transformations of Matter (Theory of Affinity I): chemical statics[34]

However, the thermodynamic reference frame in which Nernst operated cannot serve as a rigorous disciplinary demarcation. For example, Berthelot, Thomsen, van't Hoff, Arrhenius, Planck, and Ostwald exploited thermodynamics in its application to chemical phenomena. What seems most characteristic of Nernst, in contradistinction to his predecessors in physical chemistry, was his consistent adherence to physical methods and instrumentation. The tradition of Kohlrausch's and Boltzmann's institutes in which Nernst had been trained left a significant imprint.

An overview of Nernst's work on electromotive forces from 1888 to 1892 shows that he derived the theoretical potentials for the case of a completely dissociated electrolyte in two ways: In 1888, this derivation was based on the diffusion of ions; in his 1889 Habilitation, however, the potentials were derived from the "identity of the electromotive forces of a reversible process with the maximal external work which can be obtained through this process."[35] In the published version of the Habilitation in 1889, Nernst wrote that he had "already obtained the above equations through significantly different considerations," and referred the reader to his 1888 paper.[36] He proposed the following equations:

$$A = P_1 - P_2 = [(u - v)/(u + v)]p_0 \ln(p_1/p_2) \qquad [1]$$

34. Nernst never quite addressed the fundamental question of what accounts for chemical affinity, and explicitly eschewed "hypothesizing" on the subject of affinity in his most important book, the textbook of chemistry, *Theoretical Chemistry*. Nernst, *Theoretische Chemie*, 7th ed. (Stuttgart: Ferdinand Enke, 1913).
35. Nernst, "Über die Potentialdifferenz verdünnter Lösungen," p. 361. See Nernst, "Die elektromotorische Wirksamkeit der Ionen," 1889, p. 136.
36. "Obige Gleichungen habe ich bereits durch wesentlich andere Betrachtungen erhalten." Nernst, 1888, p. 635. See also Drennan, pp. 178, 201.

where $P_1 - P_2$ is the potential difference. The work necessary to bring the quantity of energy $+e'$ from P_1 to P_2 and at the same time of $-e''$ from P_1 to P_2 is

$$A = (P_2 - P_1)(e' + e'') \qquad [2]$$

In the case of two different solutions of the same electrolyte that are in contact, the electricity is transported with the "imponderable mass, the ions," and the work is calculated for the transport of $(e' + e'') = 1$, given by equation [1], in which u and v are the ion mobilities of the cation and anion (the positively and negatively charged ion), respectively, while p_1 and p_2 are the pressures of the ions in the two solutions. By expressing p_0 in electromagnetic units, equation [1] becomes:

$$E_1 - E_2 = 0.02347 \, [(u - v)/(u + v)] \, \ln(p_1/p_2) \text{ volts} \qquad [3]$$

or, as a function of temperature:

$$E_1 - E_2 = 0.860 \, T \, [(u - v)/(u + v)] \, \ln(p_1/p_2) \times 10^{-4} \text{ volts} \qquad [4]$$

Although this final expression [4] for the theory of the electromotive forces thus had little to do with thermodynamics, Helmholtz's work and thermodynamic considerations played an important role in Nernst's *methodology*. Nernst sought to find positive correlations between kinetics and thermodynamics and insisted on always checking one against the other. This insight is essential to an understanding of the context in which innovative scientists in the late nineteenth century succeeded in overcoming conceptual and experimental difficulties. The tools were often unsatisfactory, but by continually playing with alternatives, most innovators – including Helmholtz, Planck, Ostwald, and Nernst – succeeded in constructing a more complex, less reductive mode of doing science and representing reality.

The chapter on "Electrochemistry II, Thermodynamic Theory" in Nernst's chemistry textbook *Theoretische Chemie,* first published in 1893, established a direct continuity between his 1889 paper and Helmholtz's thermodynamic treatment of 1877: "I substantially simplified Helmholtz's theory for dilute solutions, by using the van't Hoff laws, and by using reversible electrodes. . . . I have made possible a quite exact experimental proof." This published claim of a direct link to Helmholtz later came to plague Nernst. His considerable originality in electrochemistry was denied by some of his peers. They charged that he had only improved on Helmholtz's work.

In "Electrochemistry III, Osmotic Theory," Nernst wrote on the mechanism of the production of a current in solutions:

The considerations presented so far rely primarily on thermodynamic foundations; it is inherent in the nature of this research method that, if carefully applied, it provides undoubtedly correct, but not intuitive [*anschauliche*] results. In particular, the mechanism of the production of the galvanic current has been left out of these considerations. It becomes evident that here, too, the recent views of the ionic theory are capable of leading us one significant step further.[37]

He devoted this chapter to the presentation of a "more special theory of the electromotive action of ions, which [he] had developed in 1888/89 and which nowadays seems to be generally accepted." He also provided the derivation of the equations for the potential gradient obtained "thermodynamically," as shown above in equations [1] through [4]. Nernst was convinced that he had provided the derivation of the equations for the electromotive force of ions in solutions both thermodynamically and on the basis of the ionic dissociation theory. In his thermodynamic derivation, Nernst did not use the term "osmotic pressure" and only implicitly made the analogy between solutions and gases, relying on what he termed the "van't Hoff laws." Thus, Nernst believed that his paper of 1889 did make explicit use of thermodynamics. But although he used and accepted the validity of the thermodynamic derivation, and although he considered it adequate, yet he nonetheless hoped to provide added "*Anschaulichkeit,*" – or "intuition" or "physicality" – to the underlying processes.

This is a significant point because it constituted the core of his later debate with Planck. In addition, this reading confers consistency on Nernst's scientific investigations. While not at all indifferent to thermodynamics, he was expressing in his own work a certain dissatisfaction with the explanatory power of thermodynamics not uncommon among physicists and chemists at the time. Preoccupation with thermodynamics and kinetic theory, coupled with ionic dissociation and osmotic studies, constitutes the matrix of references and methodologies with which a modern scientist in the physical sciences had to operate at the turn of the century. These avenues, individually incomplete and to a great extent irreconcilable, led to a multilevel approach and the necessity to work with internal contradictions.

This must be emphasized for a number of reasons. First, it underscores the continuity of perspective and concerns in Nernst's life. It explains his dissatisfaction with the nonintuitive or nonphysical aspects of thermodynamics, starting with the interest he showed, as a young Boltzmann student, in the seemingly unavoidable heat death of the universe as a consequence of the Second Law of Thermodynamics. Second, it fills the

37. Nernst, *Theoretische Chemie*, 1913, pp. 777–8.

apparent hiatus in Nernst's thinking and career, which some historians have posited. Some claim that he moved towards the formulation of the heat theorem only after a long period of electrochemical work divorced from thermodynamic preoccupations. But this is not the case. He demonstrated a persistent concern with thermodynamics both as a methodology and *"Problemstellung"* during the thirteen years prior to 1905. And third, it sheds new light on the question of whether thermodynamics did or did not play a role in the formulation of a viable electrolytic solution theory. Thermodynamics, although not "crucial" to the successful development of an electrolytic solution theory, was nonetheless an important constituent of electrolytic solution research as early as the 1880s. Therefore, Nernst's and Planck's work explains the important role played by thermodynamics in electrolytic solution theory. While retrospective analyses of these events almost always focus on one individual's particular contributions, the picture of thermodynamics, coupled with electrolytic solution theory and electrochemistry, emerges as quite complex. And while priority disputes and substantive debates, such as those between Arrhenius and Planck, tend to stress the "right" version of the facts, it is fruitful to emphasize the varied, and often discordant, atmosphere in which "the facts" first come to light.

5

The Nernst-Planck Exchange

*Und es ist für die Wissenschaft unter allen Umständen nützlich, nach-
dem man die Dinge so lange aus einem Fenster betrachtet hat, sie auch
aus einem anderen anzusehen.*   W. Ostwald, 1893

In any circumstances science will benefit if, after having watched
things for a long time from one window, we look at them from another. Intensive experimental treatment of [theoretical] conclusions is
the only means of uncovering the weaknesses and imperfections of the
chosen point of view, after having brought it to the stage at which it
can not be refuted. And we will make it our business to perform this
very endeavor.[1]

The letter in which Ostwald makes this commitment is part of a long series of private and published exchanges between Planck and Ostwald on
the foundations of energetics, the problem of irreversibility, and the various
new forms of energy postulated in the previous work of Mach, Helm, and
Ostwald. It illustrates the deadlock which the two men came to acknowledge: Their positions on the Second Law of Thermodynamics had reached
a stage at which they could be neither proved nor refuted. Their major
disagreement, as is well known, centered on Ostwald's insistence that
the Second Law of Thermodynamics is not a statement of fact on the irreversibility of natural processes. While Planck acquiesced somewhat
facetiously that Ostwald's new ideas on energetics had engendered more
experimental research than previous, even more worthy, competing thermodynamic theoretical speculations, he remained unwilling to change
thermodynamics at the expense of reconciling inconsistencies. He concluded that the efforts expended on correlating thermodynamics with the
kinetic theory of gases had failed, and that the atomic hypothesis, although successful, had become a stumbling block to progress in the field
and should better be abandoned altogether, at least for the time being.[2]

1. W. Ostwald to M. Planck, Leipzig, 2 July 1893, Körber, vol. 1, p. 52.
2. See Ostwald to Planck, 26 June 1893, and Planck to Ostwald, 1 July 1893, Körber, vol. 1,
   pp. 46–50. Heilbron, *The Dilemmas of an Upright Man*, p. 15.

The exchange between Ostwald and Planck had thus been reduced to differences in the appropriateness (*Zweckmässigkeitsfragen*) of thermodynamic concepts and their applications, an essentially teleological dispute, which sets the stage for our present concerns. These disputes, and Planck's ambivalence on whether the atomic hypothesis is merely useful or indeed fundamentally true, have previously been discussed as if Planck had only one or at most two meaningful partners in conversation, namely Boltzmann and Ostwald. And because Ostwald was generally moving in a "losing" direction with his unqualified energetic view of the world (a world in which even happiness and bibliographic classification systems could be subject to an energetic calculation), Planck's work in physical chemistry has tended to be seen as a youthful – albeit useful – prologue to the real drama that was to unfold around 1900 with the publication of his epoch making papers on the quantum theory of radiation.

And yet in the early 1890s, Planck was embroiled in other exchanges that were significantly to shape developments both in physics and in the work of Ostwald's gradually expanding circle of physical-chemical adherents. Among them, Nernst came to play an important role, as revealed in his debate with Planck in a series of papers published in late 1891 and early 1892.

Between 1888 and the late 1920s, Nernst was to cast a wide net encompassing theoretical and experimental physics and chemistry. By 1892, he felt that he had carved a niche for himself; as he later said, the field of electrolytic theory and experimentation was the domain to which he had contributed most in those early years, and he began feeling possessive about these achievements. He thus began to complain about misrepresentations and engaged in a small and subtly conceived priority dispute. During 1890 and 1891, Lothar Meyer and Friedrich Paschen had challenged the theoretical predictions of the indirect measurement of osmotic pressures by measurement of electromotive forces.[3] In addition, W. Negbaur, in an article on the potential gradient of dilute solutions, not only had contested the validity of Nernst's formulae for the electromotive force but also had attributed to Planck's elaboration of Nernst's work a greater "generality."

Nernst retaliated with a concerted effort, publishing two articles con-

3. Lothar Meyer's criticism of the Ionists prompted rather derogatory remarks from Nernst, occasioned by the publication of Meyer's latest book, *Grundzüge der Theoretischen Chemie*, Leipzig, 1890, in which Nernst decried the lack of understanding for the new developments, in particular that van't Hoff's "brilliant formula $0.02\ I = I^2$ has probably passed by [Meyer] without any trace." It was on this occasion that Nernst also mentioned his critique of Paschen. Both were published only a year later. Nernst to Ostwald, 15 November 1890, Ostwald Papers, AAW, Berlin.

secutively in the same issue of the *Annalen*.[4] The first, cosigned with R. Pauli,[5] a graduate student working in Nernst's laboratory, was a rebuttal of Paschen's claim that an experimental verification of Nernst's theory with the aid of a drop electrode had shown wide deviations from predicted values. In that paper, Nernst was still perfectly comfortable with using Planck's theoretical work as substantial confirmation for his own research. He acknowledged that his initial treatment of the electromotive forces in 1889 had encountered considerable "mathematical difficulties." He argued that at the time when the equations[6] were set up, they were employed in two directions, deriving the diffusion coefficients of salts from the gas laws and from the Kohlrausch ion mobilities; and on the other hand, the electromotive forces were deduced from the same magnitudes. He considered that the general differential equations had by then, in principle, solved the problem of finding the electromotive forces of liquid cells that are produced by completely dissociated electrolytes, and that this problem was entirely due to mathematical difficulties.[7]

He stressed, however, that Planck himself had encountered similar difficulties and had nonetheless "unambiguously accepted" the equations in a paper, which is "also novel, insofar as it proved in a strict, previously only-hinted-at and incomplete proof, that the forces acting in solution are *unambiguously* determined by these differential equations." This, his previously only-hinted-at and incomplete proof, was soon thereupon improved

4. Friedrich Paschen, *Wied. Ann.* 41 (1890): p. 184, quoted in Nernst and R. Pauli, "Weiteres zur elektromotorischen Wirksamkeit der Ionen," *Ann. Phys.* 45 (1892): 358. L. Meyer, *Ber. d. Berl. Akad.*, 12 November 1891, and W. Negbaur, "Über die Potentialdifferenz verdünnter Lösungen," *Wied. Ann.* 44 (1891): 737ff, quoted in Nernst, "Über die Potentialdifferenz verdünnter Lösungen," *Wied. Ann.* 45 (1892): 360ff and 367, respectively. The identity of titles between the papers by Negbaur and Nernst is not a mistake; it was common practice to rebut articles by publishing under the same title.

5. Nernst sent Pauli to study with Ostwald in October 1892. Nernst wrote to Ostwald that he had recently published some measurements on liquid cells with Pauli, and recommended that Ostwald might assign Pauli a project where "there is a little to calculate, since Pauli is trained more in mathematics. Of course until now he was more preoccupied with student affairs than with pure science." So in order that Pauli might finally free himself of the "Allotriis," Nernst advised him on a change of air. (Allotriis was a pet word in the correspondence among Ostwald and his students, signifying the pleasures of extrascientific life.) Pauli arrived in Ostwald's laboratory at the end of October 1892. Ostwald, however, reported laconically that upon arrival, Pauli had immediately left for a wedding. Nernst to Ostwald, 6 October 1892, and Ostwald to Nernst, 23 October 1892, Ostwald Papers, AAW. Berlin.

6. Differential equations that made it possible to calculate the transport of ions under the influence of osmotic and electrostatic forces from the friction resistances, which were obtained from the combination of hydrodynamical and electric current equations.

7. Nernst, "Über die Potentialdifferenz verdünnter Lösungen," *Wied. Ann.* 45 (1892): 353-4.

by Planck, who, in Nernst's words, succeeded in obtaining "in a manner characterized by elegance and simplicity the general integration for the case of an arbitrary number of binary and totally dissociated electrolytes in solution. By overcoming mathematical difficulties, it is now possible to calculate the absolute electromotive force of a liquid cell constituted of any number of binary electrolytes from the gas laws and the ion mobilities."[8] Nernst recalculated the electromotive forces, according to the formulae provided by Planck, from the sum of the forces active at the contact of four solutions, compared them to the experimental results obtained through the method he had proposed in 1889, and found a good correspondence.

Yet Nernst engaged Planck in a priority dispute in 1892 over what he perceived as an undue appropriation of ideas expressed by Planck in a 1891 paper on the laws of electrochemical equilibria.[9] Nernst wrote to Ostwald that he felt he was in a "difficult position" regarding Planck. As we know, Planck had by then contributed significantly to the thermodynamic elaboration of physical-chemical theories, and although less well known or recognized for work in this field, Planck was in 1891 still active mainly in physical chemistry.[10]

Why did Planck expend so much effort in the direction of electrochemistry and solution theory? What was there in the scientific environment of the time that would lead a man of Planck's background and aspirations as a rising theoretical physicist into this line of research? One might find it plausible that Planck engaged in the extension of the applicability of the Second Law of Thermodynamics to various physical and chemical phenomena. His choice of field indicates that the perspectives opened up by physical-chemical research constituted what we would term a "progressive agenda" at the forefront of research.

8. Ibid., p. 354. Refers to Planck, "Über die Erregung von Elektrizität und Wärme in Electrolyten," *Wied. Ann.* 39 (1890): 161–86, and "Über die Potentialdifferenz zwischen zwei verdünnten Lösungen," *Wied. Ann* 40 (1890): 561–76, in Planck, *Physikalische Abhandlungen und Vorträge,* vol. 1, pp. 330–71.
9. Max Planck, "Über das Princip der Vermehrung der Entropie. Vierte Abhandlung. Gesetze des elektrochemischen Gleichgewichts," *Wied. Ann.* 44 (1891): 647–56, submitted in July 1891; also in Planck, *Physikalische Abhandlungen und Vorträge,* vol. 1, pp. 382–425.
10. For Planck's work in this domain, see Hiebert and Root-Bernstein. Kuhn and Klein devote relatively little space to Planck's work between 1880 and 1893 when he almost exclusively explored the application and consequences of the Second Law of Thermodynamics to equilibria in gases and solutions, the molecular constitution of dilute solutions, osmotic pressure, vapor pressure, thermoelectricity, diffusion, and electrolysis and published some twenty papers on these subjects. See Planck, *Physikalische Abhandlungen und Vorträge,* 1958.

Upon his arrival as professor of theoretical physics in the capital of the empire, Planck had based his inaugural address to the German Physical Society in Berlin in April 1890 on the experimental and theoretical work of Nernst. The lecture elicited silent rejection from the audience and critical remarks from the chairman, Emil du Bois-Reymond: "On the basis of new measurements which Nernst had communicated to him, Planck felt very secure, but it happened otherwise. Du Bois-Reymond . . . attacked Planck's considerations, deemed the correspondence between measured and experimental data as a coincidence and some of Planck's assumptions as totally unacceptable for any who can think chemically. Planck went home quite depressed, but comforted by the thought that a good theory would prevail even without deft propaganda. That happened in this case as well, although it took some years in Berlin."[11]

On 24 September 1891, Planck lectured at the joint session of the physical and chemical sections of the 64th Congress of German Scientists and Physicians in Halle on "General Considerations Regarding Recent Developments in Thermodynamics."[12] He began by describing the two major developments in thermodynamics: the first, based on general assumptions set forth in the two laws by Clausius; the second, relying on the kinetic theory of gases as represented by Maxwell and Boltzmann. His attitude toward the latter was critical, and even though he stressed the fruits borne by the analogy between gases and solutions, Planck considered that deeper insights were to be gained by avoiding special molecular notions and applying only the two laws of thermodynamics to solution processes.

Afterward, Nernst complained to Ostwald:

> [Planck] proceeds with me exactly in the same manner as he had earlier done with van't Hoff, where in one paper he derived van't Hoff's formulae and in the next he proclaimed them as all his own. At that time you [Ostwald] reprimanded him. . . . I'm not much disposed to quietly watch this thing and will probably give him a thorough reminder [*Denkzettel*]. This whole manner of capitalizing by using foreign ideas and embellishing this foreign property with mathematical flourishes which are only there to blind those not close to the subject – and doing it often – is in the long run highly repulsive. What

11. Planck, "Über die Potentialdifferenz zwischen zwei verdünnten Lösungen binärer Electrolyte," *Wied. Ann.* 40 (1890): 561–76, in Planck, *PA*, vol. 1, pp. 356–71. Planck, *Erinnerungen*, ed. W. Keiper (Berlin: Keiperverlag, 1948), cols. 137–40, also quoted in J. L. Heilbron, *The Dilemmas of an Upright Man*, p. 12. Hans Hartmann, *Max Planck als Mensch und Denker* (Basel, Thun, Düsseldorf: Ott Verlag, 1953), p. 22.

12. "Allgemeines zur neueren Entwicklung der Wärmetheorie," in Planck, *PA*, vol. 1, pp. 372–81. For a discussion of other aspects of this paper, see T. S. Kuhn, *Black-Body Theory and the Quantum Discontinuity*, pp. 22–3, and n.51, p. 264.

do you think about it? The Halle lecture showed how much he lacks ideas of his own. . . .[13]

This reflects a disturbing pattern of behavior on Planck's part vis-à-vis not only Nernst but also van't Hoff himself. Nernst's letter to Ostwald refers back to Planck's first paper published in the inaugural volume of the *Zeitschrift für physikalische Chemie*.

While professor of physics in Kiel, Planck had submitted in October 1887 an article on the molecular constitution of dilute solutions. In it he had attempted to construct a rigorous treatment of the molecular constitution of solutes, for which anomalous vapor tensions and freezing point depressions, deviating from the gas laws, might provide information on the degree of dissociation and chemical changes undergone by a substance in solution. Already then, Planck's severe comment on the pervasive "lack of any rational basis" had been "reprimanded," in Nernst's words, in a footnote added to the paper by Ostwald, the journal's editor, in which he referred readers to a systematic paper published by van't Hoff a few months earlier in the same journal.[14]

Planck had indeed been well acquainted, if not with van't Hoff's most recent publication, then at least with the main thermodynamic applications of the gas laws to solution theory. In a previous long article published in the *Annalen* and completed in July 1887, Planck had made numerous direct references to a series of van't Hoff's papers. There he derived essentially the same formulae as van't Hoff for the lowering of the freezing point of dilute solutions.[15] In the introduction he wrote: ". . . we

13. Nernst to Ostwald, Göttingen, 1 January 1892, Ostwald Papers, AAW. Berlin. Ostwald replied:

> Your anger toward Planck has somewhat surprised me, since I didn't get the impression that he claimed to have done anything else but to derive your results by a different route. I have reviewed the paper thoroughly, and would be glad to insert, in your interest, the comment that his results are identical with and do not go beyond yours. . . . In any case the journal is at your disposal if you wish to publish a complaint, which would however rather belong in the Annalen, where P. has published.

Ostwald to Nernst, 4 January 1892, Ostwald Papers, AAW. Berlin. Nernst, never the conciliatory person and at the time less concerned with academic politics than the seasoned Ostwald, nevertheless sent his article to the *Annalen*.

14. Planck, "Über die molekulare Konstitution verdünnter Lösungen," *Z. f. phys. Chem.* 1 (1887): 577–82, in Planck, *Physikalische Abhandlungen und Vorträge*, vol. 1, pp. 274–9. See also Root-Bernstein, p. 497, for analysis of differences among Planck and Arrhenius and van't Hoff.

15. Planck, "Über das Prinzip der Vermehrung der Entropie. Dritte Abhandlung. Gesetze des Eintritts beliebiger thermodynamischer und chemischer Reactionen," *Wied. Ann.* 32 (1887): 462–503, in Planck, *Physikalische Abhandlungen und Vorträge*, vol. 1, pp. 232–73.

will discuss those quite numerous apparent deviations which have been observed experimentally from the Guldberg-Waage theory and which, as is known, have recently led van't Hoff to give a substantially modified shape to that theory."[16]

Nernst's reprimand referred primarily to Planck's claim that his researches on the application of the principle of the increase of entropy to the various kinds of electrochemical equilibria, in particular to dilute solutions, "lead . . . to a new confirmation of the theory of electromotive forces of ions . . . established by W. Nernst, independent of the conventional views."[17] In the final paragraphs of his 1892 reply to Negbaur, Nernst inserted his disagreement with Planck:

> Concerning Mr. Planck's latest investigation, in which he systematically analyzes certain theories by Lord W. Thomson, von Helmholtz, Lippmann, Warburg and me, I feel obliged to express doubts about a point which could possibly be misleading to those less knowledgeable about the subject in order to avoid in time the necessity of further similar objections. Mr. Planck notes (p. 386) that "his investigations lead . . . to a new confirmation of the theory of electromotive forces of ions . . . established by W. Nernst, independent of the conventional views." I cannot find that the thermodynamic proof in the derivation of the potential difference between two different solutions of the same electrolyte, or between a reversible electrode and the solution, is in any way – other than in its form – distinct from mine, since Mr. Planck – according to my method – equates the maximal external work which can be obtained from the transfer of ions with the potential difference which is sought. Even though I cannot therefore agree that Mr. Planck has brought a new point of view in the treatment of the problem with the help of the entropy principle, I obviously would not negate the progress which such a systematic compilation represents for the clarification of the unsettled questions.[18]

As far as Nernst himself was concerned, Planck had published two papers related directly to Nernst's work on the electromotive force, the first completed in December 1889, the second in May 1890.[19] They constituted

16. Ibid., p. 233.
17. Planck, "Über das Princip der Vermehrung der Entropie. IV.," *Wied. Ann.* 44 (1891): 386, in Planck, *PA*, vol. 1, p. 383.
18. Nernst, "Über die Potentialdifferenz verdünnter Lösungen," pp. 368–9.
19. Planck, "Über die Erregung von Elecricität und Wärme in Electrolyten," *Wied. Ann.* 39 (1890): 161–86; "Über die Potentialdifferenz zwischen zwei verdünnten Lösungen binärer Electrolyte," *Wied. Ann.* 40 (1890): 561–76, also presented in abstract, with an introduction to the theory of solutions, in the *Verhandlungen der Physikalischen Gesellschaft zu Berlin,* Meeting of 18 April 1890. Both in Planck, *Physikalische Vorträge und Abhandlungen,* vol. 1, pp. 330–71.

the only predecessors on this subject to the paper which upset Nernst in late 1891, and were the first published fruits of Planck's new career as associate professor of physics in Berlin. This series of electrochemical and thermochemical investigations was to constitute the basis for Planck's *Grundriss der allgemeinen Thermochemie (Outline of General Thermochemistry)*, published in 1893. Planck's purpose in the *Thermochemie* was to present "the concepts and laws of thermochemistry in the closest possible connection with its basic facts, independent of more special views, including atomistic views." There he formulated what essentially amounts to a positivist credo: Even though the atomic hypothesis was capable of conveying an "encompassing picture of a wide range of individual facts (*Tatsachen*)," the task of separating "unchanging facts" from the "changeable notions" to which they have led becomes even more significant. "History has repeatedly shown," Planck wrote, "that the best hypothesis, once it has fulfilled its utility, becomes the most dangerous enemy of progress which extends beyond [that hypothesis]. Then difficult crises can arise in science, and these will be overcome that much better and with fewer losses of what has been achieved if careful criticism has preceded such crises."

In the brief introduction to the book, Planck wrote: "Thermochemistry treats of the relationships between chemical and thermal phenomena, and together with photochemistry and electrochemistry constitutes the field of physical chemistry. . . . [T]he importance of thermochemistry extends to all chemical processes, and in many cases one cannot even decide conclusively whether a problem belongs to the domain of chemistry or thermochemistry. The relationships of thermochemistry to the physical theory of heat [i.e., thermodynamics] have lately proven to be so intimate that a presentation of the current state of thermochemistry becomes impossible without a closer examination of the principles of the theory of heat. Therefore, the following material is organized in such a manner that, after a historical introduction, we discuss the various branches of thermochemistry in relation to the two laws of thermodynamics."[20] Four years later, Planck published the first edition of his *Vorlesungen über Thermodynamik*, or *Lectures on Thermodynamics*, a work that was to become the famous *Vorlesungen über Thermodynamik*.[21] This edition was, in essence, a slightly enlarged and edited version of his *Thermochemie*, but the change in title indicates that by 1897, Planck presumably

20. Planck, *Grundriss der allgemeinen Thermochemie. Mit einem Anhang: Der Kern des zweiten Hauptsatzes der Wärmetheorie* (Breslau: E. Trewendt, 1893). Planck to Ostwald, 1 July 1893, in Körber, vol. 1, p. 50. Planck, *Grundriss der allgemeinen Thermochemie*, p. 1.
21. Planck, *Vorlesungen über Thermodynamik* (Leipzig: von Veit, 1897).

wished to be known exclusively as a thermodynamicist, severing his ties with two decades of physical-chemical work – although most scientists would by then not have seen thermodynamics as foreign to thermochemistry or vice versa.

The title of the earlier work is revealing, however, since it uses the term *themochemistry*, coined by Julius Thomsen, the leader of an important Danish school of physical chemistry. Thomsen had published his major work in four volumes, *Thermochemische Untersuchungen* (*Thermochemical Researches*), in Leipzig during the years 1882 to 1886. This seems to indicate that Planck was closely engaged in exploring very recent developments in this field. The *Thermochemie*, which in hindsight seems a quite unusual project coming from Planck, included a historical sketch on the development of thermochemistry, from Lavoisier to van't Hoff. One might safely assume that a textbook on thermochemistry commissioned by the press from Max Planck squarely places him in a quite different constellation from that in which we have habitually seen him.

Planck's priority disputes with colleagues were beginning to look like a habit. In his exchange of letters with Ostwald in July 1893, quoted at the opening of this chapter, Planck had written that in fact his "'Thermochemistry' has been ready for a year" and that he was at the time busy with corrections to the page proofs. In the foreword to the publication, signed September 1893, Planck mentioned that he had "regrettably" not been able to refer in his book to the other "detailed" works that had appeared since the writing of his essay, namely Ostwald's *Chemische Energie,* Nernst's *Theoretische Chemie,* and J. J. van Laar's *Die Thermodynamik in der Chemie.* This was due to the fact that Planck's *Thermochemie* had originally appeared as a chapter in the *Handwörterbuch der Chemie* – a dictionary of chemistry, of which eleven volumes edited by the chemist A. Ladenburg had been published by 1893. They covered, in alphabetical order, a wide range of chemical reactions, descriptions of organic and inorganic compounds, chemical elements, etc. In addition, specific chapters were devoted to theoretical subjects (the atomic theory, stereochemistry, thermochemistry, etc.), experimental methods (absorption, spectroscopy, analysis, electrolysis, capillarity), and other related topics (the atmosphere, light, nutrition, fertilizers, fermentation, explosives, etc.).

Helmholtz's support of both Planck and Nernst to key positions in the German academic establishment, over and against local opposition to physical chemistry in Berlin, enforces the view that both scientists, backed by the vision of the doyen of German science at the time, were clearly perceived as pioneers, if not as outright outsiders. In fact, considering the dearth of good theoretical physicists in Prussia at the time, Nernst and Planck were probably among the best bets.

Thus, while Nernst and Planck confirmed each other's work and found internal support from these external confirmations, they disagreed on the priority of formulation and on the generality of its theoretical validity. In his paper, Planck had derived the formulae for the calculation of the potential on the basis of ion concentrations, ion mobilities, and valence of ions. He compared his new formula with that which he had previously developed, from Nernst's work, on the basis of osmotic pressure. He then argued that the agreement with the laws derived earlier on the basis of osmotic pressure was "the natural outcome of the fact that osmotic pressure itself is a consequence of the principle of the increase of entropy." However, Planck added that the introduction of osmotic pressure "leads us somewhat further, since it teaches us about the movements in the interior of an unequally concentrated solution. We have to forgo this application since the entropy principle says nothing about the development in time of a change of state."[22]

Nernst acknowledged the intrinsic value that a broad, well-founded thermodynamic treatment of other similar and connected phenomena may bring to problems of the electrochemisry of solutions, but refused to allow Planck's programmatic stress on the primacy of entropy and the second law to interfere with his own priority in treating the subject of electromotive forces. Nernst recapitulated what he considered to have been the chain of thought in Planck's work:

> [Planck] theoretically obtained and experimentally verified formulae, which essentially coincide with my own, and has then performed the integration of my differential equations also for the case in which two different electrolytes are in contact. I had repeatedly indicated that with the aid of my views on the electromotive forces of ions one could also obtain the theory of liquid cells; and even if I fully appreciate the merit of Mr. Planck's extremely complete and elegant solution of the problem for the case in which the electrolytes are totally dissociated, I still cannot see it as more than an extension of my theory in a special direction, and in no case as a generalization.[23]

Planck's answer to Nernst's objections followed in the next issue of the *Annalen*. His paper was not only a specific response to Nernst but also a lesson for those skeptically inclined toward the Carnot-Clausius principle of thermodynamic cycles. Planck stressed in his reply the independence of his derivation from considerations of osmotic pressure. He wrote:

22. Planck, ibid., pp. 403–4.
23. Nernst, "Über die Potentialdifferenz verdünnter Lösungen," pp. 367–8.

I completely acknowledge that Mr. Nernst operated in his theoretical derivation of the potential difference of unpolarizable electrodes, and between two solutions of the same electrolyte, with the Carnot-Clausius principle. And I gladly add that his more intuitive manner of using the second law, without naming it, rather increases than diminishes the merit of my discovery. But, to the same extent, I have to hold on to the contention that the second method, independent of the conception of osmotic pressure, is different from the first. Only if one considers osmotic pressure and its laws as a consequence of the second law do the two methods (*Wege*) become identical; otherwise not, and at the current state of research I still consider this point fundamental.[24]

The chain of events that led to the publication of this final installment suggests that, in fact, the above paragraph could have been fully understood only by those who had followed the publication of a large number of articles by Nernst, Planck, and others in the course of the previous three to four years. In this respect, Nernst had probably been correct in writing to Ostwald about the misleading nature of Planck's statements. This is particularly true since Planck himself had first approached the above problems by examining osmotic pressure, and only later attempted a completely independent generalization.

Planck had criticized Nernst's use of thermodynamic cycles as early as 1891, in his Halle lecture, mentioned earlier. The Second Law of Thermodynamics had developed in several directions, Planck argued. The general content of the law was usually reduced to the solution of a maximum or minimum problem: Gibbs stressed entropy, Helmholtz free energy, and Duhem the thermodynamic potential. But all these forms of the second law were only different expressions of the same essence, he argued, and hence have to lead to coinciding results. They differ only in the processes to which they are applied. Planck criticized the approach of

> physicists and chemists who in the derivation of new consequences of the second law do not rely on these [above mentioned] assumptions but use certain special thermodynamic cycles, devised for the particular purpose at hand, to which they apply the primitive law that heat can not be transformed into work without compensation.[25]

Planck's reference to the direct use of the Carnot-Clausius principle as "primitive" must have irked Nernst, who surely saw himself as a rather enlightened, mathematically and physically oriented thinker.

24. Planck, "Bemerkungen über das Carnot-Clausiussche Princip," *Wied. Ann.* 46 (1892): 162–6, in Planck, *PA*, vol. 1, pp. 426–30.
25. Planck, *PA*, vol. 1, pp. 373–4.

From this exchange we may conclude that the primary difference between Nernst's and Planck's methodologies in approaching some specific problems in electrochemistry was one of purpose: While Nernst was primarily concerned with providing not only a correct but also a "concrete," visualizable theory of the processes related to the production of an electric current inside an electrolytic cell, Planck – the theoretician par excellence – pursued his thermodynamic program in relation to the generalizations and applications of the second law he had begun during his student years.

Planck's thermodynamic treatment of ionic dissociation, developed on the basis of Arrhenius's theory of 1884–7, subsequently led to fruitful research and, as Planck himself had hoped, entropy measurements now provide guidance in the laws of physical and chemical equilibria. However, we recognize that the ideas developed by Planck, in particular with regard to the electromotive forces, relied heavily upon the work of van't Hoff, Arrhenius, and Nernst.[26]

To summarize: Nernst applied a number of methods and presuppositions to his work in electrochemistry. His thermodynamic derivation of the electromotive force relied, to a great extent, on Helmholtz's similar treatment of the galvanic cell, and was instrumental in furthering Planck's work in this domain. His kinetic views were applied to electrochemical reactions under the stimulus of Ostwald and the broadening interest in ionic dissociation theory during the 1880s. The experimental and theoretical bases of Nernst's investigations lay, however, in his training as a physicist. He developed a characteristic style of reasoning and research that coupled thermodynamic and atomic conceptions, an approach which was criticized by Planck, who was then still antiatomistically inclined. The efforts of Nernst and Planck formed the starting point for the further development of a thermodynamic electrolytic-solution theory that was later extended by G. N. Lewis, P. Debye, E. Hückel, L. Onsager, and H. Falkenhagen. However, these developments, which came after the elaboration of statistical and quantum mechanics, obliterated the sharp contrasts between the original approaches: Nowadays we do not question the peaceful coexistence of thermodynamics and the atomic theory.

Physicists preferred to see their monopoly in providing the theoretical apparatus for physical-chemical phenomena unbroken. This is clearly expressed in Helmholtz's slow and even reluctant acceptance of the ionists'

26. See H. Falkenhagen, "Die Elektrolytarbeiten von Max Planck und ihre weitere Entwicklung," in *Max-Planck-Festschrift*, 1958, ed. B. Kockel, W. Macke, and A. Papapetrou (Berlin: VEB Deutscher Verlag der Wissenschaften, 1959), pp. 11–34.

armamentarium. Helmholtz did, however, consider both Planck's and Nernst's work sufficiently important to promote vigorously their appointments to Berlin professorships. In an 1889 letter, quoted by several authors but whose origin is unknown, Helmholtz had described the apparent opposition between the rigorous thermodynamicists and the physical chemists. The physical chemists' approach had been "fruitful on many occasions" and resulted in a number of correct consequences, although it contains "some arbitrary assumptions":

> The chemists, however, need such assumptions [on the partial dissociation of compound molecules of dissolved salts] in order to form a conception of the processes . . . since one has to concede that the whole of organic chemistry has developed in the most irrational manner, always hanging on to material pictures [*sinnliche Bilder*], which cannot possibly be correct in the manner in which they are presented. This whole direction of the application of thermodynamics to chemistry has a healthy core, and also occurs in much purer form in Planck's work. But the thermodynamic laws in their abstract form can only be grasped by strictly trained mathematicians and are, therefore, accessible only with difficulty to those who would have to perform the experiments on solutions and their vapor pressures, freezing points, heats of solution, etc.[27]

Already in 1892, Helmholtz seems to have expressed this confidence in the usefulness of physical chemistry for chemists in general to Ostwald himself. Thus Ostwald wrote to Nernst:

> I was recently in Berlin and visited Helmholtz. It had been said that Helmholtz intended to destroy the dissociation theory, or energetics, or both, in a kind of Nürenberg talk.[28] To my question he denied this intention and explained to me that from the beginning of his electrolytic studies, he was convinced of the absolute and total freedom of ions, and that therefore he was even more radical on the subject than we were. Only the chemists made him delay the expression of his views. . . .[29]

That Helmholtz's views were crucial to the further development of physical chemistry in Germany is of particular significance when looking at the parallel careers of Planck and Nernst. Their work was of sufficient

27. Ernst Cohen, *Jacobus Hendricus van't Hoff. Sein Leben und Wirken* (Leipzig: Akademische Verlagsgesellschaft, 1912), p. 314.

28. In 1894, the annual meeting of the German Society of Scientists and Physicians was to meet in Nürenberg. This is also a wordplay on Martin Luther's talk in the same city.

29. Ostwald to Nernst, 22 November 1892, Ostwald Papers, AAW. Berlin.

importance to convince Helmholtz that a "modernization" of what he considered a science dominated by a staid organic chemistry could be accomplished by promoting these two scientists. Nernst evidently entertained a vision of physical chemistry different from that held by his mentor Ostwald and some of his colleagues, being somewhat closer to Helmholtz and less chemically militant than they might have wished.

6

Electricity and Iron: The Electrolytic Lamp

Some believed in the 1880s that except for Richard Wagner, Germans of European reputation were found mostly in the applied sciences of chemistry and electromechanics.[1] Immersed in both, Nernst was to devote a decade of intensive work to the improvement of an incandescent lamp. In 1897, he began work on the "glow worms," as they came to be known somewhat disparagingly, for which he obtained many patents in Germany, the United States, and other countries between 1898 and 1904.[2] This was the time during which he and his collaborators became especially well versed in two major scientific and technological areas: electric and electrolytic conductivity, and the behavior of solids and gases at high temperatures. A thorough knowledge of conducting and nonconducting materials at elevated temperatures was essential to the successful development of any electric lamp. In addition to its purely practical significance, work on radiation phenomena, on the relationships among light, heat, and electricity, was the most widely studied field in the decades surrounding the turn of the century.

Nernst's early lamp design was not well known outside his circle of friends and collaborators. In November 1898, *Science* magazine described in brief Nernst's "new electric lamp" on which information had "heretofore been so difficult to obtain." His announced discovery was that magnesium oxide, "which at ordinary temperatures is a non-conductor," was found to be a "perfect conductor" at sufficiently high temperatures, emitting "a brilliant white light" when employed as an illuminating filament.[3]

Thus Nernst's work renewed interest in research on metal filaments, which ultimately led to the modern lamps, while he himself remained less successful in the ultimate refinement of his dielectric glowers as

1. Egon Friedell, *A Cultural History of the Modern Age*, trans. Charles Francis Atkinson, vol. 3 (New York: Knopf, 1933), p. 320.
2. For instance D.R.-P. 117041, 1900. See Nernst, "Einiges über das Verhalten elektrolytischer Glühkörper," *Zeitschr. f. Elektrochem.* 1900: 373–6. For a good discussion of this work, see Bartel, pp. 44–9.
3. *Science* (18 November 1898): 689–90.

illuminating marvels. However, the "Nernst glower" or "Nernst globar" has survived as an important photometric instrument. It has been a standard teaching spectrophotometer in university laboratories. And even as recently as 1994, scientists have incorporated bright Nernst glowers into powerful, time-resolved, infrared spectroscopic methods for the investigation of short-lived intermediates produced in photochemical reactions. One such recent instrument, which uses a high-intensity Nernst glower originally manufactured around 1969, pleasantly surprised chemists in Nottingham, since it was found to be "so bright that there are instances of the glower initiating photochemical reactions in the UV-visible region!"[4]

Nernst's work on the lamp came at a time of explosive developments, initiated and facilitated by the striking expansion of the electrical industries. If the 1870s had epitomized Bismarck's rule of gold and iron, the fin de siècle was becoming the era of electricity and iron – not the iron of swords and cannons but the iron and steel of the Eiffel tower, of 40-foot power generators, of cable poles and railroad tracks, and of the world's largest naval fleet. The broad availability of electrical energy facilitated by the invention of the dynamo, as well as the improvement of batteries, electric cables, power stations, engines, and illuminating devices, brought about the need for a complete reorganization and overhauling of many industrial, economic, and consumer enterprises. Easily available and affordable electricity was the major impetus for the construction of electrochemical plants that produced metals and salts, themselves raw materials necessary for the heavy and electric industries. Changes in one area prompted novel applications, which in turn raised fresh problems in need of solution and eventually led to scores of new inventions and products. The last decades of the nineteenth century were characterized by an intensifying interdependence among all industrial branches at all levels: research, production, marketing, and financing. It was a particularly important era for the consolidation of the large concerns of Germany, leading to the emergence of such well-known industrial giants as Siemens, AEG, Krupp, Zeiss, Bayer, and others. But Germany was not alone: Battles for markets and over patents were also waged by the Edison Company in the United States, competing against General Electric at home and the British and Germans elsewhere. Much was at stake in the process of electrification worldwide, where capital investments could produce impressive financial returns. No one had doubts that in the long run, illumination, communication, industry, construction, and transportation would consume huge quantities of electrical power.

4. Michael W. George, Martyn Poliakoff, and James J. Turner, "Nanosecond Time-resolved Infrared Spectroscopy: A Comparative View of Spectrometers and Their Applications in Organometallic Chemistry," *The Analyst*, 119 (April 1994): 554.

The difficulties in implementing such large-scale changes were formidable, involving an unprecedented harnessing of enormous financial and human resources in research and education. We should, therefore, view Nernst's work between 1895 and 1905 as an integral part of the larger cluster of studies initiated and sustained by the rapidly expanding electrotechnological industries of the day. Nernst's work on his lamp was contemporaneous with similar work carried out at the Physikalisch-Technische Reichsanstalt in all its three divisions (heat, electricity, and optics), at many other universities, and in the nascent industrial research laboratories.

The development of electricity, electric motors, cables, dynamos, and power stations was concurrent with the expanding range of their applications, a bilateral relationship of supply and demand that provided ample opportunities for connections between many fields of science and engineering. One of the most popular uses of electricity was illumination by means of the carbon arc lamp. In the late 1870s, a small power station supplied the carbon arc lamps that were set up on the Victoria Embankment and in several London shops. With the advent of Thomas A. Edison's first central power station located on Pearl Street in New York City, private consumers could for the first time illuminate their homes with carbon filament lamps. In 1887, the South Kensington Museum of Science installed 860 16-candlepower glow lamps, which purportedly used 57 percent less power than equivalent gas lighting.[5]

While it is generally held that electric power stations made the introduction of electric light a viable alternative to gas lighting, it seems that the reverse was also true. Industry and commerce, and later the general population, spurred the construction of power stations by demanding the availability of electricity for electric lights – safer and easier to use than gas. The gas illumination industry was not, however, immediately superseded. Street lighting with gas improved dramatically, automatic switching stations made the "gas lighter" superfluous, and gas-production methods were significantly simplified.

While it could be safely argued that one of the main purposes of the early electrical power industry was lighting, until the early decades of the 1900s dissatisfaction with almost all available illumination devices induced myriad improvements, inventions, and the investment of substantial capital. In the mid- and late nineteenth century, the only viable electrical illumination device had been the electric arc, which emitted a brilliant, bluish-white light from the production of an electric spark between two carbon electrodes. In 1893, the open arc light was replaced by the enclosed arc

5. W. T. O'Dea, *Handbook of the Collections Illustrating Electrical Engineering*, Part I, *History and Development* (London: Science Museum, South Kensington, 1933), p. 69.

flame that flickered – albeit with a somewhat decreased luminosity – inside a glass shield. By diminishing the carbon electrodes' contact with the oxygen in the air, the enclosed lamp lasted longer and the carbon electrodes required less frequent trimming. The lamp came to be operated on 110-volt constant-potential circuits. Since the arcs operated at 80 volts, a resistance was used in series and acted as "ballast." By 1898, it was found that different colors of light could be produced by impregnating the carbon electrodes with different substances: Calcium fluoride gave a brilliant yellow light, barium salts emitted a white light, and strontium salts gave a red light of lower efficiency, used for advertising purposes. These lamps could operate for some fifteen hours, and they were cleverly redesigned to have the carbon tips tilted downward and an electromagnet built into the lamp that "blew" the arc's flame downward, thus becoming suited for direct overhead lighting.[6]

The powerful arc lights were employed in lighthouses, train stations, factories, and the larger department stores, but it was only with the various incandescent filament lamps that houses could be lit. These incandescent bulbs had problems of their own, their main fault being the disintegration, or melting, of the filaments at the high temperatures produced by the passing electrical current. It was clear that an evacuated bulb, a "vacuum lamp," might solve this problem by reducing the degradation of the filament in air and the vapors released by the glass bulbs. The possibility of using electricity for illumination had been explored in the 1800s, when Sir Humphry Davy used the newly discovered voltaic pile to produce the first electric arc. It was also Davy who first brought a platinum wire to incandescence by passing electric current through it and observing the intense light emitted. W. J. Starr had experimented in 1841 with a carbon filament, but it was the advent of dynamos and generators that spurred Edison's lamp research in the 1870s. With the invention of the mercury vacuum pump, as well as the carbon filament bulb produced by Edison and Joseph Wilson Swan in the late 1870s, the modern light bulb in its essentials already existed and was in common use by 1900.

Yet upon close inspection, one finds that the practical development of an incandescent lamp was primarily a *chemical* problem, rather than a physical one. Edison's incandescent lamp consisted of a fine synthetic carbon filament inserted as part of the electric circuit into a glass bulb. The best filaments were made by cutting Japanese bamboo fibers. These were inserted into a mold that was placed in a furnace, where the filaments were carbonized. The tips of the fibers were then cut flat and squeezed into cop-

6. Henry Schroeder, *History of Electric Light,* Smithsonian Miscellaneous Collections, vol. 26, nr. 2 (Washington: Smithsonian Institution, 1923), p. 68.

per clamps, which were welded together. These copper clamps were soldered to platinum leads, which were in turn sealed through the glass and
connected to the electric wire conductors. It was a remarkable and useful
coincidence that the thermal expansion coefficients of glass and platinum
were nearly identical, preventing the bursting of glass or the rupturing of
the wires.[7] Eventually, Swan developed the parchmentized cotton thread
and the squirted thread of cellulose "which was destined to become the
universal process." This filament was improved by "flashing" in an atmosphere of hydrocarbon gas. Later, the "metallized" carbon filament increased the efficiency of the lamp, but around 1902 the much more efficient tantalum lamps came into use.[8]

Against this backdrop, Nernst managed in 1897 to build a lamp that cir-
cumvented the labor-intensive process of evacuating carbon-filament
lamps and sold the patents to the AEG. The lamp, which operated in open
air rather than in an enclosed bulb, consisted of a burner – the glower –
composed of a mixture of rare earths. Its main ingredient was zirconium.
This lamp had an increased luminosity and a more efficient fuel consumption, but it suffered from other disadvantages. The Nernst glower was
a nonconductor at ordinary temperatures, and hence required preheating.
A platinum electric coil located above the glower served as a preheater.
Once the glower reached the temperature at which it became electrically
conducting, the platinum heating wire circuit was automatically cut off
by an electromagnet. In addition, Nernst's lamp also used a ballast in order
to avoid excessive increases in current, which would lower the resistance
of the glower. This ballast was a complicated piece of microengineering:
An iron wire was installed into a bulb filled with hydrogen gas. Since iron
has a marked increase in resistance with the increasing current flowing
through it, it acted as a counterweight to the zirconium glower.
The problems associated with designing all the component pieces of this
complex lamp were addressed by Nernst and his pupils and collaborators
from a variety of viewpoints. For example, it was necessary to take into
account the precise temperature range at which the lamp was operating
and to correlate it with the most suitable "ballast," one which in a certain
chemical environment would exhibit an appropriately dramatic increase
in electrical resistance. For this purpose, exact knowledge of many physical and chemical properties of conducting materials in a range of high

7. Th. du Moncel and Wm. Henry Preece, *Incandescent Electric Lights* (New York: D. Van
Nostrand, 1882), pp. 64–6.
8. Dwight T. Farnham et al., *Profitable Science in Industry* (New York: Macmillan, 1925),
pp. 43–4.

Two Nernst lamps (Deutsches Museum, München)

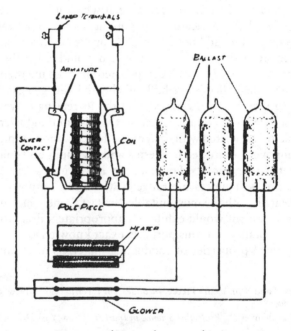

Diagram of Nernst lamp mechanism

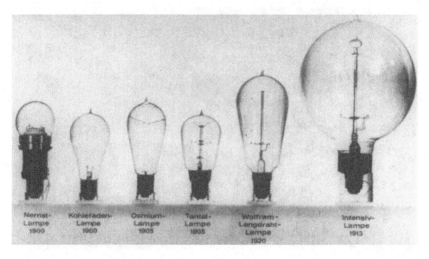

Lamps from the AEG catalog

AEG factory assembly room for Nernst lamps

temperatures had to be investigated. The reactivity of heated wires in hydrogen and other gases had to be considered as well. This research would provide Irving Langmuir with information needed for the design of the modern electric bulb. But even by 1912, specialists in the field agreed that

München department store with Nernst lamps

Nernst's lamp had exercised a "stimulating effect upon the makers of carbon lamps" and had also constituted the "cause for the renewal of experiments with the rare metals for lamp-making purposes." It might well have been that, had certain design problems been solved earlier, the introduction of metal filament bulbs would have been delayed for a considerable time by a successful Nernst glower. As it was, the modern filament bulb was the result of experiments with Nernst lamps.[9]

Nernst at first sought to obtain financial support for the development of his lamp from the Siemens company and even spent some time in their Berlin laboratories. The already well-known Siemens & Halske company, which had been founded by Werner von Siemens (1816–92) and Johann Georg Halske (1814–90) in 1847 as a telegraph factory, was by then producing not only cable lines but also dynamos, telephones, motors, and illumination equipment, in both Germany and England. Besides his celebrated discovery of the dynamo, Siemens had built the first electric motor-driven locomotive, shown at the Berlin exhibition of 1879. Two years later, Siemens introduced the first electrical tramway in Berlin.

Nernst's trips from Göttingen to the Siemens laboratories often included visits to the new director of the PTR, his erstwhile advisor, Friedrich Kohlrausch. But Siemens was not particularly interested in Nernst's designs, and eventually he turned to the AEG's general director, Emil Rathenau,

9. G. Basil Barham, *The Development of the Incandescent Electric Lamp* (London: Scott, Greenwood & Son, 1912), p. 10.

and its director Paul Mamroth (1859–1938), who visited his Göttingen laboratory.

Oskar von Miller (1855–1934) and Emil Rathenau (1838–1915) had founded the Deutsche Edison-Gesellschaft in 1883. In 1887 it was renamed the Allgemeine Elektrizitätsgesellschaft (AEG), soon to become renowned. During the next two decades, the AEG's production of a variety of electrical light sources, motors, appliances, and illumination devices became a major component of the German electrical industry. Oskar von Miller was one of the most visible and influential Bavarian engineers and entrepreneurs. He was an effusive and energetic promoter of the electrical industry in Germany and founder of the Deutsches Museum of science and technology in Munich in 1903. After having been the organizer of the first electrical exposition in Germany in 1882, von Miller became the central figure responsible for the construction of Germany's first electrical power system and for the design of Berlin's first central power station. Von Miller was particularly interested in creating networks capable of carrying alternating currents over large distances, and he was also a great proponent of hydroelectric power. In 1882, he had helped construct the first direct-current overhead line, which ran between Miesbach and Munich and operated at 2,000 volts. Almost a decade later, Miller built the longest three-phase overhead conductor, spanning 178 kilometers between Lauffen and Frankfort and supplying 25,000 volts.

Like other contemporary leaders of industry and finance, Emil Rathenau – father of the future Weimar minister Walther Rathenau – was a man of many qualities and interests. His family had resided for five generations in Berlin. They were part of the so-called *Kaiserjuden*, Jews who had succeeded in obtaining the business – and often the protection – of the imperial household, although they were far from being truly accepted by the Prussian elite.

By 1900, Rathenau's AEG's offices and factories, the AEG's hexagonal trademark, its stationery, fans, and street lamps were being designed by modern artists and architects. First among them was Alfred Messel, who built the new AEG offices between 1905 and 1907 in a Greek revival style, followed by Peter Behrens (1868–1940), who became AEG's artistic adviser. Between 1909 and 1912, Behrens built the AEG factory complex, which, with its famous turbine assembly works with glass curtain walls, became one of the most remarkable and architecturally influential buildings in Germany at the time. During these years, the future Bauhaus group of Walter Gropius, Ludwig Mies van der Rohe, and Le Corbusier collaborated with Behrens. But Behrens was, in his own right, a highly versatile and avant-garde architect, designer, decorator, and painter, whose work reflected the functionalist ideals of clarity of line. His light fixtures, some of which were exhibited in Paris in 1900, "show his predilection for clean, straight lines . . . evident in his metal and opalescent glass night light, silver candlesticks, ceramic table lamps, and arc lamps designed for the AEG." His work presaged the transition from objets d'art, which characterized Art Nouveau, to the spareness of Art Deco, which was to benefit immensely from technological advances in the years following World War I. No longer would lamps be an excuse for countless flowers, pistils, and women glowing from within through tinted and heavily ornamented glass. Instead, "the 1900 *art* of lighting gave way to the 1920 *science* of lighting,"[10] no longer concerned with decorating lights, but rather with directing light.

In the mid-1890s, the AEG was producing 20,000 carbon lamps per day, and its ever-growing output required the construction of a new lamp factory on the Schlegelstrasse in Berlin. In 1898, the AEG concluded a collaboration agreement with Nernst that was financially quite advantageous to him.[11] In its annual report for the year, the AEG wrote that since the middle of March 1898, it had been investigating the Nernst lamp in their laboratory. The major difficulties encountered had been the time it took to heat the filaments, which at ordinary temperature were nonconducting, as well as obtaining a stable filament. In addition, the AEG was attempting to improve the efficiency and versatility of the lamp, hoping it would be capable of functioning at variable degrees of luminosity. The AEG was well aware that the Nernst lamp was superior in convenience and ease of operation to the arc lamps, but that it lagged behind the carbon filament bulbs. From the beginning, therefore, the company did not expect the Nernst lamp completely to replace existing incandescent carbon lamps.

Nernst entered into a close relationship with the director of AEG's lamp manufacturing section, Bussmann, whose two sons-in-law were Nernst students. The group continued work on the development of the lamp in the AEG laboratory itself. But although the prototypes seem to have been ready by 1899, a new factory section needed to be readied before production was finally started in 1900. The next year's report stated: "After years of long and strenuous work, the Nernst lamp has finally become a complete success. The beautiful and economic light can be found in hundreds of thousands of lamps already in use. It has gained an ever-growing reputation because of its very good performance and the extremely satis-

10. Alastair Duncan, *Art Nouveau and Art Deco Lighting* (New York: Simon and Schuster, 1978), pp. 121, 139.
11. Mendelssohn quotes the sum of 1 million RM, which is not verifiable. Nernst's daughter denied the correctness of this figure in correspondence with Mendelssohn, but the folklore among physicists has persisted.

factory results of measurements carried out by the Physikalisch Technische Reichsanstalt."[12]

The AEG's investment in the Nernst lamp was spurred not so much by a desire to compete with Edison's carbon filament lamp as to provide an alternative to the gas industry's lamp, which seemed to threaten the "energy devouring" electric lamps. The competing industries "labored mightily" in order to better understand the underlying principles of illumination, improve existing lamps, and devise new ones.[13] Nernst himself was seen by foreign commentators as being engaged in a "great race between gas and electricity," as many of Nernst's friends and colleagues were eager to mention in private correspondence. The rapid progress that Edison's light had achieved was soon threatened by an invention of Auer von Welsbach, who made an improvement on his famous eponymous gas lamp – the "Auer mantle" of 1886 – by employing acetylene. In 1892, the acetylene blower gas lamp was introduced in Berlin, and within a year, the investors in the Auerlicht-Gesellschaft had netted an extraordinary profit of 65 percent, followed by similar profits over the next years. Many believed electricity would after all have to yield the supremacy to gas. "Nernst now reclaims the palm to electricity," gushed a writer in 1898, "for he expects that the cost of his light for a whole evening will be no more than that of the Edison for an hour."[14]

The AEG's hope of obtaining a serious advantage over its competition was misplaced; the Nernst lamp never became a commercial blockbuster due to the need for preheating, which could last several minutes. The AEG tried to circumvent the disadvantage of the "preheating" of the Nernst lamp by producing a hybrid filament, consisting of the Nernst heating coil and carbon filaments. This *Expresslampe* was a masterful engineering feat, but its production costs were high. They also created a Multi-Lampe, a combination of many Nernst lamps, which would require a relatively short preheating period.

Upon embarking on conductivity studies, Nernst had thought that among the metal conductors, none would be suitable for economically viable electric illumination at high temperatures. Hence, he had chosen the problematic magnesium oxide, a combination that became conducting only at higher temperatures, unlike the pure metal filaments. Auer von Welsbach, however, did not share Nernst's skepticism, and discovered in 1886 that osmium filaments would emit a steady white light, that they

12. Felix Pinner, *Emil Rathenau und das elektrische Zeitalter* (Leipzig: Akademische Verlagsgesellschaft, 1918); reprint (New York: Arno Press, 1977), p. 190.
13. Bartel, pp. 45ff.
14. H. C. Cooper, of Heidelberg, to *Science*, N.S. (203): 710 (1898).

had a rather long life, and that they could be manufactured much more easily than the Nernst mixture. In addition, the prices of electric current began to steadily decline. Although Emil Rathenau had invested huge sums and manpower in the Nernst lamp, he soon recognized the superiority of Welsbach's discovery and decided to cut his losses. Although some four million Nernst lamps had been sold, the AEG never recovered its investments and discontinued production altogether in 1909. Only spare parts and projector lamps were sold, and the AEG turned toward an intensive research program into the rare earths filament production.[15]

In September 1906, the Auer Gesellschaft succeeded, after laboriously experimenting over a period of several years with brittle osmium and tungsten filaments, in producing the first successful metal lamp. The foreign-sounding Osram lamp – a name created by the conjunction of osmium and wolfram, the latter being the German name of tungsten – became an instantaneous success. It was a devastating blow to the AEG. Emil Rathenau, who had feverishly driven his technicians to succeed, soon thereafter abandoned the Nernst lamps and moved to the production of the new bulbs and other products. By 1909, the company had established subsidiary factories in London and Paris. The original company moved to new quarters in a building that was to remain into the 1920s Berlin's only "skyscraper," having three subterranean and eleven above-ground levels.

By 1911, the tungsten lamp had been significantly improved. Many other electrical companies were experimenting with tungsten filaments – which had been mentioned in patent applications connected with improvements to the Nernst lamp – and a large number of scientists and technicians were working on the subject, as for example the chemists of the Siemens & Halske company. The preparation of viable tungsten filaments was a task assigned to the chemists, since the process involved chemical reactions of tungsten powder in an environment of various gases in order to remove impurities, such as carbon, from the metal wires.[16]

After 1913, the Nernst "glower" was completely displaced from use by his former student's invention of the inert gas–filled tungsten lamp. Irving Langmuir (1881–1957), then at the General Electric laboratories in Schenectady, New York, invented the gas-filled light bulb and thus reclaimed the supremacy first achieved by Edison for the United States.

Research into Nernst lamps did not completely disappear despite the commercial failure, however, and the details were relevant to those scientists and engineers working in conductivity studies and in the very important

15. Pinner, p. 283.
16. Barham, pp. 57ff., 121ff.

field of heat radiation. Broadly defined, incandescent lamps involve processes of heat and light radiation; that is, they contain an opaque substance strongly heated to incandescence. In 1897, Lummer had boldly articulated the common goals of pure research and of research into the industrial applications of artificial illumination: to separate light from heat, and to investigate the properties of various substances and the physical laws that controlled the two phenomena. The PTR scientists embarked upon research on black-body radiation, which, only two years later, was to provide sufficient puzzles and results to lead ultimately to a complete reformulation of the laws of physics. Max Planck's papers of 1899–1901, which drew upon the experimental results of work on black-body radiation at the PTR, produced the unheralded advent of quantum theory and eventually forced fundamental changes into the work of physicists and chemists alike.

By 1898, Lummer and Kurlbaum's radiating cavity or black body at the PTR was found better suited than previous devices for electrical and heat standardization work.[17] Although an ideal black body – one that absorbs all radiation incident upon it – exists only in theory, definitions of its properties constitute the foundation upon which all research on incandescent, radiating solid substances was based. For an ideal black body, the maximum of the total energy emitted will fall in the visible spectral region only for temperatures between 3500° and 9000°K. The two options for the material of the hot filament considered by technologists in the 1880s and 1890s were carbon and several temperature-resistant metals. Among the metals known before World War I, tungsten withstood a maximum of 3655°K, while carbon in certain forms was capable of withstanding even higher temperatures. While not ideal, carbon and tungsten could be considered to behave in ways that approximate the properties of a black body.

The efficiency of tungsten filaments was up to nine times more light per watt than carbon. Carbon also had other disadvantages: Its practical operating temperature was lower, and its radiation intensity at any wavelength (from the infrared to the ultraviolet) would therefore always be lower than that of tungsten at its normal operating temperatures; tungsten lamps were always brighter. In addition, carbon filaments vaporized fast, and hence had a shorter life. Moreover, carbon did not have the radiation selectivity of tungsten; its emissivity is pretty much constant throughout the light spectrum. With this approximately constant (flat) emissivity, carbon lamps approached the radiation patterns of a "gray" body. Tungsten's selective radiation falls in the visible range and is closest to that of a black body in the visible spectrum. Its emissivity at the visible wavelengths is

17. Cahan, pp. 147–9.

thus very high. For example, at 2100°K, the emissivities are 0.46 and 0.38 at wavelengths of 0.5 micron and 1.0 micron, respectively. (Emissivity is the ratio of the energy radiated from a substance to that radiated by a black body at a given temperature.)

However, one advantage of the carbon lamps was their robustness, which enabled them to withstand mechanical shocks. In addition, carbon has a negative temperature coefficient of resistance: Its resistance decreases with temperature. Therefore, the carbon filaments were not prone to the excessive current "rush" that occurs when tungsten filament bulbs are initially switched on.

Besides issues regarding the candlepower, or luminosity, and the efficiency of the lamps, one important area of concern was the effect of electrical frequency on the quality of any lamp. The fluttering of the electric lights available on low frequency circuits produced a fatiguing effect on the retina of the eye, and electrical engineers were interested in improving illumination because they wished "to furnish a light that will be pleasing to the users." For example, as late as 1907, researchers at the University of Wisconsin's engineering department compared incandescent lamps with the Nernst glower. They found that the incandescent bulb was "influenced by frequency very much more than the Nernst lamp," a result that was not due to the materials of which the glowers were made but to the thickness of the filament and the sensitive ballast that is placed in series with it. Ballast, or preheating devices, are still used in all modern fluorescent light tubes. Nernst's ballast introduced a resistance in series with the filament at the "slightest increase of current." Because of the lower heating temperature of the filament, and the lesser cooling during slight current variations, Nernst's lamp emitted a more steady and uniform light. However, the lamps had to work at two frequencies, and the Nernst lamp was found to have a shorter life at low (25 cycles) than at high (60 cycles) frequencies.[18]

The Nernst lamp's finest hour should have been the German electrical pavilion at the Paris International Exhibition of 1900, which was illuminated by thousands of Nernst lamps. The Paris exhibition was an unprecedented triumph for German science. In 1892, Germany had proposed hosting a great exposition in Berlin in 1900 but lost its bid. At the Chicago exhibition of 1893, Germany had done well, but in 1900, millions of Germans flooded Paris, lured by enthusiastic newspaper reports. The Paris exhibition was a decentralized affair, having taken over entire sections of

18. Frederick W. Huels, "A Comparison of the Effects of Frequency on the Light of Incandescent and Nernst Lamps," *Bulletin of the University of Wisconsin* 157 (1907): 401–3.

the city, all along the Seine from the Tuileries to the Eiffel Tower. The banks of the river were home to spectacular floral displays, to the army and navy, to hunting, fishing, and pleasure boats. Exhibiting countries had the opportunity to build their own pavilions on the Quai des Nations. The Germans built a Rathaus – a city hall of the sixteenth century – complete with the tallest spire among all other pavilions; it evoked the Hanseatic city that had become immensely popular with the publication that year of Thomas Mann's first novel, *The Buddenbrooks*.

No Teutonic symbolism, however, was necessary on this cosmopolitan occasion. The splendor of German manufacture and industry, spread over more than fifteen locations, spoke for itself. Germany, unlike France and other nations, refrained from any ostentatious military displays. The exposition itself had broken with precedent, in that its international and cosmopolitan flavor was a conscious decision: Rather than allowing each country to exhibit separately, the organizers decided on thematic clusters, concentrating on heavy and light industry, manufacture, art, chemistry and pharmaceutics, textiles, glass, and other categories. In addition, it had been decided that only those industrial finished products be shown which would "enter into the general use of all nations."[19] Hence, the machine exhibits so popular at earlier exhibitions were not the centerpiece at this one. Also, the Germans had decided that the chemical section would be organized as a national entity; instead of presenting the products of individual chemical companies, the Germans labeled only their products' names and applications. In not specifying their manufacturers, they thus underscored the total impact of a national industry, rather than emphasizing its competitive basis.

Most interestingly, the Germans had decided not to separate electrotechnology from chemistry. Therefore, the high-temperature extraction methods for aluminum, electrochemistry, the liquefaction of gases, and electrical power production were grouped together, reflecting the opinion of many – among them Nernst and Haber – that they all formed an integral whole. A section of the chemical exhibition was devoted to chemicals needed in industrial production. Among them, the rare earths took a prominent and literally colorful place, and it seemed that the technology developed between 1894 and 1900 for the extraction of rare earths had been "occasioned by the needs of the illumination industry." Although rare earths development had been born in the United States, and although French companies apparently excelled in producing other rare earths that accompany the extraction process of thorium, the center of production

---

19. Otto N. Witt, *Die Chemische Industrie auf der Internationalen Weltausstellung zu Paris 1900* (Berlin: Gaertner's, 1902), p. 4.

had "shifted to Germany," according to one important exhibition organizer. The purest thorium was produced primarily in Germany, "and its
quality has not suffered despite the reduction in its price from 1,500–
2,000 marks to 50 marks per kilo" in the six short intervening years.[20]

Another major and novel leitmotif of the exhibition was its historical,
retrospective character. The exhibition was "not only to express the height
to which industry had risen in the nineteenth century, but also the path"
by which it had done so. And while the 1893 Columbian exhibition in the
White City of Chicago – named for the white pavilions constructed for
the occasion – had overwhelmed visitors with the riches of the earth and
the promise of unlimited natural resources, in Paris it was the manufacturing and industrial potential of such materials, themselves now taken
for granted, that was highlighted.

The technical exhibits were located on the Champs de Mars. They "revealed a veritable explosion of method . . . the greatest instance of commercial encirclement the world has ever seen."[21] Among the exhibits was
the world's largest dynamo, built by the Helios company of Köln and gen-
erating 5000 horsepower, and a demonstration of liquid air and the production of cold. The British journal *Nature* extolled the impressive instrument section of the German pavilion. The sight was not "one which
brings great pleasure to an Englishman, and if he moves on to examine
the English exhibit his thoughts cannot fail to be very grave." Germany's
exports of scientific instruments had tripled within the decade between
1890 and 1900. The British saw the "flourishing state of the instrument
trade in Germany" as a direct consequence of the "unity of its aims, which
is traceable to the history of its development and to its ultimate connection to pure science."[22]

And not only the British marveled at the beautiful 250-page catalog, issued free of charge to visitors, which depicted – in German, English, and
French editions – the history, development, and applications of the various instruments, products, and machines. As with the chemical exhibits,
these were placed not according to the "usual grouping under various
firms," but in "sections embracing certain classes" of mechanical and optical instruments and the "art" of their manufacture. The catalog was an
artistic and literary event. Printed in special type used by German print-
ers during the Gothic period for texts in foreign languages, it was ac-

---

20. Witt, p. 130.
21. Paul Morand, *1900 A.D*, Trans. Romilly Fedden (New York: W.F. Payson, 1931), p. 77,
    quoted in Richard D. Mandell, *Paris 1900* (Toronto: University of Toronto Press, 1967),
    p. 82.
22. "Instruments of Precision at the Paris Exhibition," *Nature* 63 (1620): 61–62 (1900).

companied by colored illustrations and decorations in the highly fashionable Art Nouveau style of the day. The book was bound in woven beige muslin, and was accompanied by a detailed index and eighty pages of advertisements, all designed in the same style, with photographs and illustrations printed on high-quality paper. Most interesting was the historical and sociological introduction to the volume, which provided a detailed demographic survey of Germany, including data on women's employment, divorce, racial and religious groups, and emigration. Among the more remarkable data given were these: In only four years, between 1895 and 1899, the German population had risen from 52 to 55 million, and it had doubled from 1816 to 1900; 61 percent of Germans were under the age of 30, with an average of 4.7 children per family; and 1 in 80 marriages ended in divorce.

New uses for electricity were a major theme on the Champs de Mars. A moving sidewalk saved visitors time and effort, while telephone instructions and automobiles were used to help set the 606 tables and serve the food at the extravagant banquet given by the French mayors. As for the Germans, the AEG had hoped that the big event of the illumination industry at the Paris exhibition would be the Nernst lamp. They wished to light the thousands of lamps from a pavilion built especially for this purpose in the Hall of Honor.

The demonstration of a successful prototype of the Nernst lamp was being anxiously awaited in the highest German circles. When sometime earlier that year Walther Rathenau had given a lecture on "Electrical Alchemy," both His Majesty the Emperor Wilhelm II and his secretary, Eulenburg, repeatedly asked whether the Nernst lamp would soon be available. Rathenau reputedly replied: "We will have a great demonstration at the Paris Exposition. The French government has approved for us the only pavilion in the Cour d'honneur." Rathenau expected the demonstration to bring great honor to Germany, a hope with which Wilhelm concurred. "It will!" he declared, but reconciled himself to the humorous thought, "Well, if the Parisians come first, then we'll have to wait."[23]

The plan faltered, to their great chagrin. The French organizers had set aside the hours from 5:00 to 7:00 P.M. for these demonstrations, but because no new visitors were admitted onto the grounds after 6:00 P.M., and since the Nernst lamps required time and elaborate attention, "not many were allowed the privilege of actually seeing the lamps" which had delighted the daily newspapers and professional journals. The lamps had to

---

23. "Äusserungen Sr. Majestät des Kaisers gelegentlich des von Walther Rathenau im Postministerium zu Berlin gehaltenen Vortrages über Elektrische Alchymie," no author, typescript, 3 pages, 1900. Bunsengesellschaft Archives, Karlsruhe.

The Nernst Pavilion at the Paris Exhibition, 1900

be preheated with alcohol burners, gas burners, electric current, or electric arc, a procedure which was quite cumbersome; preparations could last as long as a minute for each lamp, and thus the advantage of electric light seemed to disappear. When the lamp was preheated with an incandescent platinum wire, time was saved, but with the disadvantage of additional complications and higher costs. The automatic preheating was not quite reliable, and therefore the lamps in Paris were operated with open flame heaters. The resistance of the electrolytic conductors used in the Nernst lamp decreased with rising temperature, contrary to the behavior of normal conductors. And although a counterresistance was wired into the bulbs' sockets for this purpose, as many as 600 Nernst lamps burned out simultaneously at one Paris demonstration. For all these reasons, the Paris exhibition was a "heavy disappointment."[24]

Although no completely new lamps had been displayed in Paris, the entire city had been bathed in thousands upon thousands of arc and carbon

24. Hans Kraemer, *Die Ingenieurkunst auf der Pariser Weltausstellung 1900*, reprint (Düsseldorf: VDI Verlag, 1984). Originally published as Hans Kraemer, *Das XIX. Jahrhundert in Wort und Bild*, vol. 4 (Berlin: Dt. Verlag.-Haus Bong, 1900), pp. 104–7.

lamps, projectors, lanterns, and illuminated advertisements. The exhibition grounds were lit by precisely 21,749 lamps, 90 percent of which were electric. Another 7000 nonglare lamps were installed in simple straight lines to a height of 300 meters on the Eiffel Tower.

And the aftermath of the Paris "disappointment"? Nernst's eighty international patents, his years of intense labor in the laboratory and with industry, his comparative neglect of "purely scientific publications"[25] were to engender mixed, often rancorous, feelings among his colleagues and students. Some were thrilled, however. In late 1902, his old mentor and close friend, Ludwig Boltzmann, ordered several Nernst lamps to be displayed at his home at a party for 55 colleagues and friends – they were a rarity in Vienna. A cold buffet and fifty liters of beer were prepared, but only seven gentlemen showed up.[26]

It is evident from all the above that Nernst's research interests were profoundly enmeshed with the development of electrical and electrochemical instrumentation, and with technological culture in general. What Nernst learned to do best, and what he thought he had shown others how to do, was systematically to improve illumination by raising the temperature of the conducting filament in the new lamps. Between 1896 and 1905, Nernst was located squarely within both the practical and the theoretical environments of electrotechnological research.

25. J. Henricus van't Hoff nomination letter to the Prussian Academy of Sciences for Nernst, *Physiker über Physiker,* p. 162. See also Bartel, p. 46.
26. Boltzmann, *Leben und Briefe,* p. I 260.

7

High Temperatures
and the Heat Theorem

The problem of achieving a steady illumination was probably the central question Nernst had to address in his work on an improved electrolytic lamp. It was from these practical concerns, rather than a theoretical interest in elucidating the chemical equilibrium problem, that the heat theorem will be seen to emerge.

Nernst's acumen was to realize the significance of work carried out in related fields for the domain of chemical transformations at the extremes of physical conditions. An avid automobilist, owner of the first motorcars in Göttingen, Nernst knew about engines, electrical and otherwise. In the 1890s, he purchased his first battery-operated car, and developed a powerful accumulator battery in his institute. After visiting the Paris exhibition in 1900, Nernst purchased his first gasoline-powered automobile, to be followed by many more during the next years. These not only provided relaxing and attention-arousing excursions in the environs of Göttingen but also contributed mightily to Nernst's interest in combustion processes. Unlike his mentor Kohlrausch, who had almost no relations with industry, Nernst absorbed the Göttingen spirit of Felix Klein and paid attention to mathematics, physics, and industry. A physicist by training and inclination, he never really oriented himself toward the purely "chemical" way of "seeing things."

In his frequent trips to Berlin, Nernst had seen a metropolis in the making. Compared to placid, academic Göttingen, life in Berlin was positively tumultuous. The entire city celebrated the dawn of the new century, although no one knew exactly when it would officially begin. Notwithstanding the Pope's decision that 1900 would be the last year of the previous century, Wilhelm, with support of parliament, opened the official ceremonies of the new century on 1 January 1900. The emperor ordered the expansion of his naval fleet and built the Pergamon museum of antiquities in Berlin. The main boulevards were expanded, stalls and beer halls were demolished, and Wertheim's, the first gigantic department store, was built where once a bazaar of a hundred little rooms had stood in the Leipziger Strasse. With some 16,000 factories and workshops, Berlin was becoming one of the most industrialized cities in the world. But although thousands of electric lamps illuminated the city, only the wealthier private households benefited from the use of the plentiful electric current feeding the city's industry and commerce. Almost 2 million people lived in the inner city, and another million moved into its 23 suburban neighborhoods within several years, most of whom commuted daily into the city. By 1900, Berlin could boast of being the first city with a fully electrical public transportation system. Thirty-five tramway lines crisscrossed the main square, the Potsdamer Platz, elevated trains raced through imposing stations, and 120 trains per hour were assured safe passage over intercalated bridges and roads. By 1903, they could travel at 150 miles per hour.

Nernst's passion for automobiles was unusual in Göttingen. But in 1901, the first Berlin-to-Paris rally was held, for which the ladies were provided with adequate face masks. An automobile was, of course, an expensive hobby: A small "Passe partout" model, with a 7-horsepower engine, cost 2700 marks, while the monthly salary of a metal worker was about 120 marks. Just before his triumphal car ride to inaugurate his professorship in Berlin in 1905, Nernst gave a paper at the 46th convention of the Organization of German Engineers in Magdeburg on the topic of "Physical-chemical considerations on the combustion process in gas engines." There he discussed the results of experiments on gas equilibria that were to provide the main body of data leading to the enunciation of the heat theorem that same year.[1]

The intricate path that led Nernst from work on an improved incandescent lamp, via high-temperature investigations of the various properties of metals and other compounds, to the enunciation of the heat theorem in late 1905 merits detailed examination. Such scrutiny reveals how, in practical terms – that is, in terms of everyday problems in need of solution – Nernst and his collaborators integrated, improved, and invented various methods, concepts, and instruments in use and of use to many scientific and technological pursuits.

We shall now focus on a particular, longstanding, and well-localized problem, that of specific heats. Nernst came to devise experiments related to specific heats because

1. He was working on burning filaments;
2. He needed to understand the behavior of matter at very high temperatures;
3. At very high temperatures the problem of specific heats may become paramount.

1. Bartels, p. 64.

	NERNST'S	
WORK ON ELECTRIC LAMP		WORK ON ELECTRIC MOTORS
	involves theoretically	
CONDUCTIVITY, GASES, EQUILIBRIA, RADIATION LAWS, CHEMICAL LAWS		COMBUSTION, GASES, EQUILIBRIA
	and encounters practical problems with	
ELECTRICAL CONDUCTIVITY AT HIGH TEMPERATURES		MEASURING AT HIGH TEMPERATURES
	that require	
a. STUDY OF CARBON, SPECIFIC HEATS		a. DESIGN OF MATERIALS FOR HIGH TEMPERATURES
WEBER → NERNST, EINSTEIN → LOW TEMPERATURE STUDIES		OVEN LIQUEFIER
b. OTHER FILAMENTS →		b. CALORIMETER
STUDY OF GASES AT HIGH TEMPERATURES → NERNST		
c. STUDY OF RADIATION LAWS →		c. DESIGN OF PHOTOMETERS THERMOMETERS
BLACK-BODY STANDARDIZATION → PLANCK, EINSTEIN, NERNST		

Let us schematically represent at this point what we have seen so far and in which direction we are headed. In the accompanying diagram I have reconstructed the work of Nernst (in the right-hand column) for the years 1895 to 1914 in the context of the better known theoretical developments (in the left-hand column). What I argue in the preceding and in the following chapters is that the three domains a, b, and c have to be seen as taking place both synchronically and diachronically, that all required attention simultaneously.

The Carbon Filament

In a process similar to high-temperature electric conduction accompanied by the emission of light, the combustion – or burning – of any substance or fuel will be accompanied by the emission of energy as both heat and light. The combustion process, viewed as a chemical reaction of coal and air, for example, is subject to certain chemical laws governed by particular properties and characteristics, such as the amount of air and coal available, the temperature and pressure at which the burning takes place, the speed with which the resulting gases are evacuated, and the quality of air available to the process.

For instance, coal burning in pure oxygen will produce a higher temperature than when burning in an equal quantity of oxygen in air, since in the latter case, the air's nitrogen also needs to be heated. In industrial combustion processes, a preheating of reactant gases raised their heat content in advance in order to release maximal amounts of heat upon combustion. The temperatures reached in practice, however, were much lower than predicted by the calculated theoretical ratio of reaction heat and heat capacity, and in most cases corrections had to be introduced due to several factors: the strong increase in the specific heat of gases with temperature – a feature, which we shall shortly see, is at the core of the heat theorem; the dissociation phenomena of combustion products at high temperatures; and the heat losses through conduction and radiation.

These phenomena, both chemical and physical, also occur in the heating of a carbon filament enclosed in a bulb. One thus understands that, although apparently far apart, the fields of combustion, chemical equilibria, and electric conduction are linked in that they all necessitate a thorough understanding of the laws of chemical equilibrium and the heat and light radiation theories.

It is important to introduce here a discussion of the relationship of specific heats to electricity studies, for it must have been evident to Nernst and others examining substances for their illumination and spectroscopic work that specific heats *must* change with temperature: Specific heat (c) is the ratio of the change in energy (dE) to the change in temperature (dT), that is,

$$c = dE/dT$$

The wide variations of energy (or emissivity) in a range of temperatures surely pointed to the possibility that the magnitude of specific heats increases with the rise in temperature. And indeed, a wide range of high temperatures was attainable, for which substances in the gaseous state, such as the evaporating filaments, were amenable to investigation.

Nernst had paid very close attention to this problem. As early as 1896, in his lecture course on "The Physical Methods of Chemistry" held during the spring term between 24 April and 3 July, Nernst was teaching his students methods for the determination of "1. atomic weight; 2. molecular weight; 3. the construction of the molecule." He began by introducing the three most important methods for molecular-weight measurements devised by Dumas, Hoffman, and Viktor Meyer. He singled out the determination of specific heats "as an independent method of the determination of atomic weights." This was an alternative to Avogadro's method, whereby only the weight of molecules in a given volume can be established, rather than that of individual atoms – especially for the case of unknown molecular constitution. Nernst alerted his students to the notorious and numerous "marked exceptions" to the Dulong and Petit Law of the Additivity of Specific Heats – the law that expects specific heats of atoms always to add up to the same constant value. He argued that if the specific heats vary with temperature, atomic weight determinations derived from their measurements will not provide accurate data on the constitution of molecules. Nernst discussed in detail the case of the constitution of mercury chloride, since it was "known that mercury is monatomic, i.e., its atomic and molecular heats coincide." As we saw above, he had learned of these exceptions as early as high school.

He emphasized in lectures, however, that "[n]o other known uniatomic gas has been investigated, for temperatures required are very high." Nernst taught skepticism. Although Lord Rayleigh had determined that argon and helium have the expected specific heats for a monatomic gas, Nernst countered that since they "do not crystallize, no other method for determining atomic weight is applicable . . . [but this assumption of monoatomicity was] not absolutely certain."[2]

Nor did Nernst's skeptical seeds fall on barren ground. Immediately upon his return to the United States late in 1897, Millikan, who had attended the lectures, embarked upon the examination of the ratio of specific heats at constant pressure and constant volume by employing the free-expansion method for gases.[3] Eventually, this work would, after many digressions, come to be relevant to Millikan's major contributions to modern physics, namely, the determination of the electric charge of the electron and the experimental proof for a particulate, quantized theory of matter and energy.

The possibility, however, that specific heats might not only rise but also decline dramatically at low temperatures (that is, at room temperature or below) occurred to other investigators, in particular H. F. Weber (1843–1912) at the Zürich Polytechnic. He had been examining the radiation from a carbon-filament lamp, as well as the properties of solid carbon.[4] His papers and monographs on the subject were well known. In the late 1880s, Weber proposed improvements on Kirchhoff's distribution function for the radiation of a black body, suggesting that the wavelength at the maximum intensity of the radiation distribution is proportional to the inverse of the temperature multiplied by a constant. He thus anticipated Wien's formulation of the "displacement law," which eventually prompted Planck's quantum hypothesis papers of 1900.

According to Thomas S. Kuhn, it was only during the first decade of the twentieth century that "the displacement law rapidly became a stan-

2. Robert A. Millikan Collection, California Institute of Technology Archives, 4.2. Millikan Notebooks, Göttingen 1896, pp. 1–13.
3. Millikan, *Autobiography*, p. 59.
4. H. F. Weber "Untersuchungen über die Strahlung fester Körper," *Berl. Ber.* (1888): 933–57, in Kuhn, 1978, p. 8.

dard [theoretical] tool. But it could scarcely have had that status at the time of its announcement in 1893. . . . [T]he radiation from hot bodies was primarily the province of expermentalists."[5] When Wien published his paper on the displacement law, "his only reference to experiment was through Weber's law," which required that the wavelength at which the "intensity function reached its maximum be governed" by the equation stating that the product of temperature and wavelength be constant (or that the ratio of temperature and frequency be constant).

This was the same Weber with whom both Nernst and Einstein studied in Zürich, the former in the 1880s, the latter in the late 1890s. He was also the same Weber whose experimental data on the specific heat of solid carbon Einstein was to publish in his 1907 pathbreaking paper on the quantum theory of solids, the only graph of experimental data Einstein would ever include in any of his publications.[6] In the fall of 1900, after unsuccessfully attempting to find a position as assistant to Professor Adolf Hurwitz, Einstein began studying thermoelectricity in the hope of finding work in Weber's laboratory, which had "the best facilities" at the ETH. His fiancée, Mileva Maric, was also studying with Weber and was trying to intervene on Einstein's behalf. But Einstein's rather skimpy attendance at classes and his often arrogant demeanor toward Weber had not endeared him to the powerful professor. Einstein even came to suspect Weber of writing unfavorable letters on his behalf. Maric became equally resentful. As a result of her pregnancy with Einstein's child and Einstein's insensitive treatment of her, Maric was incapacitated to such an extent that she failed her doctoral examination.[7]

H. F. Weber devoted many years of intense labor to studies of the incandescent filament. This work stimulated the researches of those experimentalists most significantly connected to black-body radiation: Heinrich Rubens (1865–1922), Wilhelm Wien (1864–1928), Otto Lummer (1860–1925), Ernst Pringsheim (1859–1917), Ferdinand Kurlbaum (1857–1927), and Friedrich Paschen (1865–1947). All the heat-radiation experimentalists and theoreticians, including Planck and Nernst, were born within a few years of one another, between 1857 and 1865. They belonged to the same generation, all physicists connected to Berlin in one way or another,

5. Kuhn, p. 7.

6. See Albert Einstein, *The Collected Papers*, vol. 2, ed. John Stachel (Princeton, N.J.: Princeton University Press, 1989), p. 135. It is interesting to note that in his first discussion of the black-body radiation, Einstein used Weber's results "without citing a source," as was his habit in many publications. Albert Einstein, "Zur allgemeinen molekularen Theorie der Wärme," *Annalen der Physik* 14 (1904): 354–62.

7. Roger Highfield and Paul Carter, *The Private Lives of Albert Einstein* (New York: St. Martin's Press, 1993), p. 65.

all trained in some aspect of radiation phenomena, whether heat, electromagnetic, or optical. They were all entering the most productive years of their careers at a time when the relationship between heat and electricity was the most important and "fashionable" topic in physical research, when the electrical and illumination industry experienced explosive growth, and when metrology and instrument design became a highly sophisticated field of specialization in and of itself.

It is remarkable, however, that Weber – or indeed any of the experimental roots of black-body radiation – is omitted from "standard" histories of the period. Thus Hans Kangro, in his very detailed prehistory of Planck's quantum theory of 1900, insisted in 1970 that "the stimulus to investigate" the radiation laws stemmed "almost entirely from scientific interest in problems of physics and astronomy." Although acknowledging that Weber "had in mind the establishment of a physical theory of the electric incandescent lamp," Kangro firmly insisted that "the participation of members of the PTR in the matter at a later date had nothing to do with the needs of the lighting industry."[8] It is true that among the major contributors to black-body research, only Lummer and Kurlbaum were actually members of the PTR; yet to see their work as somehow unconnected to the original impetus – which was, in effect, of quite immediate and recent nature – or to the surrounding idiom and direction of research seems naive at best. Kangro himself states that as early as six months after "its inception," "certain gas specialists asked the PTR in March 1888 ... to test and approve the Hefner-Alteneck amyl acetate lamp as a standard light unit." He continues: "The testing of light standards was another function of the PTR in 1891," thus effectively undermining his own argument, adding that "photometry," the search for a suitable light unit, was from the beginning an important goal of the PTR scientists.[9]

In addition to Weber's continuous and prolonged work on the carbon filament, Friedrich Paschen's work in the years 1892–5 constitutes another important prelude to Nernst's research on solid substances. Paschen published indefatigably on the spectra of incandescent solid substances, struggling with the anomalies encountered. In July 1895, Eduard Riecke, Nernst's superior at the time, communicated to the Royal Science Society in Göttingen Paschen's important paper on the regularities in the spectra of solid bodies and on a new determination of the sun's temperature.[10] Nernst, most probably, attended the meeting, and must have been ac-

8. Hans Kangro, *Early History of Planck's Radiation Law* (London: Taylor and Francis, 1976), p. 40. German edition *Vorgeschichte des Planckschen Strahlungsgesetzes* (Wiesbaden: F. Steiner, 1970).
9. Kangro, p. 151.   10. Kangro, p. 72.

quainted with the topic in any case through his contacts with the Göttingen physicists. After presenting the data obtained in numerous black-body experiments, Paschen outlined his conclusions – which W. Wien had also reached – that the frequency of thermal vibration is proportional to the temperature of the radiating body, and that the shape of radiation curves presents certain asymmetries at the extremes of the spectral range. With increasing temperature, the maximum of the radiation emission curves was shifted toward the shorter wavelengths. Paschen had also examined the properties of the incandescent carbon filament of an electric lamp, in addition to platinum and other substances. From an experimental point of view, Paschen's observational curves coincided extremely well with the theoretical extrapolation curves derived by Weber a decade earlier.

Here we close a circle of tantalizing biographical and scientific significance; Nernst's path to the heat theorem passed through experimental and theoretical work on the lamp: Much like his teacher Weber, they also both came to place an emphasis on specific heats, which became central to the formulation of a quantum theory of solids by Einstein, another Weber student; Einstein was registered for a course given at the ETH in the winter term of 1898–9, in which Weber discussed the results of his work on the radiation law;[11] and both Einstein and Nernst were to contribute – albeit in different ways – to the development of modern quantum theory in its post-Planck doldrums. The latter topic we will analyze in a separate chapter.

On the basis of all that we have so far learned, I would argue that the relative "silence" concerning Planck's quantum theory between the years 1900 and 1911 (which will be further discussed in the following chapters) was in part due to the fact that it was largely irrelevant to the ongoing experimental and technological application of radiation laws. Wien's proposed distribution law had been found to be inadequate in the infrared region of the spectrum (above 0.75 micron). This failure constituted a powerful motive for Planck's work on the topic and his ensuing formulation of a new radiation law at the end of 1900. But the theoretical and experimental research on the standardization of light measurements central to the Physikalisch-Technische Reichsanstalt's concerns was focused less on the infrared region than on the rather narrow visible spectrum (0.4 to 0.75 microns). It was, of course, essential to find acceptable laws for the behavior of a black body – the quintessential standard radiating body – but the major problems in need of solution were located in the visible range of the spectrum. In addition, the practical problems in need of solution for

---

11. Einstein, *Collected Papers,* vol. 2, p. 135, n. 11.

the electrical industry and the optical industry, in general, were located mostly below the high melting point of carbon at more than 3500°C, which, even by 1913 – the year of Langmuir's breakthrough with his tungsten bulb – had not yet been measured.

High Temperature Metal Extracting

Difficulties and issues identical to those dealt with by the physicists and engineers involved in illumination work were also faced by the metal-extracting industries, which witnessed a spectacular growth in the latter part of the nineteenth century. It was because of the high temperatures available in powerful electric arcs that aluminum, once the most expensive industrial raw material, came to cost a mere 30 cents per kilogram by the beginning of the twentieth century. It was only with the advent of electric "furnaces," and the work of Hall and Heroult in 1886, that the industrial production of aluminum became commercially viable. The various improvements in the electric arc furnaces contributed vigorously to the expansion of the manufacturing industries of steel, glass, and other products requiring extremely high temperatures. Thus, without the powerful electric arcs first employed by Elihu Thomson in 1886, the towering engineering feats of the fin de siècle would have been unthinkable: Buildings, bridges, and ships were "welded instead of riveted and bolted" and their joints were examined with the newly available X rays.[12]

Electrochemistry gradually came to supplement and even replace heat as the method of extracting metals, producing alloys, and other intermediates necessary in the rapidly changing industries. In Germany, a completely new industry was founded in the famed Essen area, where metallurgy was the main industrial backbone. This industry focused exclusively on the exploitation and uses of aluminum, and the science of "*Aluminothermie*" was a prominent feature at the Paris exhibition in 1900. The chemist Hans Goldschmidt (1861-1923) in Essen, had found that the extremely high combustion temperature of aluminum facilitated the extraction of metal oxides from natural ores. He produced the necessary high temperatures through the elevated combustion heat of electrically produced aluminum, to which the necessary oxygen was added in the form of oxides, such as iron oxide and chromium oxide. These oxides were immediately reduced, and by inserting the aluminum and the oxides into regular vessels, they burned like fireworks without the addition of external heat, producing temperatures of 2000° to 3000°C in the space of a few minutes.

12. O'Dea, p. 71.

Goldschmidt produced pure metals, such as chrome and manganese, which were used in chrome-steel and manganese-copper alloys, and also ruby and the famed carborundum discovered by Acheson in 1891, the cooled aluminum oxide used for polishing even diamonds.[13] This process was also used for the soldering of railway tracks, and by 1900, the inexpensive aluminum was already coming into use in the building of automobiles and trucks. By the same year, the "Chemical Thermoindustry" of Essen was also displaying samples of pure chromium, manganese, ferrotitanium, cobalt, niobium, and vanadium oxide, uncontaminated by the carbon that had earlier made metals brittle.

By the turn of the century, electrical furnaces had become the most appealing and practical method for heating metals. For example, 1 electrical horsepower could heat 5 grams of platinum in 1 second by 1000 degrees without heat loss. Since ease of operation was paramount, the electrical oven became the method of choice, its use being limited only by the melting point of the oven materials. Essentially, two kinds of electrical ovens were employed in metallurgy and other industries. In resistance ovens, the current was sent through a coil, through a carbon rod, or directly through the conducting installation – such as aluminum ore – itself. In others an electric arc transmitted its immense heat through conduction and radiation to a well-insulated oven. The temperatures of such ovens were too high to make them accessible to precise measurement, since even the most resistant materials – such as chrome, calcium oxide, and magnesium – melted, while gold, silver, platinum, iron, silicic acid, and even coal and magnesia evaporated.

Only from the energy of the radiation emitted in these ovens, however, could scientists approximate the temperature of the light arc: According to Rossetti it was 2500°C at the negative pole and 3900°C at the positive one, while Le Chatelier had found the temperature range to lie between 3000° and 4100°C.[14]

Thus we observe that black-body radiation, spectroscopy, and the study of chemical decompositions at very high temperatures were intimately connected conceptually and practically. It was imperative to come to an understanding of the heat (and light) radiation laws in order to measure any high temperatures correctly and precisely. This was an issue in need of resolution for the determination of the point at which equilibrium set in in chemical reactions. However, if the radiation laws failed, as they did, an understanding of the reactions taking place in these ingenious ovens was thrown into disarray as well, and no proper thermochemical calculations

13. Goldschmidt, *Zeitschr. f. Elektrochem.* 4 (1898): 494; ibid. 6 (1899): 53.
14. Le Chatelier, *Journ. de Phys.* 1(3): (May 1892).

could be carried out. Thus, *all* problems needed solving in order that *any* could be solved.

Conductivity and Electrical Motors

These developments coincided with the introduction of electrical motors and the rapid replacement of steam engines, prompting the massive high-temperature research programs at the Physikalisch-Technische Reichsanstalt. Under the direction of Max Thiesen, four researchers were active in the heat laboratory of the PTR by 1903: Karl Scheel, Ludwig Holborn, Friedrich Henning, and Louis Austin. Holborn performed research into methods for keeping high temperatures constant, using electrical currents in heating coils or vapor jackets with boiling zinc, while Max Bodenstein added liquid thermostats with oil or tin-lead filling.[15] Among other projects, Thiesen conducted an investigation of the specific heats of gases in collaboration with Helmut von Steinwehr from the PTR's own electricity laboratory. These studies were initiated at the behest of the Organization of German Engineers, which was interested in the "specific heats of gases in motors operating at temperatures and pressures up to 2000°C and 50 atmospheres."

Nernst himself was engaged in precisely the same kind of work. As we have seen in great detail in chapter 6, the heating element of the Nernst lamp, which was made of a mixture of zirconium and yttrium, became conducting when heated to incandescence. An area of much interest to researchers in chemical kinetics, particularly with regard to the ionic dissociation theory, was the measurement of electrical conductivity. Many chemists and physicists intensively pursued the electrical conductivity and the free energy of formation of melted salts at temperatures up to 1100°C, since many electrolytes known to be almost nonconducting at room temperatures were good conductors at high temperatures. Gases, too, were found to become conductors at very high temperatures, as had been shown by Thomson and Arrhenius, who attempted to prove that conductivity took place through ions in the incandescent gases of flames and in electrical discharges. Conductivity measurements had shown that the electrical resistivity of metallic conductors increases with temperature, and it was expected that there must exist an extremely high temperature domain where metals take on the characteristics of dielectrics. Mercury vapor, for example, was known to be a dielectric, and the question arose as to how the conductivity of mercury close to its critical temperature would change. This was the

---

15. M. Bodenstein, *Zeitschr. f. physik. Chem.* 30 (1899): 113, quoted in Bredig, p. 11.

temperature point at which it is possible to move continuously from the metallic-conducting liquid state into the dielectric gaseous state, a problem which both Georg Bredig and Nernst were investigating at the time.

In what has been considered separate and derivative work, Nernst and his assistant, F. Dolezalek, devoted an article published in May 1900 to a detailed analysis of polarization in a lead accumulator (or battery). One of Nernst's former students, Dolezalek was going to replace him in Göttingen upon Nernst's eventual move to Berlin. At the time, Langmuir reported that "many people seem surprised that a young man of such comparatively small reputation should be chosen to take the place of Nernst, the great." Dolezalek had spent some time at the Physikalisch-Technische Reichsanstalt "in a poor position, then got into Siemens & Halske where he stayed several years." His most important work "has been on the theory of the lead accumulator . . . [and] among others he has devised a very sensitive form of electrometer which is of great practical use in accurate electrical measurements." This electrometer was part of Dolezalek's doctoral research with Nernst in 1896, at the time when Nernst became more and more interested in electromagnetic and radiation measurements.[16] Since the electrolysis of a weak sulfuric acid solution between platinum electrodes dramatically accelerates in the range of 1.7 volts to 1.9 volts, it was expected that such a battery would not be capable of storing electricity above 1.7 volts, where the solution would completely decompose. However, Nernst and Dolezalek found that in fact lead (Pb) electrodes were much better suited for a battery operating at voltages higher than 2 volts. Instead of decomposing the sulfuric acid solution, the process of electrolysis first produced lead and lead oxide, and only above 2.3 volts did it produce hydrogen and oxygen. Therefore, "electrolysis produces from among the possible products not those necessitating the lowest amount of work, but those with a much higher energy content." This "abnormal" case of lead electrodes could be explained, Nernst argued, because the free energy of formation (the voltage necessary) has to be supplemented by another energetic component, namely a characteristic property of the electrode material. This property had to do with the "occlusion," or absorption ability, of the electrode: It had been found that lead absorbs fewer hydrogen atoms than other materials; therefore, its occlusion ability being low, the decomposition of the sulfuric acid solution requires higher energies than would be the case with other electrodes, such as platinum.[17]

16. Langmuir to his brother, Arthur, 7 April 1905, in Irving Langmuir Papers, Library of Congress, Manuscript Division, Washington, D.C.

17. W. Nernst and F. Dolezalek, "Über die Gaspolarisation im Bleiakkumulator," *Zeitschr. f. Eleketrochem.* 6 (45): 549–50.

This piece of research showed that in electrolytic reactions, the free energy is not the only quantity that drives chemical reactions, nor does the ability to execute work determine whether a chemical reaction will or will not take place. This was an important insight, pertaining to the validity of Berthelot's principle of maximum work, which we shall examine in the next section. But we here find an additional piece in the puzzle that Nernst eventually assembled in 1905.

Instruments for Conductivity, High Temperatures, and Radiation Measurement

Nernst devoted special attention to the construction of instruments for the production of high temperatures in order to investigate these conductivity problems. Already in May 1899, Nernst had presented preliminary work on electrolytic conduction in solids at high temperatures at the meeting of the German Electrochemical Society in Göttingen. It was there that he first explained the remarkable asymmetry between electrical conductivity in metals and conductivity in solid electrolytes. He mentioned that work "recently" carried out in his laboratory had shown that "at very low temperatures metallic conductivity increases enormously, while it decreases with rising temperature." With very few exceptions, electrolytic conduction, however, takes place in an exactly opposite manner: "at very low temperatures, for example that of boiling atmospheric air, conductivity barely exists, while molten salts at high temperatures can be very good conductors." Nernst noted that during his research on his lamp, he had found that the electrolytic conduction in solids "is capable of reaching quite surprising values." In addition, he had found that small mixtures of metals – rather than pure metal oxides – raised the conductivity of such solid electrolytes quite rapidly, whereas the pure oxides, which were stable only at incandescence, increase their conductivity only minimally with the rise in temperature.[18]

Nernst's paper is remarkable in several respects. While generally explaining conductivity in metals as following "the electromagnetic theory of light," in which "conductivity is not traveling with the transport of inertial mass," and where, therefore, metallic conductors are "capable of transforming light vibrations into Joule heat," he pointed to certain discrepancies. He had found that variation with temperature transforms one form of electrical conductivity into another; that is, at higher tempera-

---

18. W. Nernst, "Über die elektrolytische Leitung fester Körper bei sehr hohen Temperaturen," *Zeitschr. f. Elektrochem.* 6 (1899): 41–3.

tures electrolytes become metallic conductors, "a theoretically remark-able phenomenon, extremely useful for the praxis of light production."[19]

This paper also demonstrates that by 1899 Nernst and his coworkers were already doing some experiments at temperatures down to the boiling point of liquid air, some eight years before low temperatures became a major concern of the laboratory.

The following year, Nernst presented to the Electrochemical Society an improved model of his electrical oven, in which temperatures of up to 1400°C were reached. The "small oven" with a "pleasing exterior" was constructed of a platinum-iridium coil wound around a fire-resistant tube; at its bottom a thermocouple served as a thermometer.[20] This oven was remarkably similar to the black body constructed during exactly the same year by Lummer and Kurlbaum at the PTR. Three years in the making, their device was "an instrumentational breakthrough": It consisted of a "thermocouple placed inside a completely closed platinum cylinder whose blackened interior walls could be brought by electrical means to thermal equilibrium."

In order to provide accurate instrumentation for the determination of a valid radiation law, the PTR scientists had long been concerned with producing high-temperature measuring devices. In addition to designing his instruments, Nernst himself was interested in exploring such properties as conductivity at high temperatures. After 1900, however, the PTR scientists continued their research with a view to developing ever more accurate and uniform standards for illumination and electrical units. For instance, Steinwehr and Jaeger labored at the PTR for several years on the design of a platinum-resistance thermometer of exquisite precision, which "helped secure the foundations" of thermochemistry, a project which Emil Fischer said "was little known in public . . . and so often underestimated."[21]

At the time, Nernst was employing methods similar to those generated at the PTR and elsewhere for the determination of the temperature of the new conductors (metal oxides of zirconium, thorium, yttrium, and others) needed for the development of a novel lamp. After measuring the intensity of the current, and determining the light emission per square millimeter for the iridium vessel by applying the black-body radiation laws, Nernst calculated the temperature of the incandescent iridium and thus the temperature of vaporization for the substance under investigation. He used the data obtained by Lummer and Pringsheim – the same data that provided

19. Ibid., p. 43.
20. Nernst, "Ein elektrischer Platinofen," *Zeitschr. f. Elektrochem.* 7 (1900): 253.
21. Cahan, *An Institute for an Empire*, pp. 143, 149.

the basis for and the subsequent verification of Planck's distribution law during those same years. Nernst further calibrated his measurements by comparing the emission at the (known) melting temperature of platinum (by introducing a platinum wire at the lower end of his instrument) and found good correlations between his temperature/emission data for iridium and the data of Lummer and Pringsheim. In his work on the lamp, Nernst had used optical standardized lamps produced by the PTR, calibrated for photometric measurements.[22] As a result, he compiled a table of temperatures and corresponding light intensities that would serve as a thermometric chart for future experiments.

Thus, by 1900, Nernst had of necessity developed his own photometric standards, since the almost completely white light of the electrolytic lamp could not be compared against the strong reddish light of normal lamps. For this purpose he used an intermediate, standard electrolytic lamp. Evidently, Nernst attempted to avoid such errors as those that would lead to the infamous "N ray" affair that shook the scientific community in the years 1903 to 1906. "For greater certainty and in order to avoid subjective instances," he wrote, "these [comparative optical readings] were generally performed by several persons."[23]

Eduard Riecke and Paul Drude, the physicists who had come to listen to Nernst in his "Apostolic seminar" on the ionic theory in the spring of 1890, made significant contributions during the following years to the understanding of the electron's role in electrical conductivity. Nernst's work with electrolytic conduction lamps was superseded by metallic conductors precisely because the behavior of metal conductors had become better understood by 1904. At least as important was the fact that I. Langmuir had successfully exploited a suggestion made by Nernst himself; namely, that bulbs filled with an inert gas might contribute to better conductivity, lowered resistance, and expanded lifetime. At higher wattage, filaments operate at higher temperatures, but the presence of such gases as nitrogen or argon in the lamp can help reduce evaporation of the filament. In 1905, Langmuir worked in Nernst's laboratory in Göttingen, engaged at the time in studies of dissociation of hydrogen iodide gas in a gradient of high temperatures. When Nernst traveled to Berlin during the spring of 1905, Langmuir, a rather anxious student, became overly concerned about his ability to complete his research and pass the doctoral examination. By then, Nernst's upcoming appointment as full professor in Berlin had been announced.

22. See Nernst, "Einiges über das Verhalten . . . ," 1900, p. 373.
23. Ibid., p. 373. See M. J. Nye, "N-rays: An episode in the history and psychology of science," *HSPS* 11 (1980): 125–56.

Until 1903, Nernst did not publish much on the chemistry of high temperatures. Then at the meeting of the Bunsen Society for Applied Physical Chemistry held in conjunction with the 5th International Congress of Applied Chemistry in Berlin in June 1903,[24] Nernst presented a paper "On the determination of molecular weight at very high temperatures."[25] On the one hand, measurements of molecular weights were a classical domain for investigations into the structure of matter; on the other hand, predicting chemical equilibria of gaseous systems was a traditional topic of thermochemists in the late nineteenth century. But for Nernst, the determination of molecular weights could give him a very good indication of the degree of dissociation of molecules at the elevated temperatures. In his paper, he presented information about a new apparatus for the determination of vapor densities of gases up to 2000°C. Until then, the melting point of platinum or porcelain vessels had been a limiting factor in high-temperature experiments. Nernst strove to overcome the technical difficulties. Because in the temperature ranges attainable until then, the vapor tension of certain substances was still too low for vapor density determinations, Nernst had, by his own account, been "involved with this problem for a number of years; after many unsuccessful preliminary experiments in this direction I can now describe a method which enables measurements up to about 2000°C and maybe higher." The new instrument was, in Nernst's opinion, "a first class work of art which has been put at our disposal by the [Heraeus] company with greatest generosity," an instrument with which much better information could be collected.

As in most of his publications, Nernst paid close attention to a thorough description of the apparatus, contrary to his students' perception that he was interested only in "numbers." The novel iridium vessels, a project of great interest and pride to Nernst, on which he collaborated with his assistants Siemens and Riesenfeld, were produced for him by the Wilhelm Carl Heraeus Firma in Hanau.[26] The center of the platinum industry in Germany, Hanau was famed not only for metal extraction but also for the production of extremely sensitive instrumentation made of platinum. The original specialty of the Heraeus company had been the manufacture of platinum, but in recent years, it had begun to excel in the production of very large instruments constructed of high-grade silver that were needed by the acetic acid industry. Their most important achievement, for our

24. Tenth Meeting of the Deutsche Bunsengesellschaft, 10th Section of the 5th International Congress for Applied Chemistry, 3 to 8 June 1903.
25. Nernst, "Über Molekulargewichts-Bestimmungen bei sehr hohen Temperaturen," *Zeitschr. f. Elektrochem.* 32 (1903): 622–8. Also in abstract in *Gött. Nachr.* 1903, vol. 2.
26. Nernst, 1903, p. 622.

purpose, had been the development of pyrometers – thermometers for the measurement of extremely high temperatures in combustion ovens and the only means for measuring the heat of radiation produced in a black body. As mentioned earlier, this type of instrument had been used and calibrated by the PTR scientists and by the technicians working for the imperial Normal-Aichungskommision, the standardization facility in Berlin. In addition, the company had successfully introduced new methods of oxy-acetylene welding in the production of metallic aluminum, an extremely important procedure previously employed only for metallic lead.[27]

In his introductory remarks at the congress, Nernst noted that much progress in the study of chemical reactions could be expected from investigations of equilibria at high temperatures since at low temperatures, research is hampered by "insurmountable difficulties," including the fact that "chemical equilibria cease in the vicinity of absolute zero, in that according to theoretical considerations, all reactions at absolute zero will take place exclusively according to the laws of thermochemistry."[28] At high temperatures, by contrast, research on the vapor densities of substances (in gaseous systems) will provide data on equilibria.

Nernst then described in his paper how, because of the high manufacturing cost of iridium vessels, an iridium tube was introduced into a magnesia packing, isolated with asbestos, to which circular platinum electrodes were fused. These were connected with six bolts to flexible copper sheets in order to allow for the heat expansion of the iridium tube. The whole apparatus was approximately 30 cm tall, a small and ingenious instrument indeed. Within the heated iridium oven, very small iridium vessels, resembling crucibles, were introduced. These contained the substances whose vapors were to be investigated. Nernst employed optical methods for comparing the light radiated from the heated iridium tube with that emanated by a standardized light source.

Nernst announced that over a period of several years, he had been improving the photometric method developed by Holborn and Kurlbaum between 1900 and 1901 at the PTR. Early on, the Organization of German Engineers had also requested data on improved light-measuring instrumentation, and had set the stage for the continuing search for standardized units of luminosity and temperature scales. Nernst hoped that the use of "electrolytic incandescent" filaments, that is, his own lamps, would improve the photometric measurements. In order to determine the temperature of the iridium vessel, Nernst compared the intensity of the

27. Hans Kraemer, *Die Ingenieurkunst,* pp. 132–5.
28. Nernst, "Über Molekulargewichts-Bestimmungen bei sehr hohen Temperaturen," *Zeitschr. f. Elektrochem.* (1903): 622.

light it emitted with that of a standardized electrolytic conductor. By using colored pieces of glass, he "regulated," or compared, the emitted light with that of the electrolytic standard, heated by an electric current and previously calibrated for various electrical intensities. (Several years later, Nernst induced the company of Franz Schmidt & Haensch in Berlin to produce a spectral photometer that employed two light sources instead of only one as previously used.)[29]

He described how he vaporized the substance under investigation (such as water, carbon dioxide, potassium chloride, sodium chloride, or sulfur dioxide) at temperatures of 2000°C, obtained by passing a current of 2000 to 3000 watts through the copper strips fused to the iridium furnace.[30] The vapor densities method of V. Meyer, mentioned earlier, consisted in measuring the displacement of a very thin mercury thread enclosed in the calibrated glass capillary of a drop-bulb, in which the substance to be investigated was evaporated. The dimensions of Nernst's instrument were extremely small. Thus, the capacity of the bulb when heated was at most 2.5 cm^3, which meant that only 1 cm^3 of substance could be analyzed. At room temperature, the volume of air displaced was at most 0.13 cm^3. The amount of vaporized substance was often of the order of a fraction of a milligram. The iridium vessels were introduced into the bulb and could be reused for several consecutive runs of the measurements, being emptied after each run by blowing the remaining substance out with a platinum capillary tube.

The extreme attention to detail and the beauty of the instrument are a confirmation of many contemporaries' assessments of Nernst's exceptional experimental dexterity and ingenuity, bespeaking, in particular, his ability to design small-scale instruments capable of functioning in conditions of extreme temperatures (and pressures, as evidenced by an additional calorimetric bomb devised during the same years).

This ability impelled Nernst to devise an additional measuring instrument, the microbalance, "an exceptionally sensitive device." Its scale was "similar to an ordinary letter scale," and functioned according to the torsion principle, being suspended in a forklike support from which the scale itself was hung by glass and quartz threads. The readings were made with the aid of a telescope from a microscale calibrated at 0.5-mm intervals on a glass scale. This instrument still remained, many decades later, a miniature marvel of laboratory measurements. The 0.5-mm interval

29. Joel Hildebrand, "Das Königsche Spektralphotometer in neuer Anordung und seine Verwendung zur Bestimmung chemischer Gleichgewichte," *Zeitschr. f. Elektrochem.* (1908): 349–53.

30. Nernst, *Zeitschr. f. Elektrochem.* (1903): 622, and Nernst, *Silliman Lectures*, pp. 21–3.

corresponded to 0.0380 mg; Nernst claimed an accuracy of 1 to 2 thousandths of a milligram. The calibration took place with the aid of a precisely weighed, thin platinum wire, which was "cut into a number of equal pieces of approx. 1 mg each." Nernst pointed out the utility of the new instrument for analytical purposes, in particular atomic weight determinations. He had collaborated with the Göttingen company Spindler & Hoyer, who were taking on the manufacture of the microbalance in 1903. The new instrument became widely used in chemical laboratories.[31]

By 1906, the Meyer vapor-densities method, which Nernst had so successfully used in 1903 for his research on dissociation phenomena at elevated temperatures, proved to be unsatisfactory for the "exact determination of the degree of dissociation, and of chemical equilibria in general" (that is, in the gaseous phase), since it could not afford the accurate measurement of partial pressures. Nernst replaced it by a "streaming method," in which the gas mixtures were funneled through a long tube. Between two points of the tube, the equilibrium to be studied and the temperature t were maintained at a constant, while in the following section of the tube, the temperature was lowered rapidly to a temperature t' at which the reaction is practically stopped.

Nernst and his coworkers experimented over a period of more than eight years with a variety of catalysts and instruments. The dissociation of water over a heated catalytic platinum wire was studied by Nernst and Wartenberg according to a method developed by I. Langmuir; K. Jellinek investigated the formation and decomposition of nitric oxide; and a different method involving semipermeable membranes was developed by another Nernst collaborator, Loewenstein, who used the diffusion of hydrogen through platinum and iridium to study the dissociation of water vapor, hydrochloric acid, and hydrogen sulfide.[32]

31. Upon Nernst's death, the *New York Times* wrote:

> Nernst invented the micro-scale, an instrument of such marvelous precision that it turned at the scarcely conceivable weight of one millionth of a milligram. It could detect the presence of gold in sea water and was so fine that it could be read only with the aid of a strong magnifying glass.

"Walter H. Nernst, German Physicist – His Metallic Filament Lamp Paved Way for Incandescent Light – Dies at Age 77 – Taught Noted Americans – Langmuir and Millikan among Pupils – Won Nobel Prize for Chemistry in 1920," *New York Times*, 30 August 1941.

32. In his *Silliman Lectures*, pp. 23–4, Nernst briefly described the new apparatus and succinctly stated that "Of the results obtained . . . the following may be mentioned: The molecular weights of H_2O, CO_2, KCl, NaCl, SO_2 were normal at temperatures of nearly 2000; sulfur was almost fifty percent dissociated into atoms. Silver proved to be monatomic, as was to have been expected."

Why had Nernst, between 1903 and 1906, become interested in "partial pressures" and accurate determinations of equilibria? His researches on an electrolytic lamp had come to a halt. The successful rival Osram lamp had entered production, and Nernst's "glowers" no longer constituted a viable research project. But from his lamp related research, Nernst had learned an enormous amount about the behavior of a wide array of substances at high temperatures. The electrical wire with which he was best acquainted, platinum, became the paramount catalytic candidate for synthetic chemistry. Moreover, in 1905, Fritz Haber had published a remarkable book on the technology of gaseous reactions at high temperatures, which involved this use of platinum. Interested both in equilibria and in reaction velocities, Nernst and his collaborators studied the catalytic dissociation of water vapor passed over the glowing platinum wire introduced inside the streaming tube, a method worked out by Langmuir for "very accurate" determinations of dissociation equilibria of water and carbon dioxide.[33] Nernst's catalytic experiments, and his later work at elevated pressures, were performed to verify the data produced by Haber and his collaborators on ammonia formation equilibria. Nernst later abandoned this line of research, but he contributed significantly to industrial research on the improvement of the synthesis of ammonia.

Many other physical chemists at the time were engaged in studying chemical reactions at high temperatures, most of which were industrially and technologically promising or tantalizing. The "nitrogen fixation" was, as Ostwald recounted many years later, a sort of "philosopher's stone." The reaction between oxygen and nitrogen proceeds at very high temperatures produced, for instance, by an electric arc, and the resulting nitrogen oxide gas has to be rapidly cooled to prevent the decomposition of the gas. But these methods were extremely costly: The production costs for high temperatures exceeded by far the cost of the naturally available Chilean saltpeter, the traditional source of nitrogen oxide fertilizers.

In 1900, Wilhelm Ostwald had registered a patent for the production of nitrogen oxide that included the basic characteristics of Haber's process a few years later: Ostwald used nitrogen, hydrogen gas, elevated temperatures, elevated pressures, a catalyst, and a production cycle.[34] Although Ostwald's data could not be confirmed by the research scientists of the BASF laboratories in Ludwigshafen, he had pinpointed the essential features previously unexplored; in addition to high temperatures, high pressures and a catalyst were the missing ingredients for an eventual, successful production process of ammonia.

33. Nernst, *Silliman Lectures*, p. 31.
34. Dietrich Stoltzenberg, *Fritz Haber* (Weinheim: VCH, 1994), p. 142.

Nernst's experiments after the enunciation of the heat theorem helped clarify the ammonia synthesis. During 1906, Nernst and his coworkers studied reactions for which the chemical equilibrium data and specific heats had previously been available. By then, he was mainly concerned with a confirmation of his heat theorem, checking its validity against well-explored reactions. It was in this fashion that he also examined the formation of ammonia, attempting to match Haber's and Bodenstein's data against the theoretical values predicted by the heat theorem. Nernst had discussed his preliminary results on the nitrogen oxide–formation equilibrium at the meeting of the Bunsengesellschaft in Dresden in May 1906,[35] but without mentioning the heat theorem, or the work of Haber.

By the next summer, however, he had drummed up sufficient data to criticize Haber directly. In a presentation at the annual meeting of the society in Hannover, he reiterated disapproving remarks that he had already voiced at Yale, namely, that Haber's results on the ammonia equilibrium are the only ones not corroborated by the heat theorem. He insisted that the equilibrium of an efficient reaction should be reached at a temperature of 893°C, rather than 1293°C, as had been claimed by Haber and van Oordt in 1904.[36] Nernst and his coworkers K. Jellinek and F. Jost worked on the reaction for a year. They constructed a new instrument (for the streaming method described above) that was very similar to the iridium oven built earlier. An iron tube, encased in burnt magnesia, contained a porcelain tube surrounded with a platinum coil. It allowed a stream of hydrogen and nitrogen under pressure to pass at temperatures between 700° and 1000°C. With the new instrument and larger amounts of reactants, they had been able to obtain a better quantitative result for the formation of equilibria. To his credit, Nernst notified Haber of his new data, and by the 1907 Hannover meeting he was able to reach an agreement with Haber and Le Rossignol.

Nernst indicated in 1907 that an increase in pressure would be desirable in further studies, and thus Haber and R. Le Rossignol undertook additional experiments at elevated pressures. However, they did not employ Nernst's apparatus but a more sophisticated installation, which essentially analyzed the ammonia equilibrium, both through the decomposition and formation stages and by using a quartz tube, which was less permeable to the diffusion of gases than Nernst's metal tube in which the high pressure reaction was made to take place.[37]

35. Nernst, "Gleichgewicht und Reaktionsgeschwindigkeit beim Stickoxyd," *Zeitschr. f. Elektrochem.* (1906): 527–9. The expanded communication was published earlier, *Zeitschr. f. anorg. Chemie* 49 (1906): 213, 229.
36. Nernst, "Über das Ammoniakgleichgewicht," *Zeitschr. f. Elektrochem.* (1907): 521–2.
37. F. Haber and R. Le Rossignol, "Bestimmung des Ammoniakgleichgewichtes unter Druck," *Zeitschr. f. Elektrochem.* (1908): 181–96.

Although Haber had feared Nernst's intrusion into a field that he considered his own, Nernst eventually turned out to be a fair and honest competitor. When five years later the Hoechst chemical company brought suit against Haber's patent for the production of ammonia, all parties were stunned by the appearance of Professor Nernst as an expert witness for Haber and the BASF company. The hearing became a stage for a battle between the opinion of the mentor, Wilhelm Ostwald, who had written an expert testimony for Hoechst, and Nernst, who defended the originality of Haber's patent. Moreover, Nernst stated that although he himself had been intimately connected with improving the synthesis, his interest had been of a purely theoretical nature.[38] In a certain way, then, the ammonia synthesis was probably the first resounding success of a direct application of the heat theorem to chemical research.

38. Stoltzenberg, pp. 176ff.

8
Theory and
the Heat Theorem

In the published edition of his Silliman Lectures, delivered at Yale University between 22 October and 2 November 1906, Nernst wrote:

> In the hope of penetrating more deeply into the relations between chemical energy and heat, I have carried out in the last few years together with my students, a number of investigations on reactions at high temperatures in gaseous systems. . . .

Nernst related that he began to extend the "methods for the determination of molecular weights to higher temperatures, using iridium vessels in the vapor density method of Viktor Meyer."[1]

These statements have supported the standard account that Nernst was working at the time on classical chemical problems aimed at elucidating the conditions for the feasibility of chemical reactions. Broadly defined, the project that led to the formulation of the heat theorem – experimental work whose inception Nernst set around the year 1894 – has been viewed as the application of thermodynamics to chemical reactions, "the search for the mathematical criteria of chemical equilibrium and chemical spontaneity." And the heat theorem is considered to have been originated in chemistry, a "novel" solution to an old chemical problem. In this view, "between 1894 and 1905, Nernst carried out some experimental work with gaseous reactions in order to test the Gibbs-Helmholtz equation (and therefore the second law) over a wide range of temperatures."[2]

Nernst's investigation was, indeed, connected to the task of obtaining a theoretical method of predicting the feasibility of chemical reactions. However, Nernst was not "extending" the Meyer method for molecular weight determinations in order to improve these measurements. Instead, as we have seen in the previous chapter, he began this work in order to gain information about phenomena of dissociation at high temperatures,

---

1. W. Nernst, *Experimental and Theoretical Applications of Thermodynamics to Chemistry* (New York: Charles Scribner's Sons, 1907), p. 20.
2. Erwin N. Hiebert, "Walther Nernst," *Dictionary of Scientific Biography,* pp. 436–8.

about materials suitable for the improvement of the lamp, about filaments and their reactivity with gases, and about the relationship of electricity to heat capacity and specific heats.

The kinetics of dissociation phenomena were of interest in the study of the formation of endothermal, or heat-absorbing, substances during the decomposition of exothermal, or heat-emitting, compounds, and also played an important role in all research related to the phenomena inside gas-filled or evacuated light bulbs. The behavior of various filaments in a number of gaseous environments was particularly important, as we have already seen.

From a theoretical point of view, according to thermodynamic princi-ples, the equilibrium constant K of a reaction, such as any combustion process, is defined as the product of the active concentrations of the pro-duced substances divided by the product of the concentrations of the dis-appearing substances. If one considers the quantity of heat per mole thereby absorbed (q) as positive (which could be a function of T), then ac-cording to van't Hoff:

$$d\ln K/dT = q/2T^2$$

Since at high temperatures K is displaced in favor of the stability of the heat-absorbing systems, at extremely high temperatures the spontaneous formation of endothermal, or heat-absorbing, substances was often to be expected, as Nernst pointed out.[3] Therefore ozone, acetylene, and cyanide, which are endothermal substances, were easily produced at very high tem-peratures in the electrical arc flames discussed earlier. This kind of research into chemical reactions and physical transformations at high tempera-tures led Viktor Meyer to envisage in 1895 the possibility of splitting the monatomic vapors of elements into fractions of atoms. It was hoped that those substances that in their disintegration absorb heat would show higher dissociation rates at high temperatures. Gibbs, Guldberg, van't Hoff, Boltz-mann, and others had shown theoretically and experimentally that mol-ecules such as acetic acid, formic acid, sulfur dioxide, and phosphorus pentachloride absorb heat and dissociate.

At the time of Nernst's work, however, comparatively few precise ther-mochemical data were available for high temperatures; nevertheless, Boltz-mann had already calculated the approximate dissociation heat of iodine molecules into atomic iodine from the density change of iodine vapor at higher temperatures. The breakup of a number of poliatomic molecules was known through measurements of vapor densities at various high temper-atures. Bromine dissociates into atoms above 1200°C; sulfur's molecular

3. Nernst, *Theoretische Chemie*, 2nd ed., pp. 402, 590–604.

weight decreases from S_7 at 468°C to S_2 at 1719°C; phosphorus changes from P_4 to P_3; arsenic from As_4 to As_2. On the other hand, other elements, such as O, N, H, Cl, or Tl, were known to remain diatomic even at incandescence, whereas Hg vapor, Zn, Cd, and probably Na and K remained monatomic and did not indicate further splitting. For a series of compounds, such as potassium iodide or the chlorides of silver, thallium, mercury, indium, germanium, uranium, and others, the vapor density had been determined at high incandescence and a normal molecular weight was found, indicating no disintegration.[4]

Viktor Meyer's work and that of his collaborators had yielded, by the turn of the century, important instruments and techniques for determining the decomposition and dissociation of gases at temperatures as high as 1600° to 1700°C. For the purpose of measuring vapor densities at temperatures between 2000° and 3000°C, Meyer and his collaborators succeeded in building more resistant ovens in which graphite was burned in an oxygen blower. The most resistant metals and alloys, such as platinum, platinum-iridium, and porcelain, melted easily, and only magnesia resisted such elevated heat.[5]

From the entire body of this experimental and theoretical work at high temperatures, scientists concluded that the attainable maximal temperature rise was proportional to the heat of reaction and inversely proportional to the heat capacity of the system to be heated,[6] that is, that the variability of the specific heat dictated the optical and other properties of materials. In the late 1890s, the displacement of the chemical equilibrium of dissociable substances at high temperatures had been extensively studied by Max Bodenstein, in an examination of hydrogen iodide and hydrogenated selenium up to 520°C.[7] Moreover, it seemed that in order to obtain a filament that would quickly heat to incandescence, one could reasonably infer that a maximum temperature rise would be obtained with substances whose heat capacity – or specific heat – was relatively low.

From the data obtained in these studies, Nernst calculated the values for free energy (A) and total energy (U) of dissociation reactions. He concluded that in many cases A and U were indeed often equal at ordinary temperatures, in accord with Berthelot's embattled principle of maximum

4. See Windisch, *Die Bestimmung des Molekulargewichts,* Berlin, 1892, pp. 252–369.
5. Georg Bredig, *Über die Chemie der extremen Temperaturen,* Habilitationsvorlesung, 9 February 1901. Also published as *Physik. Zeitschr,* Sonderdruck II (Leipzig: Verlag Hirzel, 1901), p. 9.
6. See van't Hoff, "Vorlesungen über theoretische und physikalische Chemie," vol. 1, p. 240; Nernst, *Theoretische Chemie,* 2nd ed., p. 416; Ost, *Lehrbuch der technischen Chemie,* 2nd ed., p. 10.
7. Max Bodenstein, *Zeitschr. f. physik. Chemie* 29 (1899): 295, 429.

work. However, his investigations of dilute solutions, as opposed to concentrated solutions, had revealed the inadequacy of Berthelot's principle. It is evident from Nernst's early publications on the heat theorem that he was well aware of the suggestions made by Berthelot, Boltzmann, Helmholtz, Le Chatelier, Haber, and Richards as to the locus of the solution to the problem of predicting chemical equilibria and the relationship between heat transformations and chemical affinity. In his original communication at the beginning of 1906 and in his Silliman Lectures, Nernst carefully prefaced his remarks with a "historical introduction," giving due credit to all the above predecessors.[8]

One view of the historical origins of the heat theorem attributes to Nernst a mostly interpretive role, and sees him as a somewhat shrewd expositor and continuator of Marcellin Berthelot's researches on the principle of maximum work, which he enunciated in the early 1870s. Historians and chemists alike have seen Nernst's work mostly as part and parcel of an older, thermochemical research tradition that sought to explore the conditions under which chemical reactions take place, chemical equilibria are attained, and maximum chemical yields are obtained. Did Nernst's work correspond to any tradition linked to the development of thermochemistry? And if so, can one find concrete links between the thermochemists' agenda and the origins of the heat theorem? The preceding chapters seem to deny this view and to place him in a radically different context.

Theories of affinity were, after all, what the project of thermochemistry and chemistry were supposed to be all about. One of the main tasks of chemical investigations – nowadays primarily subsumed under the category of chemical kinetics – was the prediction of the direction and rate at which chemical reactions would proceed. During the period of Nernst's work, the possibilities of arriving at such predictions consisted almost exclusively in measurements of the chemical affinity, that is, measurement of the maximal amount of work (*A*) of a chemical reaction as a function of readily determined factors. The Helmholtz thermodynamic equation

$$U - A = T \, dA/dT \qquad [1.0]$$

provided the only formalism that linked the total energy content (*U*) and the temperature of the reaction (*T*) to the magnitude (*A*.) However, (*A*) could not be easily obtained, due to the indeterminate integration constant (*I*) obtained through the integration of [1.0].

Ignoring the new thermodynamical work of Clausius, Carnot, Favre, J. W. Gibbs, and others, the French chemist Berthelot insisted on the

8. Nernst, W. "Über die Berechnung chemischer Gleichgewichte aus thermischen Messungen," *Nachr. König. Ges. d. Wiss. Göttingen* 1(1906): 1, 7–9; *Silliman Lectures*, pp. 54–7.

primacy of his "principle of maximum work" as a mechanistic explanatory panacea for chemical reactivity based on the above known equation
[1.0]. In 1873, Berthelot formulated the definitive version of his principle
as follows: "Any chemical change accomplished without the intervention
of external energy [that is, any natural, irreversible change] tends toward
the formation of a substance or a system of substances which produce the
maximum of heat." He compared his principle to the recently introduced
concept of entropy as a measure of physical-chemical transformations.
However, by 1879, Gibbs had already published his work on thermodynamics, some of it in 1873, while Favre had shown that voltaic piles do
not follow Berthelot's principle, which equated heat and work. Favre and
Sainte-Claire Deville had also obtained experimental results that showed
that some substances are formed endothermally, that is, with the production rather than the consumption of heat.

Berthelot had hoped to lay the foundations of a new science of chemical mechanics based on his thermochemical investigations begun in 1864
and destined to transform chemistry by grounding it on "rational concepts and the laws of mechanics."[9] In a 1913 obituary of Berthelot, Emile
Jungfleisch[10] explained that eventually, physicists and chemists succeeded
in giving a more "precise value previously lacking" from Berthelot's formulae. Berthelot himself witnessed the acknowledgment of the value of
his much-controverted principle: "Towards the time of [Berthelot's] death,
Mr. Nernst, illustrious Berlin physicist, arrived at an exact evaluation of
the principle of maximum work as a result of a series of very elaborate researches on the measurement of specific heats at low temperatures pursued by his school."[11]

Jungfleisch provided a genetic account, according to which:

> Starting from the long-known fact that at absolute zero maximum
> work is rigorously equivalent to the heat of reaction, [Nernst] has
> shown that the same relationship will exist at all temperatures if the
> heats of reaction are invariable [with temperature].[12]

It could, of course, not have been known as a "fact" that the above relationship would hold. No such "fact," in the conventional sense of the
word, could have been observed, this being precisely the essence of Nernst's
later heat theorem, which denies the possibility of directly attaining absolute zero temperature. However, by mathematical integration of the

9. Medard and Tachoire, pp. 235–6.
10. E. Jungfleisch, "Notice sur la vie et les travaux de Marcellin Berthelot," *Bulletin de la Société Chimique de France* (1913): I–CCLX.
11. Ibid., pp. c–ci.   12. Ibid., p. ci.

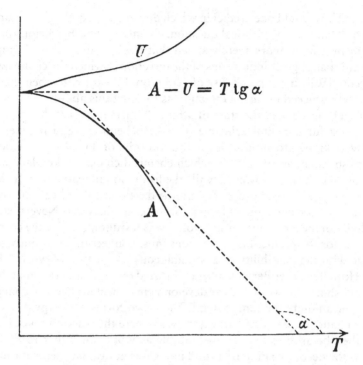

$$A - U = T \, \mathrm{tg} \, \alpha$$

The heat theorem energy curves

Second Law of Thermodynamics, this was known to be "expected" from equation [1.0] if indeed thermodynamics were rigorously applicable to chemical reactions, a debatable proposition.

Van't Hoff's Formulation of the Equilibrium Problem

Nernst's insights were anticipated by van't Hoff, who, in his 1902 lectures at the University of Chicago, submitted Berthelot's principle to careful scrutiny and suggested avenues toward a solution of this imbroglio.[13] Van't Hoff's conclusions were based on a deeper understanding of phase transitions and of the relationship between chemical reactions in solutions and electricity.

Where could one find a "fundamental principle which enables us to predict from other data whether a given chemical change will take place or

13. Van't Hoff, Jacobus H. *Physical Chemistry in the Service of the Sciences,* trans. Alexander Smith (Chicago: University of Chicago Press, 1903).

not," and how could one predict in which direction an equilibrium change will go? That was the classical question. Thomsen's and Berthelot's principle of maximum work merely stated that the heat developed during a chemical change gives indications of the direction in which the change will proceed: "[W]hen the possibility of evolution of heat exists, then the reaction will proceed in such a direction as to bring this about."

Van't Hoff sketched the state of affairs: "For many years this conception was a fundamental principle of thermochemistry, and numberless facts were known to support it . . . [but] in spite of this it is not difficult to furnish examples of cases in which chemical changes take place with absorption of heat." However, while the formation of water from hydrogen and oxygen indeed proceeds with the development of heat, the reaction between nitrogen and chlorine, for instance, does not. Nevertheless, the choice of the term "maximum work" was fortuitous, "since the correct principle for the prediction of reactions must connect the possibility of a change with the possibility of a simultaneous production of work." But van't Hoff, like other Berthelot critics, insisted on the distinction that "the accomplishment of work and the development of heat in chemical changes do not mean quite the same thing." These two forms of energy often go hand in hand, as in the case of explosives, where the reaction takes place with the liberation of energy measurable as work. But other reactions, such as the decomposition of a solid into gases at ordinary temperatures, "absorb heat . . . yet . . . the products may excercise a pressure." As van't Hoff emphasized, here one has "the possibility of accomplishing work [which] does not coincide with the capacity to develop heat, AND YET WHERE IT IS OBVIOUSLY THE CAPACITY TO DO WORK WHICH CONTROLS THE DIRECTION OF CHANGE."

The difficulty, therefore, was "finding a method for determining from other data the existence of the possibility of accomplishing work and the amount of this work in any given direction." Berthelot had investigated heats of reaction, but not "work" per se. Therefore, one needed to investigate work in addition to heat. With a barely disguised sarcasm, van't Hoff remarked that "another magnificent lifework has been suggested," that of repeating all of Berthelot's experiments with a calorimeter and determining the "ability of each reaction to do work." This project, however, was much more difficult to execute because this ability to perform work depended on the variation in the physical conditions, such as temperature and concentration, at which transformations take place. But certain examples seemed capable of furnishing information in this direction. The transformations undergone by such substances as carnallite as one varies the temperature and pressure, and in particular the extensive application of graphic methods illustrating the progression of a chemical change in relation to temperature and pressure, were expected to yield significant insights.

By changing the temperature, an equilibrium between two different solutions of carnallite is obtained at –21°C. However, between –21°C and 168°C carnallite is formed, a composite made up of three substances, $MgCl_2 \cdot KCl \cdot 6H_2O$.

> At the transition temperature, the possibility of doing work (E), of which the formation of carnallite is capable, is zero $E = 0$. However, above –21 the reaction proceeds, and can therefore overcome resistance, and since the reaction is accompanied by increase in volume this resistance might be pressure. . . . The maximum work will obviously be obtained if the resistance is so great that the formation of carnallite is just able to take place and no more, while any increase in the pressure would cause a reversal of change.

What van't Hoff was suggesting is that application of a measurable amount of pressure would give us the quantity of "work" and not "heat" characteristic of a chemical or physical transformation.

If at equilibrium one applies the principle of reversibility, then

$$dE = -W\,(dT/T)$$

or, for finite values,

$$E = -W\,(dT/T)$$

That is, the known amount of work E can be correlated to heat W.

Now the argument becomes significant for its relationship to Nernst's theorem:

> This means that at a temperature of dT degrees above the transition point, which in the present case is situated at 252°K, an amount of work dE can be accomplished. In applying this formula, since W is the heat of formation of carnallite, dE and W must be expressed in the same units, for example calories. At the transition temperature, where $dT = 0$, that is, where there is no change in temperature, E has the value zero. Above and below this point, the sign of E changes. Here the sensitivity to changes in temperature of the capacity to do work is most pronounced, and the sensitivity of the direction of the chemical change to the same influence is most noticeable.

Van't Hoff here showed that the point of greatest interest should be the point of transition, in fact the only point where a concrete relationship between heat, work, and temperature can be established. And he pointed out that if one operates at the absolute zero of temperature,

$$dT = -T$$

then $E = W$.

What van't Hoff was in effect saying is that for a reaction in equilibrium at the absolute zero of temperature, heat and work become equal, i.e., $E = W$. Therefore, Berthelot's principle is strictly correct only at the absolute zero of temperature, and the "fact that Berthelot's principle under ordinary conditions so frequently gives satisfactory results depends chiefly on the fact that our ordinary temperature of experiment is relatively low, being only 273 degrees removed from the absolute zero. In the neighborhood of such temperatures as 1000° the whole circumstances are essentially different and usually the results are in conflict with Berthelot's law."

The same reasoning would apply to the relationship between chemical change and the capacity to execute work in reactions taking place in an electrolytic cell, where electricity is developed. For example:

$$Zn + CuSO_4 = ZnSO_4 + Cu$$

In the process of the chemical reaction, an electric current is produced in the solution. Conversely, this reaction could, in effect, be stopped by applying a "resistance" that will force the reaction to proceed in the reverse direction. If one applies an electrical force:

> the [chemical] change can be brought to a halt completely if the electromotive force of the opposing current is equal to that of the cell. . . . The electromotive force when electricity is produced corresponds therefore to the pressure when the chemical change tends to bring about an increase in volume. Detailed consideration from this point of view leads us to dicover in the electromotive force a measure of the capacity to do work.

This opens up a "rich field" for research in the prediction of chemical equilibria, by discovering the conditions under which the electromotive force becomes zero.[14] And this is precisely where both Nernst's past experimental and theoretical expertise, as well as his immense knowledge of processes at high temperatures, converged and stepped into the "*Problemstellung*" which van't Hoff so clearly delineated in 1902.

Nernst's Recasting of the Theoretical Problem

Two alternative genetic explanations for the origins of the heat theorem have been proposed: It has been seen as the natural end product of nineteenth-century thermochemistry or, alternatively, as a sharp depar-

---

14. Ibid., pp. 37ff.

ture from thermochemistry, a rejuvenation induced by thermodynamics and, in particular, by the new Ostwaldian physical chemistry.

Some argue that "thermochemistry enjoyed quite an independent course of development from thermodynamics. . . ." but also that it was "reconstituted . . . as an applied domain of thermodynamics . . . in late nineteenth-century Germany with the ascendance of theoretical physicists." Generally, it has been held that the "lineage" of Hess-Thomsen-Ostwald was more or less a "resource" for the "consolidation of physical chemistry as a separate discipline."[15] Yet Nernst saw himself and his laboratory as being engaged between the years 1910 and 1917 in the exploration of "thermodynamic questions," and he unvariably viewed his heat theorem as a contribution to the general understanding of natural laws, with "special applications" in chemistry. When in 1917 Nernst published his first full-fledged monograph on the heat theorem, he placed both its origins and revelance squarely in the context of theoretical physics. His book was an "an elaboration of one of the most important chapters of theoretical physics, namely thermodynamics, into whose purview fall not only physical processes, but among others also chemical processes."[16] Even though such retrospective compilations contain active reconstructions, the organization of Nernst's scientific prose remained constant and consistent. His textbook of 1893 had been divided into sections on the "transformation of matter" and the "transformation of energy," where he had only as an afterthought added in parentheses "theory of affinity" as a descriptive for the benefit of chemistry students.[17] In 1917, too, he began his exposition with general thermodynamic considerations. He had spotted, through his intimate knowledge of electrical and chemical processes, the importance of special thermodynamic cases: He was able to correlate the behavior of an electrical force potential independent of temperature with that of a temperature-invariant mixing of concentrated solutions, both of which instantiate the special case of $A = U$. And he had spotted "strange consequences" in the shape of these functions deriving from previous attempts to relate quantitatively and graphically the evolution of changes in energy U and work A.[18]

In effect, Nernst was engaged, as we have seen, in a very specific and concrete problem: the relationship between radiant heat (or radiant energy) and physical and chemical transformations in the conducting filament, as

15. Hiebert, *DSB*. Mi Gyung Kim, *Practice and Representation: Investigative Programs of Chemical Affinity in the Nineteenth Century*, Ph.d. diss., UCLA, 1990, pp. 201–2.

16. Nernst, *Die theoretischen und experimentellen Grundlagen des neuen Wärmesatzes*, Halle: Knapp, 1917. A second revised and expanded edition was published in 1924.

17. Nernst, *Theoretische Chemie*, 1893.

18. Nernst, *Die theoretischen und experimentellen Grundlagen des neuen Wärmesatzes*, pp. 9–12.

well as the atmosphere surrounding the filament. Nernst did not mention affinity, or reaction rates, or chemical equilibria, nor was his stated aim the determination of molecular weights for their own sake. Just as he had investigated the relationship between heat and electrochemistry by applying thermodynamics to electrical and electrochemical phenomena, he extended this approach to chemical reactions not limited to liquid solutions. His transition to "dry chemistry" has been mentioned by historians without any causal explanation and, in general, his work has been fitted into the framework of a physical chemistry dominated by the programmatic pronouncements of Ostwald.

I propose, however, a third alternative explanation for Nernst's transition to low-temperature work: It relies on the immediate timing of Nernst's postulation of the heat theorem and emphasizes its continuity with his high-temperature research on electrolytic conduction in solids. So far, Nernst's conduction studies, as well as his work on equilibria at high temperatures and on molecular-weight determinations, have been seen to emerge from the traditional thermochemical research program practiced since the 1860s by several highly influential chemists concerned with the problem of chemical affinities and the feasibility of chemical reactions. In this framework, Nernst's lecture of 1905 and the 1906 published paper have been variously seen both as a continuation of classical themochemistry and as the most important event in modern physical chemistry. The most poignant conclusion of that paper was that the long-standing problem of predicting whether a chemical reaction between two or more substances is feasible could be solved with one plausible assumption. For complicated reasons, this ability to predict chemical equilibria from prior knowledge of the substances' properties had not been possible until Nernst postulated that, at very low temperature, the change in the total heat content of the substances becomes equal to the maximal amount of work of a chemical reaction (or the Helmholtz free energy that a reaction can liberate). Heat content, which is not a permissible concept in thermodynamics, means either the internal energy U, or the enthalpy $U + pV$.

From this assumption – the original form of Nernst's heat theorem – he advanced the argument that if certain physical constants, namely the specific heats of solid substances, could be measured down to the lowest possible temperatures, one would be able to determine their free energy and thus establish an affinities table for chemical substances.

Both Thomsen and Berthelot had found experimentally that at not-too-high temperatures, however, the quantity of free energy, or heat of formation A, and the amount of total energy U were, in general, not very different. In the case of condensed systems and of very concentrated solutions, the differences between A and U were surprisingly small. Since in approaching absolute zero temperature, condensation increases continuously, Nernst made the *assumption* that the change in the magnitude of U and A in the vicinity of absolute zero will tend toward o, that is, that their curves will be tangential in the close proximity of absolute zero. Therefore:

$$\lim dA/dT = \lim dU/dT \qquad \text{(for } T = \text{o)} \qquad [1.1]$$

This was Nernst's simple line of thought, the fundamental assumption of the heat theorem in its initial version published in 1906. It was believed to apply to pure solids or liquids, but not to gases. As its consequence, all processes at absolute zero would take place without either changes in molecular specific heats or entropies; the algebraic sums of the molecular heats of the reacting substances at this temperature equal zero, and the Kopp rule regarding the additivity of heats applies strictly.[19] The shortest version of the Nernst theorem would result if one states that the entropies vanish at absolute zero, as would be postulated by Planck in December 1911.[20]

Previously, the constant A of any chemical reaction could be obtained only by indirect knowledge of affinities, as given by knowledge of the position of the chemical equilibrium. But this method often failed due to the slow reaction rates of organic substances. Generally, the determination of the concentration of reactants at equilibrium proved to be an extremely difficult procedure. The Nernst theorem facilitated the measurement of A by thermal measurements, that is, by evaluating reaction heats and specific heats. This required measuring the temperature-dependence functions of these variables, and examining their graphical representations. The temperature functions, rather than absolute reaction heats or specific heats, were much more easily and precisely measurable. In addition, much unexploited experimental data were already available in the literature, data which could be used for the theoretical verification of the heat theorem equation.

According to equation [1.0]:

$$A = -T\int UdT/T^2 + IT \qquad [1.2]$$

But from the Nernst theorem [1.1] one can derive:

$$A = -T\int UdT/T^2 \qquad [1.3]$$

which for

$$U = U_0 + f(T) = U_0 + \int \Sigma \, ncdT \qquad [1.4]$$

19. Kopp's rule (1864) about the additivity of specific-heats states that the molecular heat of a solid substance (the product of specific heat and molecular weight) equals the sum of the atomic heats of the elements contained in that substance.

20. Max Planck, *Uber neuere thermodynamische Theorien (Nernstsches Warmetheorem und Quantenhypothese)*, Leipzig: Akademische Verlagsgesellschaft, 1912. A lecture delivered on 16 December 1911 to the German Chemical Society in Berlin.

is transformed into

$$A = U_0 + T\!\int\! f(T)\mathrm{d}T/T^2 = U_0 + T\!\int\!\mathrm{d}T/T^2 + \int\! \Sigma ncdT \qquad [2.0]$$

where c is the molecular heat of the individual reactants, and nc is the algebraic sum of these heats in the sense of the reaction equation, that is: $nc = n_1c_1 + n_2c_2 + \ldots - n'c' - n''c'' - \ldots$ for any reaction equation $n_1c_1 + n_2c_2 + \ldots = n'c' + n''c'' + \ldots$

Therefore, for condensed systems, A could be generally obtained from U_0 and the combination of specific heats ($c = n_1c_1 + n_2c_2 + \ldots - n'c' - n''c'' - \ldots$) as a function of temperature.

Thus, equation [1.4] allows A to be determined from measured values of specific heats as functions of T (temperature).

In order to also obtain A for gaseous reaction systems, Nernst used the relationship of A to the chemical equilibrium constant K_p, that is,

$$A = RT \left(\Sigma v \ln \pi - \ln K_p \right) \qquad [2.1]$$

where $\Sigma v \ln \pi$ has the same significance regarding the $\log_{\mathrm{nat}}$ of vapor pressure π as Σnc has for the above molecular specific heats. If we now introduce the expression [2.1] for A into [2.0] we obtain:

$$-RT\ln K_p = U_0 - T \int f(T)\mathrm{d}T/T^2 - RT\Sigma v\ln\pi \qquad [3.0]$$

(at the same total initial and final pressure of the reactants, $A = -RT\ln K_p$).

Nernst introduced for each of the reacting substances a vapor pressure formula instead of π, in which the nature of each of the reacting substances is expressed by a characteristic constant i, such that:[21]

$$I = -R\Sigma ni$$

By introducing the vapor tension formula for π into [3.0], the $f(T)$ member of [3.0] disappears, and K_p can be obtained solely from the reaction and molecular heats for the gaseous state, or for the liquid state of homogeneous systems. In the case of heterogeneous systems, where solid or liquid states coexist with gaseous substances, only the vapor pressures of gaseous substances were substituted for π. Then the molecular heats for the gaseous and solid or liquid phase had to be obtained in addition to the reaction heats U. Nernst justified this method by stating that, although every solid substance possesses a low-saturation vapor tension, it remains constant at constant temperature as compared to the change in tension of added gases, such that at equilibrium the product of the tensions remains constant, a magnitude calculated according to the above procedure.

21. F. Pollitzer, *Die Berechnung chemischer Affinitäten nach dem Nernstschen Wärmetheorem, Sonderausgabe aus der Sammlung chemischer und chemisch-technischer Vorträge.* (Stuttgart: F. Enke, 1912), particularly pp. 27, 48, 121–4, and eq. 25.

Devising the Tests

In order to obtain the specific heats used in this analysis of chemical equilibrium, Nernst had to apply and improve existing methods for calorimetric measurements. By the late nineteenth century, several such methods were well in place, notably in the research projects of Berthelot and Thomsen, whose views and publications dominated the field.

Hans Peter Julius Thomsen (1826–1909) spent four decades in his Copenhagen laboratory, where the temperature was kept constant at 18°C, performing an estimated 4000 chemical calorimetric experiments. Like Berthelot and many other contemporaries, Thomsen became embroiled in sundry scientific controversies. Although he did not directly try to assemble a "school," preferring a small number of disciples to large research groups, his metaphysics undoubtedly set him in opposition to the phenomenological positivists, such as Berthelot and Ostwald. He belonged to the generation of "elder" thermochemists whose chemistry was "entirely linked to atoms and molecules."[22] Thomsen's often biting criticism was aimed against experimental errors on heats of combustion, heats of dissociation, the formation of ammonia, multiple proportions, and other work performed by Bruhl, Favre, Stohmann, Hermann, Klingen, Hagemann, Lagerlof, and others.

His most acrimonious critique, begun in 1873, was directed against Berthelot's claim to have been the founder of thermochemistry. The altercation between the two lasted for another three decades, and as late as 1905, Thomsen contended that Berthelot's famous calorimetric bomb was inferior to his own universal burner, and that his rival's heat of combustion data were useless in establishing correlations with the constitution of organic substances.[23] Thomsen performed direct combustions for the determination of specific heats and the heat of formation of chemical compounds, but "he preferred to use solutions whenever possible." Berthelot first mentioned the bomb in 1885, in a joint memoir with P. Vieille. Despite the persisting myth ascribing credit to Berthelot, Vieille had actually been the inventor of the instrument which he described in an article drafted in April 1879 but published in 1884.[24]

The calorimetric bomb allowed the measurement of the heats of combustion of solids, volatile liquids, and gases. It was, in particular, for gaseous hydrocarbons and organic compounds, where combustions were often

22. Helge Kragh, "Julius Thomsen and 19th-century Speculations on the Complexity of Atoms," *Annals of Science* 39 (1982): 42.
23. Medard and Tachoire, 1994, pp. 161ff.
24. Kim, p. 223.

incomplete, that Vieille and Berthelot's new instrument was exceptionally well suited. The burning of the substances took place in a constant volume, in oxygen compressed at what were then high pressures of up to 25 atmospheres. The calorimeter was built from gold-plated steel, with an interior volume of 247 cm^3, weighing 662 g. Berthelot and his collaborators gathered impressive data on heats of reactions and formations, which culminated in 1897 in a 1500-page work on *Thermochemistry – Data and Numerical Laws*, published the same year as Planck's *Thermochemie*, discussed earlier.

It was no wonder, then, that for his investigations into chemical processes at high temperatures, Nernst could draw on an impressive array of experimental data and instrumental designs, a body of work with which he had been acquainted from his student days in Leipzig.

9

Berlin: Low Temperatures

At the time of Nernst's move to Berlin at the end of 1905, Germany was toying with war. While the Russians were engaged in Japan, Germany prepared for an expansion into France. But in December 1905, panic broke out on the Berlin stock exchange, and an economic understanding with England became necessary. (Germany was not yet ready to engage the greatest naval power of the world.) The number of students matriculated at the Berlin University that year reached 7100, in addition to the 6300 auditors that crowded Germany's largest academic institution.

When Nernst moved to the Second Chemical Institute as Landolt's successor, the institute was immediately renamed the Physikalisch-Chemisches Institut. The other chemical laboratory, under the direction of the distinguished organic chemist Emil Fischer, was simply called The Chemical Institute of the University. Fischer's knowledge of physical chemistry was limited, but his interest in new measurement methods and devices had led him to consult Kohlrausch on the matter. As a result, Fischer's laboratory had acquired "installations and rooms especially designed for the *Grenzgebiet unserer Wissenschaft* – the border field of our science." Theodore Richards, Otto Hahn, and other physical chemists explored problems in this domain while working in Fischer's institute. Fischer did not build a dedicated section for physical chemistry in his own institute, although he had "already envisioned the Leipzig Professor Georg Bredig as section leader." Apparently, he deferred to Landolt and van't Hoff, "who reproached him that they were the ones who represented the interests of this discipline. Unfortunately, and with Fischer's great regret, the physical separation of the two sister institutes spoiled a traditional systematic preoccupation of the experimental chemists with the physical part of their science."[1]

Nernst's arrival changed matters, however. The "scientific and human personality of the new colleague, which increasingly confirmed the good

1. Kurt Hoesch, *Emil Fischer. Sein Leben und sein Werk* (Berlin: Verlag Chemie, 1921), p. 150.

choice made in his appointment," seems to have been sympathetic (*sympatisch*) to Fischer all along. "The younger colleague, versatile, many faceted, full of curiosity and enterprise, took upon himself the necessary obligations of academic chemistry with verve and to everyone's satisfaction."[2] In fact, Fischer had been instrumental in Nernst's appointment. When in 1904 Landolt announced his intention to retire as of the spring term of 1905, Althoff, Nernst's old protector, appointed a commission constituted of Fischer, Planck, and the physicist Emil Warburg. Several prominent names were included on the list of proposals, but by 2 December 1904, the appointment of Nernst was already complete. Even before his actual move, he was decorated by the emperor and made *Geheimer Regierungsrat,* or privy councillor, and was admitted to the Berlin Academy of Sciences in the fall of 1905.

Nernst and Fischer became allies in many organizational battles fought in Berlin and on a national scale. Fischer, experienced and practical, as well as physically imposing, "remained the leading and determining man in all the appointment and allocations policies" at the university. Years later, impressed by the discovery of the chemical transmutability of elements, he expressed doubts about the final, fundamental importance of his own field and avowed that, had he had the chance to start all over, he would have devoted himself to molecular and atomic structure studies, rather than organic synthesis.[3]

Despite the good impression and warm welcome afforded him, Nernst and his collaborators published relatively little during the first three years in Berlin. Nernst was preoccupied in those years with rebuilding the institute's laboratory in the Berlin Bunsenstrasse, which had been found inadequate for a large group of researchers. Despite it all, Nernst took professional trips, such as a visit to Yale University in 1906, where he delivered a series of lectures on his heat theorem to American audiences. Nernst was engrossed in the "great" problems of the day, and he took advantage of ongoing research into the instrumentation, techniques, and (industry-driven) incentives that guided work in many laboratories, including those of the PTR in Berlin. And after his lecture on the heat theorem to the Göttingen Academy of Science in December 1905, the project of refining data on specific heats experimentally and theoretically became his central preoccupation.

It was also at this time that Nernst became more and more involved in the plans for establishing the Chemische-Technische Reichsanstalt, an institution similar to its physical counterpart. The project eventually resulted in the formation of the Kaiser-Wilhelm Society and its attendant

2. Hoesch, pp. 145ff.   3. Hoesch, p. 151.

scientific institutes in 1911. In his own laboratory, he made significant contributions to the theory of the electrochemical bases of the physiology of nerve functioning, a continuation of his previous well-known theory of membranes.

The shift from high- to low-temperature research that was to occur gradually during Nernst's first decade of work in Berlin is, therefore, to be seen in the larger context of his involvement with the PTR, whose board of directors he joined in 1905, shortly before his final move. His old mentor from Graz, Fr. Kohlrausch, had by then been the president of the PTR for a decade. But not many seats on the board of directors had become vacant since Nernst's appointment as full professor due to the fact that by 1897, the emperor had agreed to grant lifetime tenures to all current and future members.[4]

At the time, discussions at the PTR reveal that not only Nernst but also Kohlrausch and other members of the board were asking for further studies on measurements of specific heats. Carl Linde, the industrialist best known for his refrigeration and low-temperature equipment company, for instance, desired measurements of specific heats at high pressures.[5] In a later session of that year's meeting, Planck also urged the "execution of a systematic experimental sequence on the thermodynamical behavior of a carefully selected substance." He strongly suggested an examination "not only of the equation of state, but also of specific heats, the Joule-Thomson effect at various temperatures, the critical temperature, vapor laws, etc." Planck was quite insistent in urging the analysis of argon, a suggestion strongly supported by Nernst. Measurements on the specific heats of oxygen, nitrogen, carbonic acid, and water vapor at higher temperatures and pressures were already on that year's proposed research plan of the PTR. Nernst explained to the PTR board that such investigations would benefit from correlations with explosion experiments, which gave data on expansion coefficients and specific heats, experiments that were notoriously difficult to execute. He volunteered to provide the PTR with "the necessary quantity of argon (20 liters.)" He would first investigate the liquid gas with the explosion method, and offered afterward to allow Holborn the use of the inert gas.[6]

It is important to emphasize here that such additional investigations were required not just by pure science but by technology as well. The PTR board discussed issues in their broadest significance, but its choice of research projects was intimately linked to the fields of specialization of its members,

4. Cahan, *An Institute for an Empire*, p. 74.
5. PTB, Kuratoriumssitzungen, Protokolle, Sign 24012/241. 11 March 1908, p. 3.
6. PTB, ibid., 13 March 1908, pp. 5–7.

as well as to the most pressing problems in need of solution. There is no evidence in their deliberations that distinctions between practical as opposed to purely academic interests were formulated or even uttered.

Nernst and Warburg were apparently competing on occasion for funds to support their respective interests. In 1909, Nernst proposed to the PTR the investigation of specific heats at very low pressures. Warburg had proposed that since absolute resistance determinations had not been carried out at the PTR for twenty years, a redetermination with new precision instruments was necessary down to a precision of 1 part in 10^4. Warburg appealed to scientific chauvinism in arguing:

> The sister institutes abroad have also approached the topic, and the Reichsanstalt cannot lag behind. First, because it wouldn't be appropriate to accept foreign data without scrutiny, and secondly, because a preoccupation with these issues furthers our ability to solve other similar tasks.

But Nernst steered the discussion to his own agenda. While acknowledging the "importance" of the electrical resistance standardization work carried out at the PTR, he asked that attention be devoted to a stabilization of the "entire temperature scale." Although Roentgen balked at Nernst's suggestion, he endorsed the fruitful collaboration. But Nernst "warned" that the ohm standardization would lead "to a battle among nations for one decimal point," since similar endeavors were already under way in "Teddington and Washington," the two national standardization laboratories in England and the United States.[7] Nernst evidently disagreed with the justification of such a large-scale project based solely on nationalistic arguments, and considered exaggerated precision measurements futile.

Nernst first published measurements of specific heats in 1909, although he had earlier investigated the topic as a by-product of experiments in chemical reactions kinetics.[8] Little was known at that time about the behavior of gases at higher temperatures, despite its great importance both for the calculation of chemical equilibria and for the future elaboration of the kinetic theory. At the time, only Holborn and Henning's work of 1907 on water vapor, carbon dioxide, and oxygen, was available. The figures obtained at very high temperatures through the explosion method had only very recently led to reliable data. Nernst had obtained data on the ammonia equilibrium, and now intended to apply his novel instruments to the exploration of other gases and vapors.

---

7. PTB, ibid., 12 March 1909, pp. 2–3.
8. "Specific heat and chemical equilibrium of the ammonia gas," *Zeitschr. f. Elektrochem.* 3 (1910): 96–102, submitted December 1909.

Nernst's redirection of his research from high temperatures to low temperatures can be seen, and has been seen by his students, as a move from chemistry to physics, from a domain in which thermodynamics and the energetic viewpoint tell us something useful and valid for chemistry, to a new region where guidance would have to come from a theory not yet created, the quantum theory, whose form could not yet be discerned. Nernst must have known that this approach was of rather limited interest to most chemists, but he was convinced that he was working on something fundamental: His new heat theorem would become a guide in studying the extraordinary state of matter when it approaches the absolute zero of temperature. As we saw in chapter 6, Nernst had experimented with a range of temperatures even prior to 1899, and the record of his research discussed so far seems to invalidate the view according to which Nernst realized the importance of low temperatures for the solid state only after the publication of Einstein's paper in 1907. Instead, this "conversion" happened much earlier, and grew directly out of work in technical and applied fields with which Nernst was concerned between 1897 and 1905.

Moreover, at the PTR, close connections were established between the specific-heats research and high-temperature research, and the PTR scientists' simultaneous studies of low-temperature production. The latter were strongly stimulated by the presence of Carl Linde on the PTR's board. Linde, the "Siemens" of low-temperature technology, argued that "experimental and theoretical work [on specific heats] must go hand in hand." An improved understanding of the operation of fuels during the combustion processes at high temperatures and pressures was necessary for the understanding of the operation of the highly efficient internal combustion engines.[9] Low-temperature investigations were initiated as early as 1896, and they provided support and incentive for Nernst's work in the areas both of high and of low temperatures. Moreover, the PTR also began studies of electrolytic conductivity soon after Nernst's doctoral adviser, Friedrich Kohlrausch, became its director in 1895 upon the death of Helmholtz.

Thus we see, again, that the heat theorem evolved in the nexus of technological and experimental efforts carried out simultaneously in a number of locales and in a network of personal, academic, as well as industrial relationships.

Prior to the publications of Nernst and Kamerlingh Onnes, research on the behavior of matter at the extremes of the temperature range and on important instrumental advances had accumulated on several fronts. By 1901, the chemistry of extreme temperatures had become a topic of sufficient

9. Cahan, *An Institute for an Empire,* pp. 137–9.

interest and depth to warrant a comprehensive Habilitation survey by
Georg Bredig, at the time a lecturer at the University of Leipzig.[10]

At the end of the nineteenth century and the beginning of the twentieth, the problem of liquefying gases had become highly important to the
refrigeration industry. Only in the post–World War II era has the production of liquefied gases and low-temperature instrumentation been exported
to outside technologies, no longer a problem of purely scientific interest –
similar to the times when chemists produced their own instruments and
physicists built their own measuring devices, or similar to the very recent
past when preparations of cultures, tissues, and enzymes were the main
scientific problem of biochemists. But in the first half of the twentieth century, millions were invested in liquefaction plants, and these required prolonged experiments that, in part, took place in academic laboratories. Even
today, "there is no hard and fast definition of where ordinary refrigeration
gives way to cryogenics." Generally, "scientific" cryogenics begins under
$100°K$, in particular the phenomena of superconductivity (below $20°K$)
and superfluidity (below $2°K$) "for which no known analogues exist in the
everyday world at room temperature."[11]

The precise measurement of extremely low or high temperatures was
perfected between 1899 and 1900 by the application of the gas laws in
the gas thermometer.[12] The measuring device, made of glass, platinum-
iridium alloy, or porcelain, relied on a displacement method, whereby the
gas from the thermometer, mostly air or nitrogen, was dislocated by another gas, such as carbonic acid or hydrochloric acid, whose volumes were
measured at room temperature. But at low temperatures, air and its components were useless because they depart from the gas laws and because
they eventually liquefy. Therefore, the gas thermometer was filled with hydrogen and, at the lowest temperatures, with helium at low pressures, according to Olszewski's methods. Another method well established by
1900 was the use of electrical resistivity measurements of metals, at a time
before the discovery of superconductivity. It was viewed as "a quite lin-

10. Georg Bredig, *Über die Chemie der extremen Temperaturen.* Habilitationsvorlesung, 9
    February 1901. Also published as *Physikalische Zeitschrift*, Sonderdruck II (Leipzig:
    Verlag Hirzel, 1901).
11. P. V. E. McClintock, "Cryogenics," in Robert A. Myers, ed., *Encyclopedia of Physical
    Science and Technology*, vol. 3, Academic Press, 1987, pp. 825–6.
12. Holborn and Day, *Wied. Ann.* 68 (1899): 817; *Ann. d. Phys.* 2 (1900): 505; Carl
    Barus, *Die physikalische Behandlung und die Messung hoher Temperaturen*, Leipzig,
    1892; Guillaume et Poincaré, *Rapports presentés au Congrés International de Physique*,
    vol. 1, Paris, 1900. Quoted by Bredig, p. 5. In what follows I rely heavily on the con-
    temporaneous account of Bredig in order to give an accurate description of the state of
    thermometry and low- and high-temperature instrumentation at the turn of the century.

ear function of temperature in such a manner that at absolute zero [resistivity] will become zero."

Thermocouples also came into general use, in particular in the form of the platinum–platinum/rhodium element of Le Chatelier. Nernst measured temperatures by immersing a body that was previously heated to a given temperature into a calorimeter of known heat capacity, such as a platinum cube, and evaluating the loss of heat in the process. Another method, developed by Daniel Berthelot, was based on the change of the refractive index of gases with temperature and pressure.

Finally, melting and boiling points were traditionally used as fixed points for given temperatures. For a long time, the porcelain industry used vessels made of a mixture of quartz, feldspar, chalk, and kaolin in order to measure oven temperatures, since these materials melt according to their composition at various but known temperatures in the range of 1100° to 1700°C.[13] For the same purpose, the known melting points of metals and their alloys, i.e., alloys of gold, silver, and platinum, heated to bright incandescence were used. In addition, the work of spectroscopists on the noncharacteristic radiation, that is, the thermal or black-body radiation that would later be intimately related to the developments of quantum theory, had provided methods of determining the temperature of a body from its radiation intensity with the help of a bolometer or photometer, methods discussed by Langley, Paschen, Violle, Le Chatelier, and others. Yet on fundamental matters, such as the temperature of the sun, the specialists disagreed: Rossetti found the temperature of the sun to be 9965°C, while Le Chatelier set it at 7600°C and Paschen at 5400°C.[14]

When Nernst entered the low-temperature research field, anecdotal evidence suggests, he had, characteristically, little patience for the construction of a big-scale liquefying plant and thus proceeded to design his own cryogenic apparatus.[15] However, practical concerns, rather than mere temperament, dictated the choice of scale and applicability. Nernst, in contrast to the "professional liquefiers," wanted to measure heats (specific heats) as function of temperature. Because the value of the specific heat decreases with decreasing temperature, he needed to measure very small quantities of heat, so he *needed* sensitivity – and he needed it increasingly the more the temperature dropped. His goal was not the attainment of record low temperatures, or record quantities of liquefied hydrogen, but calorimetric measurements, a goal which set him apart from the laboratory of his contemporary, Kamerlingh Onnes.

13. See Ost, *Lehrbuch der technischen Chemie*, 2nd ed., p. 230.
14. Bredig, p. 8.
15. See, for example, the account of K. Mendelssohn.

In his own words, for his researches on the specific heat of solids, where "the object to be investigated . . . is small and thus the heat to be removed from the object in order to cool it is negligible as compared with that required to liquefy a liter of gas, . . . it appeared evident that simpler methods could be devised."

The copper calorimeter was the first device built by Nernst and his students at the beginning of 1910.[16] A copper calorimeter had been described as early as 1840 by Heinrich Hess, but of a very different design. Hess was interested in the measurement of the quantity of heat liberated by the mixture of two solutions, generally acid-base neutralizations or dilutions. The cooling element was water, enclosed in a wooden chest, into which the copper vessels were introduced.[17] Instead of a calorimetric cooling mixture, Nernst used a copper block, which held the sample to be analyzed, and which was heated or cooled. The block was placed into a Dewar insulating vessel. Thermoelements served as thermometers, introduced into the copper block by means of glass capillaries. The substance was dropped into the calorimeter through a glass tube. The whole instrument, weighing only 400 g, was immersed in a constant temperature bath made of ice or carbonic acid, keeping the temperature of the upper end of the thermoelements constant. Determinations were made of the difference between the upper end of the thermoelements and the lower copper block, into which the substance had been dropped. These thermocouples were made of ten copper-constantan thermoelements coupled in series, off which the electromotive force was measured by a millivoltmeter.[18]

Two additional devices were constructed in order to heat or cool the samples before introducing them into the calorimeter. These consisted of very small silver vessels, which were of such dimensions that they could be easily inserted into the calorimeter. For low temperature readings, the samples were cooled in a quartz vacuum vessel through which a cooling tube passed. Alternatively, the silver crucible containing the sample was cooled inside a test tube into which dry hydrogen flowed. Either of the two devices was then positioned above the calorimeter; by opening appropriate stoppers at the bottom of the heating or cooling device, the silver vessel was dropped into the calorimeter. As Nernst reported, "the

16. Walther Nernst, F. Koref, and F. A. Lindemann, "Untersuchungen über die spezifische Wärme bei tiefen Temperaturen. I." *Berl. Ber.* 1 (1910): 247–1. Presented in the academy's session on 17 February 1910.

17. Kim, p. 215.

18. W. Nernst, *The Theory of the Solid State, Based on Four Lectures Delivered at University College, London, in March 1913* (London: University of London Press, 1914), pp. 15–16.

whole manipulation took about 3 seconds,"[19] time during which variations in the internal calorimeter took place, but which could be compen-
sated for by the fact that the method was based on relative, rather than
absolute, measurements. In order to calibrate the calorimeter, Nernst used
water or lead, whose specific heats were known over a wider range of temperature. In this device, heating and cooling experiments were used alternatively, with each lasting some twenty to thirty minutes.

This method, used in order to obtain average specific heats,[20] had to be
supplemented with a method for the exact determination of specific heats
at a given temperature. It also needed modifications in order to enable
work at lower temperatures than those provided by liquid air. For this purpose, the vacuum calorimeter was designed. Nernst's vacuum calorimeter
was thus an instrument "most superbly suited for its purpose and it may
well be said that low-temperature physics dates from its invention. It is to
low-temperature research what the spectrograph is to spectroscopy and
the Wilson chamber or the Geiger counter to nuclear physics."[21]

This quite simple instrument was developed into a rather complicated
apparatus for special purposes, and was the "method of choice" until
World War II.[22] The oldest variant of the apparatus consisted of a block
of the material to be investigated, suspended in a vacuum vessel. The
block was covered with a fine wire mesh, through which electrical energy
in known quantities was applied to the substance to be examined. The
wire served both as a heating device and as a thermometer, since the
known electrical resistance of the platinum wire indicates the change of
temperature. The block and the coils were surrounded by a cooling liquid, helium or hydrogen, which was introduced into a tight vacuum vessel. After the substance in the calorimeter was brought down to the temperature of the liquid, the vacuum vessel was evacuated, with the liquid
being pumped. Gradually, the temperature of the substance in the calorimeter was raised in successive, known increments by passing electricity
through the platinum coil. Measuring the resistance of the wire gave information on the sample's temperature, by comparing the wire resistance
to a resistance box.[23]

19. Ibid., pp. 18–19.
20. These experiments of average specific heats were performed and reported by F. Koref,
who was assigned the task of testing the usefulness of the apparatus at low temperatures.
21. Ruhemann, pp. 136–7.
22. Mendelssohn, *Cryogenics,* p. 46.
23. Nernst used the wire as both heater and thermometer, which constituted a superb technical feature. Modern calorimeters, however, use separate devices, such as a carbon re-
sistor thermometer. Mendelssohn, p. 46.

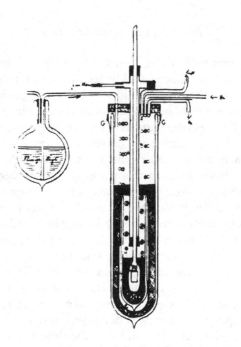

Diagram of the liquefier

The liquefier

During the winter term 1909–10, Nernst used the calorimeter for experiments at 20°K. Later, an improved platinum calorimeter was shown to the Prussian Academy of Sciences in Berlin at the beginning of 1910, on the same occasion as the copper calorimeter. Three other types of calorimeters were designed, essentially upon similar principles. Metal substances were used as a block of cylindrical shape, surrounded by a block of the same metal, around which the platinum wire was coiled. Poorly conducting substances were analyzed using the silver-vessel method, to which the wire was attached and inserted into the silver block. Alternatively, the platinum resistance was wound on the exterior.

In 1913, Nernst described this instrument to British colleagues as the outcome of more than four years of stabilizing the calorimeter. Nernst gave full credit to Eucken's innovation, which Nernst afterward improved in some details, "which really enables one to measure the true specific heat, even at the lowest temperatures." The innovative aspect of their method was that *the substance to be examined itself acted as the calorimeter* which was heated up a few degrees by means of a platinum wire through which a measured amount of electrical energy was passed. The increase in temperature was determined by using the same platinum wire as a resistance thermometer.

Nernst insisted in his lectures that he had not been solely interested in thermometric measurements or verifications: "The range of validity of the laws of Dulong and Petit, and of Neumann and Kopp, was not the chief point of interest, but rather the question of the diminution of the atomic heat below the value 3R."[24] This statement of Nernst's ties in quite explicitly with his agenda as stated in 1906 in his Silliman Lecture at Yale, at a time when he was only dimly envisaging a move toward low temperatures.

At about the same time, Nernst also asked the PTR to produce and calibrate a new thermometer for the observation of vapor pressures of liquefied gases, which allows "at low temperatures a convenient and precise reproduction of the temperature scale." He thought that, although he had been able to calibrate platinum thermometers, further work on lead thermometers should be continued, since he considered lead a more reliable substance. At the time, there were resistance thermometers, with platinum or lead wires, and liquefied gas thermometers where vapor pressure fixed the temperature. Planck's argon project of the previous year had apparently not been followed through, since Planck reiterated – again with

24. Nernst, *The Theory of the Solid State,* pp. 13–14, 21. Originally in German, with an almost identical text, in Nernst, "Untersuchungen über die spezifische Wärme bei tiefen Temperaturen II.," *Berl. Ber.* (1910): 262ff.

Nernst's "enthusiastic" support – his request for measuring its properties. He defended the proposal by arguing that an examination of the equation of state of liquid argon would facilitate a "decision" between the "consequences of van der Waals's equation of state and those resulting from Nernst's heat theorem."[25] In 1911, Nernst pointed out to the members of the PTR board that platinum resistance thermometers had a distinctly "different" behavior at low temperatures, compared to the gas thermometers then being examined for calibration (helium, argon, and hydrogen thermometers).

On the same occasion, Nernst reported to the PTR that a commercial company was planning the industrial production of liquid hydrogen, a prospect that worried v. Linde. He envisioned that the transport of liquid hydrogen on railroads and its storage would pose "great difficulties," and that therefore, the Berlin Society for Refrigeration had refrained from industrial fabrication. Nernst, however, countered by saying that the similarly dangerous transport of liquid chlorine and even gasoline had been overcome: "One will have to learn to handle liquid hydrogen," he exclaimed.[26] It was unavoidable that Nernst would be extremely interested in promoting the production of liquid hydrogen: In addition to the vacuum calorimeter, Nernst and his collaborators had to construct a suitable hydrogen liquefier.

The new liquefier was first described at the 18th Meeting of the Bunsengesellschaft in Kiel in May 1911, although preliminary results had already been presented in an informal manner by Nernst in January.[27] This liquefier was remarkably similar in construction to his high-temperature iridium oven. But instead of a surrounding heating device, Nernst now constructed a liquefier that contained liquid hydrogen obtained through precooling, by passing compressed hydrogen through successive liquid air baths. The hydrogen flowed through a double series of coils, first cooling down, then expanding back to normal pressure, thereby cooling a sample which was introduced through a silver tube into the interior of the bronze cooling vessel. The apparatus produced 300 to 400 cm^3 of liquid hydrogen per hour, at a 10 percent efficiency rate. It could be topped off with additional liquid air. In order to monitor the pressure inside the instrument, and to avoid accidents, a manometer was introduced through the silver tube, as well as a thermoelement for temperature readings.[28]

25. PTB, ibid., 10 March 1910.　　26. PTB, ibid., 8 March 1911, 10 March 1911.
27. Nernst, "Über neuere Probleme der Wärmetheorie," *Berl. Ber.* (1911): 65–90.
28. Nernst, "Über einen Apparat zur Verflüssigung von Wasserstoff," *Zeitschr. f. Elektrochem.* 17 (1911): 735–7. M. Le Blanc inquired about the availability of the apparatus, and Nernst obliged by indicating that the Firma Gebr. Hoenow in Berlin, Frankfurter Allee 118, provides the apparatus for 250Mk, a price which Le Blanc thought quite acceptable for purchase by "a number of laboratories." Ibid., p. 735.

A number of papers on related topics were presented successively in the same session of the meeting on 27 May, to be discussed in the next chapter.

Nernst repeated the experimental algorithm employed so far, introducing the calorimeter into the liquefying device, thereby avoiding the transfer of the liquid hydrogen. He found that it was easy to work with small quantities of liquid hydrogen due to the small size of the entire apparatus, enabling him "to make experiments nearly daily without very great expense." With each successive experiment, the temperature increased, and true specific heats were measured at each interval.[29]

It was in the process of calibrating his platinum thermometer that Nernst again had to pay attention to the problem of conductivity in metals at low temperatures, effectively overlapping with Kamerlingh Onnes's researches at Leiden. In order to evaluate the temperature changes correctly, Nernst needed a correlation between the conductivity and the temperature of platinum: "Difficulties were encountered while determining the resistance of the platinum wire at various temperatures, but it was found possible to fix certain definite temperatures and to find a formula giving the very variable temperature coefficients between 20–40°K with sufficient accuracy."[30]

He acknowledged, however, that the solution to this intricate problem of measurement was "difficult and lengthy." As standard points, Nernst used the melting point of ice and the boiling point of solid carbonic acid, of oxygen, and of hydrogen. The platinum wire used was also compared with a resistance of very pure lead wire. He found the "very regular curve given by lead according to the experiments of Kamerlingh Onnes" as the most useful. Avoiding the substantial changes in resistance induced by even the smallest impurities at low temperature, Nernst found that the pure lead supplied by Kahlbaum from the PTR "gave at the aforesaid standard points exactly the same resistances as the sample examined by Kamerlingh Onnes." Nernst was elated to find an empirical formula that linked the resistance curve of his platinum wires with that of the platinum specimens examined by Kamerlingh Onnes. For the resistance of two different samples of platinum, w_1 and w_2, and by reducing the resistance at 0 to 1000 degrees,

$$w_1 = (w_2 - \alpha)/(1 - \alpha)$$

29. Nernst, *Theory of the Solid State,* p. 25.
30. Ibid., p. 24. Originally in German, Nernst, "Untersuchungen über die spezifische Wärme bei tiefen Temperaturen. III.," *Berl. Ber.* (1911): 306. Presented to the academy on 23 February 1911.

and small in comparison with 1, "which is always the case with relatively pure platinum," he obtained for the temperature coefficients:

$$dw_1/dT = dw_2/dT \left[1/(1 - \alpha)\right]$$

The calibration proved burdensome inasmuch as: "this rule has been found excellent for all kinds of platinum wire freely and loosely wound; yet it is not to be applied to platinum fused on to quartz (resistance thermometer of Heraeus) and gives a slight inexactitude when the wire is fixed in a solid enamel varnish."[31] This detailed discussion illustrates only that Nernst experimented with concrete devices that were to be used as thermometers – wires that had to be practically fused to enamel on thermometric scales and had to behave reliably and sensitively over a broad range of temperatures and repeated, rapid changes.

In his original communication to the academy, Nernst had added in an appendix that one ought to point out "a very remarkable relationship" that he had established empirically between atomic heat and electrical resistivity, which "has not previously been predicted by the electron theory." That is why Nernst had to determine "as accurately as possible" the temperature coefficient of the resistivity of the platinum wire in the interval of 20° to 80°K. For calibrations he employed the data of Kamerlingh Onnes and of Travers on the vapor-pressure curves of oxygen and hydrogen, and in addition, thermometric measurements of the lead-resistance thermometer. He had been aware that the "resistance curve of lead deviates much less from a linear curve in the interval of room temperature to that of boiling hydrogen than that of platinum." And he continued:

> But it was found that the temperature coefficient of platinum, which from room temperature down to $T = 65$ varies quite considerably, diminishes at even lower temperatures extremely quickly, very similarly to the Einstein curves [for specific heats]. . . . Thus the temperature coefficient is somewhat analogous to the specific heat, and it is further noteworthy that both magnitudes tend at higher temperatures to a value independent of the nature of the metal.[32]

What Nernst had found is that as one moves toward lower and lower temperatures, the resistivity of the metal declines significantly and that, thus, conductivity increases quite dramatically.

He found an evident analogy between the validity of the Dulong-Petit rule and that of Clausius, according to which the temperature coefficient

31. Ibid., pp. 29–30.
32. Nernst, "Untersuchungen über die spezifische Wärme bei tiefen Temperaturen. III.," *Berl. Ber.* (1911): 311ff.

of resistivity and the expansion coefficient of gases are similar. Most noticeably, Nernst continued:

> The above considerations lead first to Planck's radiation formula as a simple formula for the electrical resistance, which has proven itself well in the domain of the rapid decline of the temperature coefficient, that is, *especially there where I need it urgently* [emphasis added].[33]

By assuming the "Planck formula" and fitting it to his resistance formula, Nernst calculated the resistance of the lead wire according to:

$$w = 0.1626/(e^{38/T} - 1) + 0.00070$$

and compared the obtained values with those measured for lead by Kamerlingh Onnes. He found a good correlation above 20°K, although for lower temperatures the figures disagreed. For platinum, Nernst used his own measurements and found that Planck's formula was a "useful interpolation formula for certain domains of electrical resistivity." He acknowledged that this could be a "coincidental correlation," although according to his experimental data, "this can hardly be the case anymore."[34]

In 1910 and 1911, Nernst and his collaborators had thus advanced considerably on the path toward an early "discovery" of superconductivity, and with his usual acumen had pointed to lead and platinum as the most remarkable examples of deviations from accepted theories of electrical conductivity. And although Nernst never laid claim to any originality in this domain and refrained from any serious discussion even at the Solvay Congress in October 1911, it was in 1913 that he voiced a faint reminiscence of the conductivity studies of 1911. In his Wolfskehl Lecture given at the University of Göttingen in the last week of April 1913, Nernst said: "As Kamerlingh Onnes and [I myself] have noted separately and almost simultaneously, the decrease of the temperature coefficient occurs at lower temperatures, the lower the atomic frequency of the metal."[35]

The interpretation of the calorimetric data Nernst used for determining the specific heats required several elaborations upon the thermodynamics of solid bodies. At the temperatures of liquid hydrogen with which Nernst was initially operating, the calorimetric quantity that he could measure was the specific heat at constant pressure, whereas the specific heat at constant volume as a function of temperature is the one required

33. Ibid., p. 312.   34. Ibid., p. 313.
35. W. Nernst, "Kinetische Theorie fester Körper," in M. Planck, P. Debye, W. Nernst, M. v. Smoluchowski, A. Sommerfeld, and H. A. Lorentz, *Vorträge über die kinetische Theorie der Materie und der Elektrizität. Gehalten in Göttingen auf Einladung der Wolf-skehlstiftung* [April 1913], with contributions by H. Kamerlingh-Onnes and W. H. Keesom, and Preface by D. Hilbert (Leipzig/Berlin: B. G. Teubner, 1914), pp. 61–82.

for application to the heat theorem. It was therefore imperative to determine first the data on compressibility (k) and the volume (V) coefficient of thermal expansion, which are needed to use the equation that relates the two magnitudes according to the general thermodynamic equation:

$$C_p - C_v = \alpha^2 VT/k$$

But since these quantities were not well known at normal temperatures, and even less well known at low temperatures, it was "necessary to be able to calculate the $C_p - C_v$ correction in a different manner." Nernst and Lindemann correlated the melting point of a solid with the amplitude of vibration of the atoms at that temperature, equating the latter with the atomic distance in the solid. His formula for the vibrational frequency was

$$\nu = 3.08 \cdot 10^{12} \, (T_s/mV^{2/3})^{1/2}$$

where T_s was the melting point, m the molecular weight, and V the volume. By employing Einstein's formula relating the optical frequency to compressibility and a previously obtained relationship by Gustav Mie, they obtained

$$3\alpha \text{ proportional to } C_v/T_s,$$

and hence:

$$C_p = C_v + C_p{}^2 \, (T/T_s)A_0$$

where $A_0 = 0.0214$ is a constant. Thus the ratio C_p/C_v could be calculated from melting points, and an important obstacle to the more rigorous experimental verification of the heat theorem could be removed.

By 1912, Nernst's work at the extremes of the temperature range had even become of interest to the daily press. Thus, the *Berliner Tageblatt* carried a long article on the "strange things" that were happening in a large house in the middle of Berlin, close to the banks of the river Spree, where people were busy producing "bombs, diamonds, and radium." By that time, Nernst could boast that he had been able to reach $-253°C$ with liquid hydrogen, temperatures which he employed in order to study the abnormal behavior of diamond:

> When cooling diamonds in the vicinity of this extreme temperature, the following was observed: Up to -230 degrees the diamonds showed a normal behavior. They had their specific heat, they gave heat to the surrounding colder bodies. That however suddenly stopped. They were totally indifferent when introduced into a -240 or -250 degrees environment, and no heat could be extracted or added any longer. They were, in a higher sense, frozen and dead. One conceives of heat today

as the vibration of the atoms, the smallest parts of a body. . . . Through the diamond experiment one has therefore found that the atomic vibrations can be brought to a standstill. As soon as this happens, the concept of heat does not exist any longer for the dead body. . . . One has found that the heat death [*Wärmetod*] with the given substances occurs earlier if the temperature at which they melt is higher. For diamond, whose melting has not succeeded so far, one could therefore have foreseen the strange behavior at low temperature.[36]

Nernst patiently explained to the reporter how the measurement of such extremely low temperatures, at which normal thermometers fail, is carried out with the help of a variety of instruments that supplement common thermometers: In some, one registers the expansion of gases, such as helium, which do not liquefy at low temperatures; in thermoelements, electrical currents are produced by the changes of temperature; and in yet a third method, the change in the resistivity of wires takes place with temperature changes.

Kamerlingh Onnes made the crucial discoveries of superconductivity and liquid helium, which for a decade or more were studied for their own sake, so to speak. It was Nernst who established and carried out a real physical research program of low-temperature physics for the first time. And it was the third law and its (eventual) essential connection to quantum ideas that ultimately established the fundamental character of low-temperature studies.

36. Artur Fürst, "Bomben, Diamanten und Radium," *Berliner Tageblatt,* 2? February 1912, evening edition.

10

*The Incorporation
of Quantum Theory*

Although recognized widely as the initiator and organizer of the First Solvay Congress in Physics, held in the fall of 1911, Nernst is not usually considered a contributor to the internal development of quantum theory. Despite the apparent dearth of documentation in the literature published between 1907 and 1911, he and a number of other physical chemists took considerable interest in Einstein's paper on the specific heats of solids. Their research aims were quite different from those of most physicists at the time. Thereby, an interesting and lively discussion on the relevance of the quantum theory of matter (as opposed to radiation) was introduced into the chemical community.

Einstein's paper "Planck's theory of radiation and the theory of specific heat" has been analyzed in great detail.[1] During the four years between its publication and the Solvay Congress, Nernst was virtually the only physicist-chemist who addressed the problems regarding the quantization of energy raised by Einstein's paper, and it was to a large extent due to Nernst's efforts and publications – as Martin Klein has pointed out in the earliest and most detailed discussion on this topic – that the community of physicists became cognizant of Einstein's work. Primarily during and after the Solvay Congress, a number of young theoretical physicists with physical-chemical training expanded upon Einstein's work, the best known among them being Peter Debye, Max Born, and Theodore von Kàrman. The vacuum implied by Kuhn's conclusion, however, was filled, but by a different scientific group than traditionally held responsible for advances in quantum theory.

Among the firsthand reports on the reception of Einstein's paper of 1907 is a letter from Arnold Sommerfeld, the eminent Munich theoretical physicist, to Hendrik A. Lorentz, a telling testimony not only of the

1. Martin Klein, "Einstein, Specific Heats, and the Early Quantum Theory," *Science* 148 (1965): 173–80; Thomas S. Kuhn, *Black-Body Theory and the Quantum Discontinuity, 1894–1912* (1978); rev. ed. 1987, p. 211. Abraham Pais, *"Subtle Is the Lord": The Science and the Life of Albert Einstein* (Oxford: Oxford University Press, 1982), pp. 389–401.

complex and even confused state of quantum theory and relativity at the time, but also of the deep mistrust and outright anti-Semitism that characterized some of the best physicists of the time:

> Now we are all ardently waiting for you to address the whole complex of Einstein's papers. As brilliant [*genial*] as these papers are, it still seems to me that something almost unhealthy lies in this unconstruable and nonintuitive [*unkonstruibaren und anschauungslosen*] dogmatism. An Englishman would hardly have conceived this theory; it might be that here too, as with Cohn, the abstract conceptual manner of the Semite finds expression. Hopefully you will succeed in endowing this dazzling conceptual skeleton with real physical life.[2]

In 1907, Einstein was a young and generally unknown entity, an outsider to the small but rather homogeneous elite of European physics who had been denied academic positions and did not include civil servants in the Swiss patent office. He had so far refrained from public appearances. The person and his work were indistinguishable to Sommerfeld, who, though admiring the brilliance, deferred to Lorentz to disentangle the intricate issues raised by Einstein in his pathbreaking papers of 1905 and 1907. It is nonetheless also true that as late as 1911, Einstein himself, more than anyone else at this time, knew that there would have to be a theory that accounted for "quantum phenomena." He also knew that the theory did not exist yet, that it was so far only heuristic. He was impressed by what had been learned, but sensed how far was still to go in the conceptual elaboration of the quantum theory. At that time, he prefaced a discussion of his report to the Solvay Congress with the following remark:

> We are all in agreement that the so-called quantum theory of today is a useful device [*Hilfsmittel*] but not a theory in the usual sense of the word, in any case, not a theory which can at present be developed in a coherent form. On the other hand it has also turned out that classical mechanics, which is expressed in the equations of Hamilton and Lagrange, cannot be considered any longer a useful scheme for the theoretical representation of all physical phenomena.[3]

2. "Jetzt aber warten wir alle sehnlichst, dass Sie sich einmal zu dem ganzen Complex der Einstein'schen Abhandlungen äussern. So genial sie sind, so scheint mir doch in dieser unkonstruierbaren und anschauungslosen Dogmatik fast etwas Ungesundes zu liegen. Ein Engländer hätte schwerlich diese Theorie gegeben; vielleicht spricht sich hierin, ähnlich wie bei Cohn, die abstrakt-begriffliche Art des Semiten aus. Hoffentlich gelingt es Ihnen, dies geniale Begriffs-Skelett mit wirklichem physikalischen Leben zu erfüllen." A. Sommerfeld to H. A. Lorentz, 26 December 1907, Lorentz Papers, roll 4, LPC, AIP.

3. A. Eucken, ed., *Die Theorie der Strahlung und der Quanta. Verhandlungen auf einer von E. Solvay einberufenen Zusammenkunft* (Halle: Wilhelm Knapp, 1914), p. 353.

Yet he had written a report extraordinary in its scope and structure on what must have been the most significant topic in quantum theory at the time. Entitled "On the Current State of the Problem of Specific Heat," the paper, almost fifty pages long, is a programmatic attempt at synthesizing the various strands of quantum research, and an agenda for future research. That the quantum theory had gained significant importance prior to the congress is also evident from the fact that Planck, Sommerfeld, Einstein, and Nernst all compiled synthetic reports, while James Jeans devoted his entire presentation to the treatment of the kinetic theory of specific heats according to Maxwell and Boltzmann. It is, therefore, clear that the experimental research on specific heats prior to 1911 produced sufficient interest and concrete material to elicit attention from the major theoreticians. Moreover, Einstein's paper repeatedly emphasized the compelling experimental evidence found in the research of Nernst and his collaborators. Einstein included in his paper a comparative graph that showed the theoretical predictions for the temperature dependence of specific heats plotted against the experimental data, a graph reproduced from one of Nernst's publications of 1911, and he commented:

> Even though systematic deviations exist between the observed and theoretical curves, the correspondence is an entirely stunning one if one considers that the individual curves are completely determined by one parameter, namely, the eigenfrequency of the atoms of the given [chemical] element.[4]

Einstein's formula for the specific heats of solids predicted that all specific heats will tend toward zero with the decline of temperature toward absolute zero, in contrast to the predictions of classical statistical mechanics, according to which specific heats would remain constant at a value of $3R$ or 6 calories for 1 gram-atom of a monatomic solid at constant volume.

The incorporation of quantum theoretical concepts afforded Nernst a bridge between atomism and thermodynamics. These two theoretical approaches, which are quite distinct, and which for some scientists at the time presented an either/or alternative (Wilhelm Ostwald, for example), were combined by Nernst into a powerful theoretical and experimental tool. Nernst was well qualified to incorporate quantum considerations into the complex matrix of his previous work. The tendency prevails to associate the first quarter of the century with the consolidation of the quantum hypothesis as a preparatory stage for the breakthrough in quan-

4. A. Einstein, "Zum gegenwärtigen Stande des Problems der spezifischen Wärme," in Eucken, p. 334.

tum statistical theory produced by E. Schrödinger, M. Born, and W. Heisenberg in the mid-1920s. This, however, overlooks several conceptual leitmotifs that originated in the nineteenth century and were not overthrown by what is called the quantum revolution. Nernst's example illustrates that research well under way at the beginning of the century was modified by the quantum hypothesis through gradual "working through," without, however, dramatically changing ongoing work or producing spectacular conceptual reformulations.

On 17 February 1910, Nernst presented to the Prussian Academy of Science the first two installments of the "Researches on the specific heat at low temperatures," which he had performed in the course of the preceding year with F. A. Lindemann, A. Eucken, F. Koref, and A. Pollitzer.[5] Together with other data from the literature, their measurements of the specific heats of metals and salts down to $73°K$ confirmed the predictions of the heat theorem within the limits of experimental error: namely, that specific heats would decrease with temperature and tend toward the same limit at absolute zero temperature. In the brief conclusions to the long experimental paper, Nernst wrote:

> As a general result we have found that, in accordance with the earlier work of Behn, Dewar, and others, specific heat decreases strongly at low temperatures, so that one gains the impression that, corresponding to the requirements of Einstein's theory, it tends toward zero.

Nernst predicted that investigations at still lower temperatures (that of liquid hydrogen) should provide a final decision. Besides Nernst's brief mention of Einstein and Planck in his textbook in the sixth edition of 1909, this is probably the first direct reference to the relationship between his own work and the quantum theory.[6] Simultaneously, additional publications came forth from his students. In March 1910, F. A. Lindemann and A. Magnus presented two papers on both the experimental fit with Einstein's formula and the application of the heat theorem to the thermodynamics of chemical reactions.

During the same year, T. W. Richards of Harvard University published an important paper on a series of determinations of specific heats at low temperatures. Richards (1868–1928) had visited Ostwald in Leipzig and Nernst in Göttingen in 1895, and had thenceforth devoted his career

5. W. Nernst, F. Koref, and F. A. Lindemann, "Untersuchungen über die spezifische Wärme bei tiefen Temperaturen. I.," *Berl. Ber.* 1 (1910): 247–61. Nernst presented the results in the session of 10 February 1910, together with W. Nernst, "Untersuchungen über die spezifische Wärme bei tiefen Temperaturen. II.," *Berl. Ber.* 1(1910): 262–82.
6. Nernst, *Theoretische Chemie*, 6th ed., p. 700.

primarily to the experimental determination of atomic weights. Since the Dulong-Petit rule linked specific heats to atomic weights, Richards and his students embarked on a thorough and extended program of calorimetric measurements, examining 25 chemical elements. His data became a rich trove for Nernst's future work, although Richards's temperatures never reached below that of liquid air. His data were tabulated generally in two temperature intervals, $-188°$ to $-20°C$, and $0°$ to $100°C$. However, Richards made no reference in that paper to either Einstein or to Nernst's work, but he did describe the "convex" shape of the plotted data, writing that "all tend increasingly downward toward the absolute zero. From these observations it may be inferred that at very low temperatures the specific heats become exceedingly small and it would not greatly surprise us if in the neighborhood of absolute zero many of them were less than 0.2 of their value under ordinary conditions."

The problem is that Richards plotted the difference between the atomic heats of different elements in the range of 4.14 to 6.7 calories, since he was interested in the variability of atomic heat with atomic weight. He never attempted to plot changes with temperature for one and the same element, something he could not have done since his were average relative measurements, sets of readings taken at two different temperatures always separated by the same amount (dT of approximately $200°C$). Without altering the fixed endpoints, this method could not lead to a proper graphic, visual representation of the change in specific heats, let alone detect the significance of the analytic shape of the curve. In fact, it could give only two values for any substance, one a "low temperature" average, the other a "high temperature" average. It took Richards five years for his experiments and analysis before publishing, yet, as we shall see in the last section of the book, he nonetheless attempted, several years later, a weak claim to priority for an "early" discovery of the third law.[7]

Magnus and Lindemann proceeded to interpret a number of regularities that Richards had detected in the drop in atomic heat within the groups of the periodic table.[8] They proposed to alter Einstein's formula slightly in

7. T. W. Richards and F. G. Jackson, "The specific heat of the elements at low temperatures," *Zeitschr. f. physik. Chemie* 70 (1910): 414–51.
8. A. Magnus and F. A. Lindemann, "Über die Abhängigkeit der spezifischen Wärme fester Körper von der Temperatur," *Zeitschr. f. Elektrochemie* 8 (1910): 269–72. In the following paper, submitted to the same journal five days later, Magnus applied the proposed formula for a chemical reaction, in order to test the validity of Nernst's theorem and to calculate therefrom the electromotive forces from thermal data; ibid, pp. 273–5. In May 1910 at the meeting of the Bunsengesellschaft in Giessen, Nernst again discussed the relation between the thermodynamic theory of electrolytic conduction, his heat theorem, and Einstein's work.

order to account for the fact that at very low temperatures, the specific heats seemed to decrease not exponentially but somewhat more slowly. They assumed this to be due to the fact that specific heats were measured at constant pressure, where an additional expansion work and possible contributions of electrons to the specific heats should be taken into consideration. Dewar's work of 1905 had already drawn attention to the problem of electronic contributions to the specific heat of solids in investigations that he had extended down to the temperature of liquid hydrogen. Theoretically, if one took into consideration that for n conducting electrons for n atoms a quantity of at least $3/2nkT$ kinetic energy per unit volume should be added, it would bring the total energy to $3RT + 3/2RT = 9/2RT$, which would increase the specific heat by 50 percent, or to the equivalent of 9 calories per degree per unit volume. But Dewar's experiments of 1905 showed that the specific heats at room temperature for conductors and for insulators were quite similar, and therefore excluded the possibility of electron contributions.

The problem of the electronic contributions was particularly thorny, and it is very interesting that Nernst and his students attempted to come to grips with it. In this context, the relationship of chemistry to the quantum hypothesis, then relatively new, was faced with substantial difficulties. Without one's going into the details of contemporary knowledge concerning the structure of atoms and molecules, it is evident that the electronic structure of matter, as it related to chemical and other properties, was almost totally unexplored in 1910–11. It was one of the main preoccupations of those involved in electrical conductivity and theoretical quantum problems in physics, but had not become an integrated research topic in chemistry.

Planck and Einstein exchanged letters with H. A. Lorentz about this time that testify to the difficulties involved in ascertaining the precise role played by electrons in quantum phenomena. Planck wondered why energy is transferred only in discrete portions, and avowed that "there are better prospects of holding the electrons responsible for these peculiar transition conditions than the ether, primarily because the latter has a so much simpler structure and is much better known than the electrons."[9] Thus, four years after Einstein proved the ether to be superfluous in the electrodynamics of moving bodies, the ether as a heuristic device for understanding

9. "Eine secundäre Frage ist erst: warum geht die Energie nur portionenweise über, und da glaube ich allerdings, dass es mehr Aussicht hat, die Elektronen für diese seltsamen Übergangsbedingungen verantwortlich zu machen als den Aether, und zwar hauptsächlich deshalb, weil letzterer eine so viel einfachere Structur besitzt, und viel besser bekannt ist als die Electronen." Planck to Lorentz, 10 July 1909, roll 3, Lorentz Papers, AIP.

processes of energy transfer still held a powerful grip on the mental conceptions with which Planck and others operated. And yet, apparently for entirely idiosyncratic reasons, the strange quantum phenomena were being located in the realm of the "lesser known," the electrons. The electron was a labile entity, whose mass and charge had only recently been established, yet for which no proper scattering theory through matter was available. But it was hoped that one could link the electron to a variety of unsolved problems: localization of quantum interactions, conductivity, and chemical reactivity. Thus, Einstein wrote to Lorentz in February 1911 after he had visited Holland and had learned of Kamerlingh Onnes's work:

> The relationships between electrical conductivity and temperature seem to me to become highly important. If one could only avoid the difficulty of not knowing whether the change in conductivity is due primarily to the change in the number or in the free path of electrons, or to both.[10]

Einstein was evidently well acquainted with the literature on this subject, as was to become clear from his detailed report on specific heats at the Solvay Congress later in the year. Nernst and many of his colleagues had acquired a good knowledge of conductivity problems and their correlation to other physical and chemical properties, primarily based on solution and galvanic studies. Nernst had earlier pointed to the relationship between the behavior of specific heats and electrical conductivity, before Kamerlingh Onnes's discovery of superconductivity was known. The complexity of the conductivity studies was compounded by the fact that Millikan's experiments on the determination of the charge on the electron dated from 1909. In 1911, at the Solvay conference, Jean Perrin was still compelled to discuss at great length the electron theory and the merits of both sides in the ongoing debate between Felix Ehrenhaft and Millikan. Ehrenhaft's experiments claimed to show a continuously varying electric charge, while Millikan's oil-drop experiments seemed to demonstrate electric charges consisting only of integral multiples of a unit charge – the charge of the electron.

Chemists were gradually introduced to these shared new concerns, as can be seen from a presentation on current developments in inorganic

---

10. "Hochwichtig scheinen die Beziehungen zwischen elektrischer Leitfähigkeit und Temperatur werden zu wollen. Wenn nur nicht immer die Schwierigkeit hereinkäme, dass man nicht weiss, ob man die Änderung der elektrischen Leitfähigkeit hauptsächlich auf Änderung der Zahl oder auf Änderung der freien Weglänge der Elektronen oder auf beides zurückführen soll." Einstein to Lorentz, in which he thanks him for their nice visit. Zürich, 15 February 1911, Lorentz Papers, AIP, New York, roll 4, p. 2.

chemistry at the meeting of the Bunsengesellschaft in May 1911. One of the main reports came from the distinguished Professor A. Werner of the Zürich Polytechnic, who discussed the state of the theory of valency, in particular the work of Abegg, Ramsay, and Stark. At the time, chemists recognized no difference between organic and inorganic bond structures, and this led to considerable confusion in theoretical discussions. It was accepted that chemical substances are probably donors and acceptors of electrons (Ramsay), and that in complex substances there are polar bonds; major difficulties, however, persisted due to prevailing notions of primary and secondary valencies. The relationship between valency and electro-valency (the number of electrons added or lost in electrolytic dissociation and ion formation) was not precisely known, and although some corre-lation was seen to exist, Werner questioned "whether the transition of electrons occurs already in the nondissociated state or only at electrolytic dissociation." Werner's theory of affinity postulated that the chemical bond was the sum of an infinite number of small forces, that it was not a fixed unit. And Alfred W. Stewart, noting in his influential *Recent Advances in Organic Chemistry* of 1908 that the electron theory had brought about a "rejuvenation" of physics, advocated that chemists "must go to the physi-cists . . . [and] borrow as much of their theory as seems likely to help us with our own branch of science."[11]

Nernst, however, due to his deep knowledge of the ionic theory of so-lutions and of indirect evidence from different areas of physical research, was persuaded in favor of the existence of elementary electronic processes in the solid. During the discussion after Werner's paper, the only remark on this matter to be published was that of Nernst, who stated that the question whether electrons in salts are already tied to the respective atoms, such as Na and Cl, "has been totally solved recently" by E. Rubens, who showed that the absorption bands of residual radiation indicated that ions exist in the undissociated solid.[12]

In addition to the training in electrochemistry and electricity acquired during his student years, Nernst had also been intimately acquainted with the work of Paul Drude, the most important contributor to the classical electron theory of conductivity around 1900. Drude, who had first applied

11. A. W. Stewart, *Recent Advances in Organic Chemistry*, London, 1908, p. 263, quoted by M. J. Nye, "Chemical Explanation and Physical Dynamics: Two Research Schools at the First Solvay Chemistry Conferences, 1922–1928," *Annals of Science* 46 (1989):466. Nye's paper shows that indeed before the 1920s, electronic theories of atomic structure were only minimally explored and accepted even by physical chemists, whom she shows to have been nonetheless most active at the interface between physical advances in atomic theories and chemical interpretations.

12. *Z. f. Elektrochem.* 1911 (15): 603.

the term "electron" to J. J. Thomson's "corpuscle," discovered in 1897, incorporated, much like Nernst himself, the kinetic theory of gases into his theory of electrons,[13] by treating the kinetic energy of freely moving electrons like the corresponding kinetic energy of the molecules of a gas.

For some time before his move to Leipzig in 1894, Drude had worked with Nernst in Göttingen. Nernst highly admired Drude, and facilitated his appointment in Leipzig. In 1894, Nernst and Drude published an important joint paper on dissociation processes. Here they coined the term "electrostriction" to account for the binding of the solvent molecules to the ions, showing that the process of ion "hydration" plays an important role in all electrolytic reactions. In this paper, they gave a quantitative treatment of the role of the solvent, in which the solvent was considered to be a continuum-conducting medium, rather than one composed of discrete molecular entities. They showed that solvation of the ions themselves contributes to the conductivity of a solution. This meant that the dissociation of ions was not the only factor accounting for the transport of electricity.

While providing support for Arrhenius's ionic theories, Nernst and Drude gave the solvent itself an important role to play. Arrhenius's vigorous opponent on the matter of the existence of ions, Henry Edwards Armstrong, for instance, had objected that according to Arrhenius's views, water merely played the role of a "dance floor for ions."[14] These minor "improvements" on Arrhenius's work were to have a snowballing effect on Nernst's chances for a Nobel Prize after 1906, as we shall see. Moreover, Nernst's early work on the Hall effect, and the development around 1905 by H. A. Lorentz of a model involving for the first time exclusively negative free electrons, must have drawn Nernst's attention to the relationship between thermal and electrical conductivity, the problems related to the Wiedemann and Franz rule, and the interconnection of these problems and chemical explanations of the behavior of conductors.

The problem of electronic contributions to specific heat was also discussed in 1911, a year after Magnus and Lindemann's paper was published, by Johannes Koenigsberger.[15] Koenigsberger had earlier treated conductivity as a problem of equilibrium between free and bound electrons in a metallic atom.[16] He intended to show that "experimental data speak for

13. Paul Drude, "Zur Elektronentheorie der Metalle," *Ann. Phys.* 1 (1900): 566–613.
14. Keith J. Laidler, *The World of Physical Chemistry* (Oxford: Oxford University Press, 1993), p. 214.
15. Johannes Koenigsberger, "Über die Atomwärmen der Elemente," *Z. f. Elektrochem.* 1911 (8): 289–93.
16. J. Koenigsberger and O. Reichenheim, "Über ein Temperaturgesetz der elektrischen Leitfähigkeit fester einheitlicher Substanzen und einige Folgerungen daraus," *Phys. Zeitschr.*

a quite high contribution of electrons to the atomic heat, such that in metals at high temperatures one has to assume in most cases one free electron per atom. There is a fundamental difference between the atomic heat of metals, semiconductors and isolators." Koenigsberger, in fact, anticipated the complexity embedded in entropy and suggested that at least two contributions should be assumed to exist: that of free electrons and that of the vibrations of atoms and bound electrons.

The main correlations between electrical properties and specific heats were published by Nernst and Lindemann early in 1911. The publications of the first months of the year were proof of an intensive and accelerated effort to find experimental confirmations for the heat theorem and, in particular, to contextualize new experimental data, the heat theorem, and the quantum theory of solids. Thus, in another series of simultaneous publications, presented to the Prussian Academy on 23 February 1911 and released in print on 16 March 1911, Nernst and Lindemann directly addressed the parallelism between the behavior of electrical resistivity and specific heat, and its relationship to the electron theories available at that time.[17] In his paper, characteristic for its impetuous style and quick and extensive conclusions, Nernst discussed all the issues relevant to the problem of specific heats, resistivity, the theory of electronic structure, and quantum theory.

The investigations of the behavior of specific heats down to the temperature of liquid hydrogen carried out on lead, copper, aluminum, silver, diamond, and tin showed that they could all be represented functions of temperature in the range of approximately 21° to 100°K, registering significant drops and showing that an unmistakable extrapolation would lead to zero specific atomic heats for all the materials at the absolute zero of temperature.

Prior to the Solvay meeting, Nernst published essentially the complete version of his Solvay paper in two installments in the *Zeitschrift für Elektrochemie*. The first, submitted on 21 February 1911, was entitled "Zur Theorie der spezifischen Wärme und über die Anwendung der Lehre von den Energiequanten auf Physikalisch-Chemische Fragen überhaupt"[18] –

7 (1906): 507–78; J. Koenigsberger and K. Schilling, "Über Elektrizitätsleitung in festen Elementen und Verbindungen," *Ann. Phys.* 32 (1910): 179–230.

17. W. Nernst, "Untersuchungen über die spezifische Wärme bei tiefen Temperaturen. III.," *Berl. Ber.* (1910): 306–15; F. A. Lindemann, "Untersuchungen über die spezifische Wärme bei tiefen Temperaturen. IV.," *Berl. Ber.* (1910): 316–21; W. Nernst and F. A. Lindemann, "Untersuchungen über die spezifische Wärme bei tiefen Temperaturen. V.," *Berl. Ber.* (1910): 494–501. The latter paper was presented to the Academy on 6 April 1911.

18. *Z. f. Elektrochem.* 7(1911): 265–75.

"On the theory of specific heat and the application of the theory of energy quanta to physical-chemical problems in general." The paper reflects Nernst's frame of mind at the time and shows that he was already utilizing the quantum theory as a fundamental guide for his research.

Nernst's paper was essentially of a theoretical nature, employing experimental data obtained at high temperatures. In an appendix, however, he discussed the results of investigations at "very low temperatures," measuring the true specific heats both at the temperature of liquid air and, more recently, at that of the boiling point of liquid hydrogen. From the curves for lead, silver, zinc, copper, and aluminum and their extrapolation to absolute zero temperature, Nernst concluded that the data strongly supported the quantum theoretical treatment: There was an "unmistakable tendency" of the atomic heat to diminish to zero "or at least to very small magnitudes" at absolute zero temperature.

Until 1911, the thermodynamics of solids had not been of any interest to physicists, who mainly investigated the optical properties of solids and their crystal structure, nor to physical chemists. On the other hand, once Nernst had postulated the requirement that specific heats decline and that the total and internal energy approach each other asymptotically, he had to attack the issues on several fronts: to find experimental confirmation for the decline of specific heats in solids, and to continue to examine chemical reactions and gases, while concurrently addressing the more difficult question of the thermodynamic proof of the theorem, one which he eventually failed to produce.

Like other scientists, Nernst initially took an ambivalent attitude toward the quantum. On the one hand, he referred to it several times as a rule of calculation, a fruitful working hypothesis. Such statements were echoed, for example, by Einstein at the Solvay conference. On the other hand, a strong argument can be made to demonstrate that Nernst's active use, verification, and modification of quantum theoretical constructs placed him among the active "doers." Here his advantage as an experimentalist materialized. Despite his skepticism, he attempted to make the quantum *"anschaulich,"* to find some kind of physical picture, if not of the quantum of energy itself, then at least of its localization, the properties it conveys to matter, and the locus of quantum interactions. On occasion, the reserve in Nernst's statements on the correctness of the quantum hypothesis seems to stem from a different reason: not that he doubted its validity, but that he promoted it with caution, preferring to avoid speculation.

Nernst presented his work and views to a heterogeneous audience. Physical chemistry had been accepted only gradually by German chemists, the majority of whom were steeped in the problems of organic synthesis; then, too, Berlin in particular had been resisting the applications of solution theory and thermodynamics to chemistry.[19] While Nernst in the years 1905 to 1911 was published primarily in the *Zeitschrift für Elektrochemie*, we nonetheless have to consider that at the time, he was the director of the Second Chemical Institute of the Berlin University and a member of the quite conservative Prussian Academy of Science. Since no other chemist or physical chemist prior to 1909 had considered the applications of quantum physics to chemistry, its introduction by Nernst into both textbooks and active laboratory research was, therefore, a major contribution.

In the papers of early 1911 mentioned above, Nernst began by introducing a derivation of the quantum theory of solids. It is significant that in this derivation, Nernst emphasized that Planck's radiation law, that is, the quantum theory of discontinuous emission and absorption of energy for a black body, should also apply to ions, that is, to electrically charged atoms and groups of atoms; further along in the paper he gave attention to specific chemical applications. After describing the theoretical significance of the quantum approach for specific heat measurements, he wrote:

> The perspectives which open now for the molecular theory of chemical processes are far-reaching; one could even say that here all theories, [as, for example, even Boltzmann's penetrating reflections] had to remain fruitless as long as the quantum hypothesis was not taken into consideration; this of course presupposes that the quantum theory justifies with more or less approximation the data, as can hardly be doubted. . . . [The quantum theory] has borne at the hands of Planck . . . and Einstein . . . such rich fruits, and is capable of so many applications, as I believe to have shown above, that research has the duty to take as varied as possible a stand and to submit [the theory] to experimental examination.

Nernst attempted to convey intuitive substance to what he called "this calculation rule." He expected that the solution would come from viewing the interactions between atoms as analogous to Ampère's laws, rather than Newtonian laws of gravitation.[20] With characteristic metaphoric understatement, Nernst described the results of experimental investigations performed in his laboratory during the preceding years as a *"nicht unwesentlicher Beitrag zur Quantentheorie"* ("a not insignificant contribution to quantum theory"), and concluded with a rare personal statement:

19. E. N. Hiebert,"Walther Nernst and the Application of Physics to Chemistry," in R. Aris, H. T. Davis, and R. H. Stuewer, *Springs of Scientific Creativity* (Minneapolis: University of Minnesota Press, 1983), p. 210.
20. Magnus and Lindemann, p. 274.

> I believe that nobody who, through many years of practice, has gained
> a somewhat reliable feel for the otherwise not always simple verifi-
> cation of theory through experiment, will be able to look at the above
> diagram [coincidence of Einstein and Nernst-Lindemann formula with
> experiment] without being convinced by the immense logical power
> of the quantum theory, which makes the essential features readily un-
> derstandable.

And he immediately added: "The fact that the diagram also satisfies some requirements of my heat theorem should be mentioned as an aside."[21]

This was the focal perspective for Nernst, who used this line of argumentation and significant portions of the same manuscript on three occasions during 1911: first, in an address to the Prussian Academy of Sciences
in January; next, in his two part exposition in the *Zeitschrift für Elektrochemie* in February; and later in the Solvay congress report in November.

At the Kiel meeting of the Bunsengesellschaft in May 1911, Niels Bjerrum
presented a paper on the specific heats of gases in which he calculated
from spectroscopic data the specific heats of hydrogen, nitrogen, carbon
dioxide, and water vapor, as well as ammonia. Only a short while earlier,
Nernst had succeeded in appointing Bjerrum as an assistant in his laboratory, where he put him to "work on specific heats at high temperature." Among his current assistants, Nernst explained to the government
authorities, "nobody is available for this purpose." At age 33 Bjerrum,
both of whose parents were professors at the University of Copenhagen,
had initially received funding from the Carlsberg Foundation to work
for a year with Nernst.[22] In this paper, he used both the Einstein and
the Nernst-Lindemann formulae, taking into consideration that transla-
tional, rotational, and vibrational energies of molecules are quantized,
and demonstrated very good correlations with the experimental calori-
metric-observed specific heats.

In the discussion following the presentation, two important suggestions
were made; they illustrate how incipient analogies led to further experimental and theoretical work in related, but not altogether traditionally close,
research fields. The first observation, made by Nernst, may be considered
to be a hint of his developing interest in astrophysics:

> According to Einstein's formula . . . at very high temperatures one
> should expect that the visible light bands also should contribute to
> the specific heat. We can suppose that iron vapor is mostly present on

21. Ibid., p. 275.
22. Nernst to Education Ministry, 6 February 1911, Merseburg.

the sun . . . and it is characterized by an abundance of spectral lines. Each independent line should correspond at least to one degree of freedom. We thus arrive at a very high atomic heat of iron vapor . . . and the much-discussed question as to why the sun cools so slowly can be answered according to Einstein's theory, without recourse to radioactivity, by considering that . . . gases at very high temperatures have a very high specific heat.[23]

Another conclusion, of great importance to chemists, was the fact that Bjerrum's spectroscopic data indicated ways by which to distinguish between single and double bonds in carbon compounds. These were formed due to the various vibrational modes available to oxygen atoms, for example, when comparing carbon monoxide and carbon dioxide. Most interestingly, A. L. Bernoulli from Bonn University commented that the spectra of gases would exhibit multiple bands due to the coupling of the various individual vibrations and would lead to additional eigenfrequencies, as compared with the spectral lines of solid ionic crystals.[24]

On the same occasion, E. Grüneisen talked about the relationship between the thermal expansion and the specific heats of metals at varying temperatures, taking into consideration the predictions of Einstein and Nernst-Lindemann. He estimated that a study of the variability of the proportionality of thermal expansion and specific heats at different pressures would give an indication as to whether one or more eigenfrequencies contribute to the specific heat according to Einstein's formula.[25]

The next occasion on which quantum theory was discussed among both physicists and chemists was the 83rd Meeting of German Natural Scientists and Physicians in Karlsruhe, 24 to 30 September 1911. In the physics section, the Viennese physicist Fritz Hasenöhrl presented a paper entitled: *"Über die Grundlagen der mechanischen Theorie der Wärme"* ("On the foundations of the mechanical heat theory"). He pointed out that the experimental confirmations of Einstein's formula for specific heats are the "greatest difficulties" that the kinetic theory has encountered, and despite the lack of a satisfactory theory and the failure "to adapt the element of action **h** to habitual conceptions," he concluded that "Planck's assumption is a useful working hypothesis." He further suggested the examination of Planck's theory in another domain, that of a nonharmonic oscillator, one whose frequency is not constant – as in Planck's hypothesis – but is any function of the energy.[26]

23. *Z. f. Elektrochemie* 17(1911): 731–4.   24. Ibid., p. 735.
25. E. Grüneisen, "Das Verhältnis der thermischen Ausdenung zur spezifischen Wärme fester Elemente," *Zeitschr. f. Elektrochem.* 17 (1911): 737–9.
26. *Physik. Zeitschr.* 12 (1911): 932–3.

At the meeting, Nernst presented an exposition that would be the second part of his Solvay talk, where he gave a qualitative interpretation of his results:

> Recent measurements of specific heats have doubtlessly shown that, in accordance with the requirements of the quantum theory, for any substance above the absolute zero of temperature there exists a domain in which the concept of temperature practically loses its significance. Therefore, in this domain any property which is determined by the average behavior of the atoms, such as the volume, is independent of temperature.[27]

The physical picture Nernst tried to present is indicative of his style of thought – which can be termed a *model-building by analogy* – and reflects ideas in current use at the time. He construed an analogy between dilute solutions and a solid in which the atoms of different energy content are "dissolved," much as he had been persuaded by van't Hoff's analogy of osmotic pressure and dissociation processes, or by Drude's model of a solid as a gas:

> There exists a temperature domain in which the atoms of the solid are almost at rest; only a very small number of atoms have received one single energy quantum, while the number of those atoms which have more energy quanta can be ignored. Then we can consider the solid body as a very diluted solid solution of the atoms which have one quantum in the much more numerous resting atoms; the concentration of the latter is proportional to the energy content.[28]

Nernst correlated calculations and experimental measurements that showed specific heats, thermal expansion, and compressibility to be all heavily temperature-dependent, and tending to zero at absolute zero temperature. For the internal energy content, Nernst used the analogy with solutions whose energy content will not change with the adding of small quantities of solute. This led him to the proof that $\lim dU/dT = \lim dA/dT = 0$; "therefore, this heat theorem established by me six years ago proves to be a special case of a more general law [Satz] derived from the quantum theory." In particular, after Planck's lecture to the German Chemical Society in mid-December 1911, which the two men may have first discussed, Nernst accorded "greater generality" to quantum theory and viewed the heat theorem as subsumed under this novel theory on the constitution of

27. "Über ein allgemeines Gesetz, das Verhalten fester Stoffe bei sehr tiefen Temperaturen betreffend," *Physik. Zeitschr.* 12 (1911): 976–9.
28. Ibid., p. 977.

matter and radiation. Nernst also correlated other properties, such as electrical conductivity, thermal conduction, and Peltier effects, with specific heats, thermal expansion, and compressibility, for which he drew corresponding conclusions as to significant temperature dependence.[29]

The most important aspect of Nernst's Karlsruhe paper probably was his desire to point out the relevance of the new developments in quantum theory for a series of physical-chemical properties, such as heat and electrical conductivity. He wrote: "Against the requirements of electron theory, but totally in the spirit of the quantum theory, Kamerlingh Onnes found at very low temperatures a region in which the resistance of platinum does not vary anymore." Referring to his observations of 1910, he stressed that the relationship between the behaviors of specific heat and conductivity should be pursued within the quantum theoretical framework.

In the discussion following Nernst's paper, Einstein asked whether the "lecturer does not consider it possible that in approaching absolute zero, the conductivity of pure metals becomes infinite." Kamerlingh Onnes had found that small additions greatly influenced conductivity, and that the purest metals had an extremely low resistance. Einstein inquired whether different samples of aluminum had been used and checked accordingly. Nernst replied that despite the fact that pure aluminum was hard to produce (and that impurities apparently contributed to the higher resistance), he expected resistance to "still have a characteristic finite value for every metal," a consideration which in hindsight has been generally proved correct at the lowest temperatures so far attainable.

Then Arnold Sommerfeld asked how quantum theory "pictures" (*"stellt sich vor"*) this influence of impurities. Nernst replied that Lindemann had assumed that in the vicinity of absolute zero, resistance was determined by the motion of atoms: "Therefore the resistance becomes much greater. But the impurities are in a so-called solid solution. It would be possible that the small impurities are in a kind of gaseous state, that is, that they are in motion at low temperatures, that they enter in vibrations, and then one could explain that they have a great influence on conductivity. This conception," he continued, "has much in its favor (*hat vieles für sich*), but one could not say anything certain."[30] This impromptu theorizing on current experimental data vis-à-vis gases, solutions, and solids vividly illustrates, on the one hand, the status of speculation as to the mode in which quantum considerations could be enhanced or supported by available experiments; on the other, the analogous model-building process that Nernst employed allowed him to transfer concepts from the theory of solutions

29. Ibid., p. 978.   30. Ibid., pp. 978–9.

to the then inconclusive conceptions of electronic structure and conductivity.[31]

The first communications relating to the discovery of superconductivity had been submitted by Kamerlingh Onnes at the end of April 1911 to the Amsterdam Academy of Sciences, and again in May, and had been published in two installments in the *Communications*. However, although Nernst and probably Einstein as well were informed of the developments in Leiden, the close succession of events during 1911 and the relatively low exposure of the Dutch experiments by the end of 1911 had not yet brought the Leiden investigations to the attention of most physicists. On the contrary, it seems that Nernst's studies on specific heats, of which Kamerlingh Onnes was cognizant at the beginning of 1911, spurred the acceleration of resistance measurements at low temperatures in Leiden.

The work described thus far preceded the Solvay congress, where quantum papers were presented by Einstein, Planck, Nernst, and Sommerfeld. The quantum discourse among the relatively small number of scientists, primarily German, became at this time even more complicated because of Planck's elaboration of a second quantum theory, one in which there was quantum emission but continuous absorption, and which led both Planck and Einstein to speculate on the existence of zero energy. Nernst continued to work on the quantum theory of specific heats, publishing numerous papers during 1912 and 1913, and inducing students and colleagues, such as Arnold Eucken, Niels Bjerrum, and Otto Sackur, to collaborate on or expand some of his suggestions. The proposals made by Nernst – that the quantization of atoms in the solid should be applied to rotating gas molecules – led to investigations by himself, Bjerrum, Eucken, and Eva von Bahr and were published between 1911 and 1913.

31. Ibid., p. 979.

11

"The Witches' Sabbath": Nernst and the First International Solvay Congress in Physics

"... the thing must and will take its course ...." Planck, 1910

At the end of October 1911, twenty-one prominent physicists met at the luxurious Hotel Métropole in Brussels for what was to become known as the First Solvay Congress in Physics. During the following decades, such congresses continued to take place regularly in the fields of both physics and chemistry. Although the daily press took little notice of this first meeting, apart from incendiary gossip on the alleged elopement of Madame Curie and Paul Langevin, it remained, in the memory of those present, a unique and significant gathering.

The congress was, indeed, the first instance of an international meeting devoted exclusively to a specific agenda in contemporary physics. Although international meetings in science were becoming increasingly popular, they still remained limited in number and scope. The International Physics Congress in Paris in 1900 and the various meetings of the BAAS (British Association for the Advancement of Science) attracted numerous European scholars. Specialized meetings, as for example those of the Deutsche Chemische Gesellschaft or the Deutsche Elektrochemische Gesellschaft (the Bunsengesellschaft) were attended primarily by German-speaking scientists but attracted some French and Central European scholars.

In historical literature, the first Solvay congress is depicted as a significant event in the physical sciences of the twentieth century because it enlarged the participation of physicists – in particular, British and French scientists, such as E. Rutherford (via his student Niels Bohr) and Henri Poincaré – in the quantum-theoretical discourse. Poincaré was one of the most active discussants at the meeting, after which he went back to Paris and settled an essential issue, that the Planck "energy" levels for the oscillations were necessary as well as sufficient for the Planck distribution law for black-body radiation. His papers written between 1911 and 1912 made a major impression, and they persuaded James Jeans to take quanta seriously. In addition, historians point out that Niels Bohr's knowledge of the deliberations in Brussels had a direct influence on his elaboration of a quantized atomic model. Following the congress, the number of publications on theoretical and experimental work in the new field of the quantum theory of solids increased substantially, shifting the discourse from radiation problems to other areas, such as specific heats and the quantization of the atom.

The convocation of the congress by Nernst was motivated by several intersecting considerations. The most poignant may well be the substantive one, what Planck in particular perceived as a set of contradictions and difficulties that defined the agenda for the conference. Another is the personal and public nature of the meeting, initially convened confidentially but later broadcast in the daily and scientific press. But third, the conference was also designed as a response to Ernest Solvay's personal scientific and social interests at the time, and as prelude to the formation of a more permanent scientific organization in Brussels – a detail overlooked by historical narratives. This concerted effort, begun before 1911, culminated in the official establishment of a Solvay Institute for Physics six months after the conference. Solvay, who viewed himself as a "deep thinker" about physics and other, mainly economic-political issues, distributed copies of his privately published "Étude gravito-matérialitique" to the participants at the congress, and expressed the hope that his study might be the basis of some discussion by his learned guests – a hope which was eventually frustrated. When we view the broader perspective of the various initiatives staged by Nernst and others, both in Germany and abroad, to establish scientific foundations and research institutes, and of their quite active search for adequate funding, we may better unravel the significance of the congress.

While on a trip prior to the meeting of the French Physical Society, Nernst visited Einstein in Zürich during the first week of March 1910. Having just completed an article to be published in "Guye's revue" – the *Journal de chimie physique* – Nernst had been invited at relatively short notice to deliver a paper in Paris on 31 March. Phillipe-Auguste Guye had received 25,000 francs from Ernest Solvay to found the journal. His brother and collaborator, the physicist Charles-Eugène Guye, was among Einstein's teachers at the Eidgenossische Technische Hochschule (ETH) in Zurich.[1] It seemed to him "not much of an effort to follow the friendly invitation," even though it meant postponing a planned trip to England – his host there being his colleague Arthur Schuster, to whom he wrote on 10 March:

> At the moment I am in Lausanne, partly in order to get a breath of
> fresh air, and partly to brush up my French, but tomorrow I am al-

1. See Nernst "Untersuchungen . . . ." (1911). Regarding Nernst's paper in Paris, see Klein 1965, p. 177, and Pelseneer n.d., p. 77.

ready returning home. On my trip here I visited Prof. Einstein in Zürich. It was for me an extremely stimulating and interesting meeting. I believe that, as regards the development of physics, we can be very happy to have such an original young thinker, a "Boltzmann redivivus"; the same certainty and speed of thought; great boldness in theory, which however cannot harm, since the most intimate contact with experiment is preserved. Einstein's "quantum hypothesis" is probably among the most remarkable thought [constructions] ever; if it is correct, then it indicates completely new paths both for the so-called "physics of the ether" and for all molecular theories; if it is false, well, then it will remain for all times "a beautiful memory."[2]

This early enthusiastic testimonial regarding the young Einstein provides insight both into the state of the still unformed quantum theory ten years after its formulation, and into the personality of Nernst. Unmistakable is his almost boundless admiration for Einstein, whom he sought to attract instantly to Berlin as a result of their encounter.[3] The letter is one of the earliest written assessments of Einstein's contributions by a leading contemporary scientist. In his 1909 letter recommending Einstein for an associate professorship at the University of Zürich, Professor Alfred Kleiner of that institution described the young scientist as "among the most important theoretical physicists."[4] A year later, enthusiastic reports about Einstein began circulating among European scientists. Thus, for example, physical chemist George Bredig, a close colleague of Nernst, wrote to Svante Arrhenius:

Among the local colleagues, besides [the chemists R. Willstätter and A. Werner], I am most interested in the associate professor at the university, the physicist A. Einstein, a still young, totally brilliant chap

2. W. Nernst to A. Schuster, Lausanne, 10 March 1910, Royal Society, London.
3. A postcard from Nernst to an unnamed correspondent (probably Emil Warburg), dated 31 July 1910, reads: "I have made inquiries regarding Einstein, but have not yet received any news; but the matter in question does not depend on whether E. is now in a better position." ("Wegen Einstein habe ich mich erkundigt, aber noch keine Nachricht erhalten; übrigens ist die betreffemde Angelegenheit wohl unabhängig davon, ob E. jetzt etwas besser gestellt ist.") (Staatsbibliothek Preussischer Kulturbesitz, Berlin, Autogr. F2e1898[5].) Subsequent correspondence indicates that Warburg offered Einstein a position at the Physikalisch-Technische Reichsanstalt in Berlin in the summer of 1912, and that the University of Vienna made a serious offer at about the same time. (A. Einstein to H. Zangger, summer 1912, undated, The Albert Einstein Archives, The Hebrew University of Jerusalem.) This was a period of ascendancy for Einstein, during which he received an honorary doctorate from the University of Geneva, began his teaching duties, and attended the first physics conference in Salzburg in 1909.
4. Pais, 1982, p. 185.

[*ganz genialer Kerl*] from whom one can learn a lot. He is also a Boltzmann student and has a wonderful talent for representing complicated
things simply. What is your opinion of his kinetic papers (specific
heat), relativity principle and elementary quanta? Unfortunately we are
losing him, because he has been appointed professor in Prague instead
of Lippich. I believe he has a great future, at present he lives with his
wife and children in very modest conditions. He certainly deserves a
better fate.[5]

   By 1909, Einstein's theory of relativity had certainly drawn the attention
of German scientists, but he had not yet achieved personal prominence,
and as yet enjoyed relatively few personal contacts with influential scientists. A year later, however, Einstein found himself in greater demand. At
the end of 1911, he was offered a position in Utrecht, strongly supported
by Lorentz, but he preferred to follow a call back to Zürich.[6] Ironically,
Johannes Stark, who was to be a leader in the anti-Einstein and antirela-
tivity campaign a decade later, had made one of the earliest attempts to
seek an academic position for Einstein in Germany, to whom he wrote: "I
intervene for you wherever I can, and it is my wish, that I would have soon
the opportunity to recommend you for a theoretical professorship in Germany."[7]
   The letter to Schuster is the only evidence, other than a letter from Einstein to Laub (16 March 1910) and a much-later oral comment of Einstein
to his friend Georg Hevesy (1885–1966) regarding the visit of the "great
Nernst" to the ETH.[8] However, it may well be that Nernst had met Einstein even earlier than March 1910. In reminiscences of her husband, Emma
Nernst recalled much later that Nernst first met "the genius in the attic in
Bern, while he worked at the patent office,"[9] although this seems unlikely.
   That Nernst compared Einstein to Ludwig Boltzmann, the most prominent, independent, and at the same time controversial theoretician of the
recent past, points to the uniqueness and originality that Nernst detected
in Einstein. The latter's being at the time, together with Paul Ehrenfest,
probably the most consistent elaborator of Boltzmann's statistical physics,
apparently prompted Nernst's (correct) vision of a "Boltzmann redivivus."
Who better than Nernst to hold this view, inasmuch as he had worked

5. G. Bredig to Arrhenius, 22 January 1911, Arrhenius Papers, KVA, Kingl. Vetenskaps-
Akademien, the Royal Swedish Academy of Sciences, Stockholm.
6. Pais, 1982, 239–41; Einstein-Lorentz Correspondence, roll 3, H. A. Lorentz Papers, AIP
Archives, New York.
7. J. Stark to Albert Einstein, 19 February 1908, The Einstein Archives.
8. Kuhn, 1978, pp. 214–15.
9. Transcript by E. v. Zanthier Nernst, 7 pages, Archive of Bunsengesellschaft, Darmstadt.

with Boltzmann, who must have been for him the archetypal theorist! As to his judgment of Einstein's work, Nernst in his letter is obviously referring to Einstein's proposal of light quanta, elaborated in 1905, as well as to the quantum theoretical treatment of solids, published in 1907. It is surprising that Nernst mentions "Einstein's quantum hypothesis" without reference to Planck, and yet evidently as addressing both "the ether" and "molecules." The letter reflects Nernst's recognition of the depth and scope of Einstein's thought, which complements Einstein's contemporary and posthumous evaluations of Nernst, whom he depicted as "one of the most original and interesting scientists" with whom he had ever been closely associated. Einstein paid homage to Nernst's "amazing scientific instinct" (*"verblüffender wissenschaftlicher Instinkt"*), "coupled with a sovereign knowledge of an enormous [body of] factual material always present in his mind, and a rare command of experimental methods and tricks, at which he was a master."

Einstein's obituary of Nernst was published in 1942 "at the request of the editorial leaders" of the *Scientific Monthly* at a time when the American press was generally reluctant to praise German scientists – especially one such as Nernst, who was identified in the minds of many as a contributor to chemical warfare during World War I. Einstein's obituary of Nernst was, therefore, an act of unusual devotion. Yet it reflects caution, stressing that "Nernst was neither a nationalist nor a militarist," but rather endowed with "a very far-reaching freedom from prejudice," which "set him apart from almost all his compatriots." Einstein recognized that

> the beginnings of the quantum theory were assisted by the important results [of Nernst's] caloric investigations, and this especially before Bohr's theory of the atom made spectroscopy the most important experimental field. . . . What distinguished him from almost all his fellow countrymen was his remarkable freedom from prejudices. He was neither a nationalist nor a militarist. He judged things and people almost exclusively by their direct success, not by a social or ethical ideal. . . . At the same time he was interested in literature and had such a sense of humor as is very seldom found with men who carry so heavy a load of work. He was an original personality; I have never met any one who resembled him in any essential way.

Irving Langmuir refused to contribute any memorial to Nernst, for personal as well as political reasons, and intimated that Millikan and other American students of Nernst would concur in this attitude.[10]

10. Nachruf Walther Nernst, Reader index 245, German typescript in The Einstein Archives; also published in *Scientific Monthly*, 1942. See letters of Ernst Salomon to A. Einstein,

The Nernst-Einstein relationship emerges as truly one of a kind. Both men have been credited with caustic anecdotes at the expense of the other. Einstein is reported to have hurt Nernst by raising doubts in public about the validity of Nernst's heat theorem in 1924, during the famed weekly physics colloquium at the University of Berlin. He was also said to have suggested mockingly to Arnold Berliner, the editor of *Die Naturwissenschaften,* that he publish the letters of refusal received in lieu of contributions to a planned sixtieth birthday Festschrift in honor of Nernst. However, the man who related these anecdotes acknowledged that Einstein's humor masked a "high appreciation" of Nernst.[11] On his side, Nernst was apparently "at first irritated by Einstein's Theory of Relativity and said in mock annoyance: 'Why, Planck and I engaged him just as you take on a butler, and now look at the mess he's made of physics; one can't turn one's back for a minute.'"[12] This must refer to the general theory, since Einstein was not in Berlin in 1905. There is some point to Nernst's sally. He surely hoped that Einstein would continue to be as close to experiment as he had been in the years before 1914, when he arrived in Berlin. Einstein's total absorption in creating general relativity might have disappointed Nernst. Einstein never became a "lab theorist" like himself.

Yet it is evident that from their first encounter, each entertained a deep respect for the other's abilities and peculiarities. Einstein's praise for Nernst's "freedom from prejudice" was for a trait especially dear to Einstein, who was rightly concerned with the sometimes veiled but often explicit anti-Semitism affecting his career.[13] Nernst's letter is devoid of any insinuations regarding Einstein's origins, in contrast to the early pronouncement by Arnold Sommerfeld in his 1907 letter to Lorentz mentioned in an earlier chapter. However, his misgivings about the abstractness of Einstein's thought did not impinge heavily on Sommerfeld's later appreciation of his work. And yet, in the same year that Nernst met Einstein, Sommerfeld did not voice any comparable enthusiasm after having encountered the young "genius" in Salzburg for the first time:

> Einstein, who showed himself for the first time at this year's Salzburg meeting of [the German Association of] Scientists, is a terribly likeable [*ungemein sympatischer*] and modest person.

22 November 1941 and 4 February 1942, The Einstein Archives. Also see Albert Einstein, *Out of My Later Years,* rev. repr. ed. (Westport, Conn.: Greenwood Press, 1975). Originally published 1950 by the Philosophical Library, New York.

11. H. M. Cassel to Einstein Archives, 5 March 1972, The Einstein Archives.

12. Charles Lindemann, who also studied with Nernst in Berlin, writing to the biographer of his older brother, Frederick A. Lindemann. Smith 1962, pp. 25–6.

13. For A. Kleiner's report to the University of Zürich, see Pais 1982.

This was all Sommerfeld had to report, although earlier in the letter he announced:

> I too am now a convert to the theory of relativity; especially Minkow-ski's systematic form and conception have facilitated my understand-ing. However, I am sufficiently old-fashioned to resist temporarily the Einsteinian conception of the light quanta. Stark's light quanta, against which I have recently expressed myself, must not be to your taste ei-ther.[14]

The Solvay Congress was initiated in the summer of 1910 by Nernst, who consulted Max Planck on the possibility of convening a physicists' meet-ing to discuss the latest developments in radiation theory. Nernst subse-quently presented such a proposition to the Belgian industrialist Ernest Solvay (1838–1922). Inventor of the successful soda manufacturing process that reaped him a considerable fortune, part of which he donated over the years to the founding of the Solvay international scientific institutes, Solvay was one of the major European philanthropists of science and education. Nernst's acquaintance with Solvay dated at least to 1909, when Nernst, together with Berlin's leading chemist Emil Fischer, had proposed the award of the Leibniz Medal of the Prussian Academy of Sciences to Solvay in recognition of his "services in, and rich donations for, the promotion of the sciences."[15] At the time, Solvay was extending his industrial interests into Germany, having already financed an institute of physiology in 1895 and an institute of sociology in 1909, both housed in the Parc Léopold in Brussels. He had contributed to the establishment of a commerce depart-ment at the University of Brussels and had engaged the collaboration of a substantial number of Belgian scientists and intellectuals. In addition to his own preoccupations with social reform, Solvay intended these institutes as contributions to the development of science in Belgium.

By early 1910, the quantum theory in the context of radiation questions had been the topic of discussion among only a very small number of sci-entists: Planck, Lorentz, and Rubens; followed by Einstein and Ehrenfest; and around 1909, Sommerfeld, P. Weiss, F. Hasenöhrl, and a few others. In this atmosphere, Planck was rather reserved and pessimistic about the prospects of a meeting in 1910, urging a postponement by one more year. He wrote to Nernst: "Such a conference will be more successful if you wait until more factual material is available."[16]

---

14. A. Sommerfeld to H. A. Lorentz, 9 January 1910, Lorentz Papers, roll 3.
15. Fisher 1922, p. 156; Fischer, n.d.
16. Planck to Nernst, 11 June 1910, in Pelseneer n.d., pp. 2–3; also in Klein, 1965, Her-mann, 1971, and Kuhn, 1978.

Both Planck and Nernst stressed the need for a correlation of theory and experiment. In his letter, Planck suggested that the community, in particular the theoreticians, were not yet ready to embrace quantum theory: "It is a fundamental presupposition for the convocation of the meeting (as you have duly mentioned in your proposal) that the theoretical situation created by the radiation laws, specific heats, etc. is incomplete, unbearable for any true theoretician, and that therefore necessity would prompt cooperation and jointly seeking a remedy." According to Planck, "a conscious need for a reform, which would motivate" scientists to attend the congress was shared by "hardly half of the participants" envisaged by Nernst. Planck was skeptical that the older generation – which included Rayleigh, van der Waals, Schuster, and Seeliger – would attend or would "ever be enthusiastic." In his view, only a few scientists, and not even the younger among them, were sufficiently interested in this particular problem. Since only Einstein, Lorentz, W. Wien, and Larmor were, in his opinion, possible participants in such an effort, Planck proposed to delay the meeting for a couple of years. By then "it will be evident that the gap in theory which now starts to split open will widen more and more," and the circle of concerned physicists would expand and the meeting would attract the deserved attention.[17]

For Planck, the awareness of a crisis was a sine qua non for a successful convocation. Without acknowledgment by a critical mass of prominent scientists of the widening gap between various theoretical constructs, such a modest meeting would not even take place. Planck primarily emphasized the dissonance among theories, while Nernst seemed to deplore the lack of experimental data. It seems that both Nernst and Planck anticipated that the congress would eventually draw the attention of the larger community, of "a hundred times more eyes," and despite the small number of possible participants, they were hoping for a major impact on the orientation of physics.

Despite Planck's exhortations for caution, Nernst proceeded, with characteristic optimism and a healthy measure of confidence, to engage Solvay, in whose name he had drafted the original invitation. In it, Nernst did not refrain from forceful declarations: He stated distinctly that "a revolutionary reformulation of the foundation of the hitherto accepted kinetic theory" was taking place. The existing radiation formula derived from a consistent elaboration of the kinetic theory conflicted with "all experience." In addition, many consequences of the kinetic theory, such as theories about specific heats (constancy of the specific heat of gases with changes

17. Planck to Nernst, 11 June 1910, in Pelseneer n.d., pp. 2–3; also in Klein, 1965, and Kuhn, 1978.

of temperature, validity of the Dulong-Petit rule to the lowest temperatures), were "completely contradicted by many measurements." However, the new quantum theory, which restricts the motion of electrons and atoms about their rest positions, solves some of these contradictions, despite the fact that "this view is so foreign to the previously used equations of motion of material points that its acceptance must doubtless be accompanied by a wide-ranging reform of our fundamental conceptions."[18]

Nernst's pragmatism and his expectations that the conference would bring about a convergence of theory and experiment prevailed over Planck's skepticism, because Nernst was convinced the quantum could be incorporated into experimental work and coexist with Newtonian physics at the macroscopic level, even if some of the fundamental clarifications were lacking. The convocation letter's closing paragraph, which is not well known, illuminates Nernst's beliefs, albeit expressed under Solvay's signature. Nernst strongly believed that a personal discussion among researchers, even those who were only indirectly involved in these specific topics, would, even if it did not provide a definitive answer, "open the path to the solution of these questions through a preliminary critique." He felt that it was necessary for the further progress of work if one could at least decide which views should be retained and which should be "submitted to a far-reaching revision" in light of recent experimental developments.[19]

Nernst was advocating a conscious cooperative effort, an active expansion of the "circle of belief" in the necessity of a reformulation of the kinetic theory.[20] While in his own work his focus lay entirely on establishing useful correlations with available observational and experimental data, he was genuinely seeking clarification of the quantum "theory" and its relationship to "classical" theory and not just experimental agreement. As we have seen from his previous work on the relationship between electricity and chemical reactions, between thermodynamic and molecular interpretations of electrochemical and physical transformations, and between various radiative phenomena, he was, more than many others, conscious of the interconnections across domains. As we shall see in the following chapters, it seems that as a consequence, one of the major reasons for the delay in the formal recognition of the validity of both the quantum theory and Nernst's heat theorem by the Nobel Prize committees was

18. H. A. Lorentz Papers, roll 4; also quoted in Kuhn, 1978, p. 215. I have slightly altered Kuhn's translation, substituting "view" for "conception" ("Auffassung" in the original), and "fundamental conceptions" for "fundamental intuition" (Fundamentalanschauungen in the original), since the use of the term "intuition" for "Anschauung" applies mostly in philosophy.

19. Pelseneer n.d., p. 8.   20. Galison, 1987, p. 276.

an excessive narrowness of perspective in assessing the availability of convincing supporting evidence, theoretical and experimental. As a very active participant in the meetings of the Prussian Academy of Sciences and the German Physical Society, as a vigorous member of the Berlin community, and as a close colleague of Planck, Nernst was not far removed from discussions of Planck's work, as we have seen in our earlier treatment of his electrolytic conduction experiments and his work on the electric lamp.[21]

The congress centered on Nernst's main scientific concerns at the time: It was at this conjunction – of previously available data, his own newer measurements, and Einstein's theory of specific heats of 1907 – that the validity of the quantum hypothesis emerged most forcefully. Nernst's list of proposed participants included prominent theoreticians (Planck, Jeans, Einstein, Sommerfeld, Lorentz) and experimentalists (Warburg, Kamerlingh Onnes, Rutherford, Langevin, Curie) spanning two generations. The papers were to address mainly, and equally, radiation and specific heats. The young secretaries Maurice de Broglie and F. A. Lindemann were later to become important contributors in their own right. Nernst had recommended seven discussion topics, which were all addressed by designated speakers: the derivation of Rayleigh's radiation formula; the extent to which the kinetic theory of ideal gases corresponds to experience; the kinetic theory of specific heats according to Clausius, Maxwell, and Boltzmann; Planck's radiation formula; the theory of energy quanta; specific heats and the quantum theory; and the consequences of the quantum theory for a number of physical-chemical and chemical questions.

It is difficult to gauge the success of the congress in light of Planck's and Nernst's conflicting expectations. In the immediate circumstances of the meeting, Planck was probably correct. British scientists were poorly represented at the congress. Lord Rayleigh, whom Nernst had envisaged as president of the proceedings, only sent a letter, read by Lindemann during the morning session on the first day of the congress, discussing some specific issues but conceding: "I fear I have nothing new to add to what I have so far stated on the subject."[22] This is a position consistently adopted by Rayleigh; when asked three years later at the British Association meeting to comment on Niels Bohr's atomic model, he answered: "Men over seventy should not be hasty in expressing opinions on new theories."[23] J. J. Thomson, Larmor, and A. Schuster declined the invitation for a variety of

21. See Nernst, 1903, p. 623, where he mentions that his temperature measurement device, which he had "used for a longer time, is in principle the same as the optical pyrometer described in the meantime by Holborn and Kurlbaum" in a 1901 paper.
22. Eucken, 1914, p. 41.   23. Eve, 1939, p. 233.

reasons. Wilhelm Röntgen, who had been on a provisional list as late as May 1911, did not participate.[24] The French had come in significant numbers, but by all accounts and interpretations, were poorly acquainted with the kinetic theory in light of the "new" quantum hypothesis. An exception was Paul Langevin, who several years earlier had already discussed with Lorentz an extension of Rayleigh's radiation formula to short waves.

Most papers presented at the three-day conference and published a year later contained little new information compared with previous publications. This impression is reinforced by Einstein's candid statement to his friend Michele Besso: "In general, the congress there looked like a lamentation over the ruins of Jerusalem. Nothing positive was accomplished. My considerations on vibrations elicited great interest and no serious objection. I benefitted little, since I heard nothing that was not known to me."[25] And even if this were literally true – although some of it seems to have been just bravado – no other participant could have made such a claim.

These remarks are in stark contrast to the obligatory genuflecting letter that Einstein sent Ernest Solvay at the close of the meeting, thanking him for one of the most beautiful memories of his life.[26] Regardless of his critical and cynical appraisal in correspondence with his close friends Besso and Heinrich Zangger after the conference, Einstein considered the Brussels meeting an important professional event. He had expressed pleasure and enthusiasm at the invitation and prepared a long paper on a topic suggested by Nernst. He had written to Nernst: "I accept with pleasure the invitation for the Brussels meeting and will gladly compose the report assigned to me. The whole enterprise pleases me terribly, and I hardly doubt that you are its soul."[27]

Einstein's only previous major professional exposure had taken place two years earlier in Salzburg, where he had met, for the first time at age 30, a real theoretical physicist. His position within the German-speaking scientific community changed quite rapidly during the years 1909 to 1911, and his ambivalence toward distant colleagues, their authority, and their influence, as well as the instability of his own career, is reflected in his correspondence.[28] It seems that the preparation for the Solvay congress, coupled with his visit to the meeting of the German Association of

24. Nernst to Lorentz, 15 May 1911, H. A. Lorentz Papers, roll 4.
25. Einstein to Michele Besso, 26 December 1911, in *Einstein-Besso Correspondence,* 1972, p. 40.
26. Einstein to E. Solvay, 22 November 1911, in Pelseneer n.d., p. 18.
27. Einstein to Nernst, 20 June 1911, in Pelseneer n.d., p. 12.
28. Einstein often made divergent remarks about the same event in different letters to the same correspondent (e.g., Zangger) or to various correspondents.

Scientists and Physicians in Karlsruhe in late September 1911, followed by a lecture series in Zürich, strained Einstein to some extent. He was looking forward to the time when he could again be "his own master," after the completion of the Solvay paper. He felt "harassed" by his "drivel" for the Brussels meeting. In March 1911, Einstein had moved from Zürich to Prague. During the years 1908 to 1911, he experienced frequent changes in his career and family life. These were the years in which he devoted most of his research to quantum problems. Midway through 1911, Einstein returned to problems in the theory of gravitation, and he afterward abandoned serious preoccupation with the quantum theory.[29]

A vivid account of the congress, and of Einstein himself, was given by the very young, quite talented (and characteristically immodest) Frederick A. Lindemann, who accompanied Walther Nernst and served as one of the secretaries to the congress. At the close of the meeting, Lindemann wrote to his father from Brussels that the council had been "invited by Mr. Solvay, the inventor of the Solvay soda process, at the instance of Nernst. . . ." He found the discussions "most interesting but the result is that we seem to be getting deeper into the mire than ever. On every side there seem to be contradictions." With his usual insolence, Lindemann reported to his "Pap" that his "melting point formula and my calculation of the photo effect were about the only things which were not contradicted. . . ." He enjoyed the company of Einstein, who made the greatest impression on him, "except perhaps Lorentz." To the young, wealthy Englishman, Einstein looked "rather like Fritz Fleischer en mal but has not got a Jewish nose. He asked me to come and stay with him if I came to Prague and I nearly asked him to come and see us at Sidholme. However, he does not care much for appearances and goes to dinner in a frock coat. He says he knows very little mathematics and can only set up general considerations, but he seems to have had a great success with them."[30] Lindemann and Einstein seemed to agree in their admiration for Lorentz but were otherwise quite incompatible. The letter reflects much on Lindemann's own personality. Frederick, age 25, and his slightly younger brother, Charles, were accomplished socialites, having been educated on the Continent and having worked for several years in Nernst's laboratory in Berlin. Lindemann's reference to his own work was, indeed, not contested on the

29. "Nun aber – wenn auch noch der Hexensabbat in Brüssel vorbei ist – bin ich bis auf die Kollegien wieder mein eigener Herr." Einstein to Besso, Prague, 21 October 1911, *Einstein-Besso Correspondence*, 1972, p. 32. See Pais, 1982, pp. 185–90, 201.
30. F. A. Lindemann to his father, 4 November 1911, Lord Cherwell (F. A. Lindemann) Papers A91, Nuffield College, Oxford. Excerpts of this letter are in Smith, 1962, p. 43, and in Klein, 1965, p. 179.

experimental level, although the theoretical implications were widely debated. Lindemann's conservative attitudes persisted later in his life, when he remained critical of Einstein's naive meddling in politics and in liberal and "worthless" causes.[31]

Einstein's careful, published report and comments at the congress fit well into the almost pervasive skepticism among the most ingenious physicists of the period when faced with the near "mysticism," as he said, forced on them by the quantum hypothesis. For instance, he wrote to H. A. Lorentz after the congress: "The h-disease looks ever more hopeless," referring to Planck's constant h.[32] Einstein did not foresee that Poincaré's critical remarks would materialize into a systematic and analytic examination of quantum theory several months later.[33] Nor did he give due attention to Ehrenfest's work, which he had read, published on 21 October 1911, which predated Poincaré's, in which Ehrenfest proved "that energy quantization was a necessary and sufficient condition for Planck's law."[34] Einstein's casual metaphors about "lamentations" and the "delight for Jesuits," coupled with misgivings about the correctness of Planck's "second" quantum theory, constitute a record of the development of ideas in a period of change. This record indicates that the so-called abandonment of fundamental, classical, physical concepts was a slow, gradual process, characterized by deep-ranging explorations and reconsiderations, yet also by mathematical manipulation, a certain amount of empiricism, and a large dose of intellectual playfulness.[35]

Significantly, Einstein's attitude and pronouncements during the congress reflect his deep admiration for, and sincere deference to the authority of, H. A. Lorentz – in 1911 and for years to come the chairman of the Solvay physics congresses. Einstein's veneration for Lorentz was often expressed in their correspondence. It was in Brussels that Lorentz probably decided to initiate the proposal that Einstein become his successor at Leiden, a possibility which Lorentz pursued privately quite intensely. Einstein, who had been under consideration for a position at Utrecht since

31. Smith, 1962, 26–7.
32. Einstein to Lorentz, 23 November 1911, Lorentz Papers, roll 4, p. 4.
33. Poincaré, 1911, 1912.
34. Klein, 1970, p. 252. To Klein it seems unlikely that Einstein had mentioned Ehrenfest's paper in Brussels. This assumption is verified by an examination of the original comments recorded by de Broglie, and by the participants themselves, as found in *Registre contenant* . . . n.d. For the most far-reaching and detailed exposition of Poincaré's work see McCormmach 1967.
35. For a convincing description of the ad hoc nature of the prehistory of the quantized atomic model, see Heilbron and T. S. Kuhn, 1969.

August 1911, refused the position at the end of November 1911. His letter to Lorentz was tinted with regrets and expressions of reverence and filial love.[36]

Einstein's report to the Solvay congress eschewed direct discussions of light quanta and of their physical interpretation. During the preceding years, he had struggled with and eventually abandoned an explanatory project, about which he had written to Lorentz in May 1909 in a long and detailed letter: "Regarding the light quanta I must have expressed myself unclearly. In fact I am absolutely not of the opinion that light should be conceived as consisting of independent quanta that are localized in relatively small spaces. . . . As I have said, according to my opinion one ought not construct light from discrete, independent points." Einstein went on to express his view that the "individual quantum of light is a point surrounded by an extended field of vectors, which somehow decreases with distance." After some further considerations, he added: "The essential seems to lie not in the supposition of singular points but in the assumption of such field equations as allow for solutions in which finite energy quantities propagate in a given direction with the speed c."[37]

For several years, Einstein was clearly unwilling to commit himself to a simplified physical conceptualization of light quanta. He knew that there was one theory of quanta in the time under discussion, and for that reason his characteristic mode of argument at that time was: What can we learn about the structure of radiation, about the spatial distribution of electromagnetic energy, from the combination of experimental fact (e.g., Planck's distribution law) and general principles (thermodynamics and statistical, or Boltzmann's entropy and probability relation)? His 1909 papers discuss this approach and the way in which it suggests the wave-particle, dual nature of radiation.

He therefore focused in his exposition on matters of mathematical consistency and rigor, on the internal correlation between the foundations of "molecular mechanics and electromagnetism" and the new quantum considerations. Like other scientists, Einstein exercised extreme caution in his published statements, and available records do not necessarily reflect accurately his oral comments at such meetings as the Solvay congress. Thus, for instance, Nernst wrote to F. A. Lindemann regarding the publication of the 1913 (second) Solvay proceedings: "Please send me the keynotes of

36. Pais 1982, pp. 209–10. A. J. Kox has analyzed Einstein's complex attitude toward Lorentz and has shown that his admiration for Lorentz verged on being intimidated by him and that he did not want to enter into a direct or indirect competition with him. Kox, 1993.

37. A. Einstein to H. A. Lorentz, 23 May 1909, Lorentz Papers, roll 3. The exchange of letters between Lorentz and Einstein early in 1909 refers to Einstein's article "Zum gegenwärtigen Stand des Strahlungsproblems," 1909.

my comments on the Sommerfeld lecture, I will then forward everything to Paris. With Einstein I'm afraid that, because he constantly changes his view, he unconsciously gives a somewhat different content to his comments than they had when [originally] spoken. . . ."[38] This ambivalence in Einstein's statements, public and private, underscores the generally tentative status of work in the quantum domain at the time of the congress. We may thus begin to understand that the congress did not represent a process of validation of a new physics proposed by only a few – but eventually correct – theorists. Rather, the reports indicate that the prevailing attitude was one of hope that quantum conceptions might be incorporated into classical physics.

In his welcoming remarks to the participants, the respected Hendrik Aanton Lorentz expressed satisfaction that almost all who had been invited were indeed attending the proceedings.[39] He stated that "modern investigations have shown more and more the serious difficulties" in attempting to "imagine" the motions of molecules and atoms, the nature of the forces and processes that take place in the ether, and that the kinetic theory seemed not to provide the satisfaction that physicists had felt even one or two decades earlier. Lorentz expressed the feeling which he thought to share with his colleagues, that they were "in a dead end, since the old theories do not have the power to penetrate the darkness which surrounds us on all sides." But the "beautiful hypothesis of the energy elements, which was first formulated by Planck and then extended to many domains by Einstein, Nernst, and others" seemed to disperse some of this gloom. It had "opened unexpected perspectives," and "even those who regard it with a certain misgiving must recognize its importance and fruitfulness." Lorentz's attitude was mostly circumspect. He did not believe that progress could be achieved collectively; rather, solutions would develop more easily as a result of individual efforts: "A lonely thinker in a hidden corner of the world" might find an appropriate solution.[40]

38. W. Nernst to F. A. Lindemann, 11 November 1913, Cherwell Papers D167, Oxford. The Second Solvay Congress on Physics took place between 27 and 31 October 1913. Its proceedings were eventually published after the war, eight years later (*La structure . . . 1921*).
39. Correspondence regarding the administration of the Solvay scientific institutes and congresses is in the *Paul Langevin Papers* n.d. Lorentz accepted the presidency of the proceedings in the summer of 1911, and from then on became the leading figure in the Solvay conferences and the president of the Solvay scientific institutes, which were funded by an endowment of 1 million Belgian francs in 1912. For a discussion of Lorentz's role in the context of turn-of-the-century physics and his involvement in quantum research, see A. Hermann, 1971, pp. 29–50.
40. Eucken, 1914, pp. 5–6. For a retrospective of Solvay congresses, see de Broglie, 1951.

This was quite unlike Nernst's short opening address in which he reminded the physicists of the Karlsruhe chemistry congress of 1860, where an attempt had been made to redefine the system of atomic weights. Despite the lack of agreement, that congress eventually had a distinct effect by drawing general attention to these problems, and soon afterward "complete clarity was achieved." Echoing views he had aired in the convocation letter, Nernst ended by uttering the hope that the Solvay conference would "later have an important influence on the development of physics."

It is reasonable to propose that Nernst, as an experimental scientist directing an active research institute along interdisciplinary lines, held a different conception of the success of collaborative efforts than did Lorentz. His large Berlin team had been attacking the problems from many perspectives, debating results in weekly colloquia and publishing briskly. Lorentz was certainly far removed from this style. Paul Ehrenfest, when he succeeded Lorentz in October 1912, reorganized academic life in the department according to what he had experienced during his stay in Göttingen, primarily influenced by Felix Klein.[41]

Lorentz's speech is significant not only in that it differs radically from Nernst's buoyant optimism but also because its pessimism was based on a thorough, yet different, knowledge of the state of theoretical developments at the time. Lorentz was well acquainted with current work by the major physicists working on quantum problems at the congress – Planck, Einstein, Nernst, and Sommerfeld – and had also read advance copies of their reports. He had been informed of Nernst's work through personal contacts and correspondence.

Early in 1911, for example, a critical article on Nernst's heat theorem had been published in the proceedings of the Dutch Royal Academy of Sciences, in connection with which Nernst wrote a complaining letter to Lorentz. On that occasion, Nernst sent Lorentz his papers on the specific heats and the quantum theory that had been published earlier that year. Nernst traveled to Amsterdam at the end of that week, planning to visit Lorentz in Leiden.[42] Later in the month, before publishing his scathing response to Dutch critics, he wrote to Lorentz:

> [In my paper] it is particularly stressed that at low temperatures the force potential A [soon to be called "Ergal" by Helmholtz] cannot possibly change if one assumes the quantum theory to apply; because in light of this theory the individual atoms of solid and liquid bodies in particular retain their position unchanged even a little above the

41. See Kant, 1974; Klein, 1970.
42. Kohnstamm and Ornstein, 1910; Nernst, 1911.

absolute zero [of temperature]; lim dA/dt = 0 is in fact the *whole* content of my theorem.[43]

Nernst's argument centered on the correlation between the predictions of his heat theorem and the quantum theory of solids. He continued in his letter:

> [My] second [recently published] note . . . *contains maybe one of the most convincing confirmations of the quantum theory,* together with a mild [*gelinde*] modification of the same; by the way, as I have convinced myself in an exchange of ideas with Planck and Einstein, our attempt at explaining the new formula can be understood only if one employs the representation of the quantum theory that I have recently given in the *Zeitschrift für Elektrochemie* and recently sent to you; in itself, [my] attempt to decompose the energy content into a kinetic and a second potential component is subject to certain reservations.

Nernst here referred to his joint paper with Lindemann in which two Einstein terms were used for the calculation of specific heats. With the same letter, Nernst enclosed the provisional invitation to the *conseil*, stressing that it was still only a preliminary draft but one that gave a full insight into the "enterprise." He was eager to have Lorentz chair the meeting. He suggested that Lorentz's tasks could be reduced to reading the individual reports in advance and leading the proceedings; but he urged Lorentz to present a paper as well and not to decline participating as president. Before closing his letter, Nernst again returned to his scientific dispute, asking Lorentz to submit his reply to the academy and adding: "Naturally, however, I am very curious to know your view of the whole question; in any case, I believe that whoever negates my heat theorem will have to explain thoroughly the extremely convincing correspondence, as for instance in the case of sulfur and other examples discussed in detail in my paper."[44]

Einstein and Lorentz had also discussed the problems to be raised in Brussels prior to the meeting. During Einstein's visit with Lorentz early in 1911, Lorentz had drawn his attention to a typographical error in one of Einstein's papers. On that occasion, Einstein and Lorentz had analyzed the quantum theory of specific heats. In continuation of their discussions

43. Nernst to Lorentz, 2 May 1911, Lorentz Papers, roll 4.
44. Nernst to Lorentz, 15 May 1911, p. 1, Lorentz Papers, roll 4; emphasis added. On this occasion, Nernst also presented to Lorentz an informal initial list of possible participants: Einstein, Hasenöhrl, Lorentz, Langevin, Planck, Perrin, W. Wien, Larmor, Jeans, Schuster, Seeliger, van der Waals, J. J. Thomson, Rutherford, Röntgen, Sommerfeld, and Brillouin, together with Solvay, Lorentz, Nernst, and Goldschmidt. Nernst to Lorentz, 15 May 1911, p. 2, H. A. Lorentz Papers, vol. 4.

"on quanta in the oscillation of material constructs," Einstein wrote to Lorentz that the theory has to apply to each constituent particle within the solid body under consideration, and that the thermal energy of each particle would have to be distributed among the resonators according to Planck's formula. From Einstein's letter it seems that Lorentz was, even after their meeting, not yet convinced of the validity of Einstein's arguments in favor of the applicability of the quantum theory to material substances.[45] Einstein returned to the same subject, in essentially the same terms, in his report to the Solvay congress, an argument he had already developed in his 1909 paper, discussed above.

The twelve substantial reports submitted to the congress, which had circulated in advance, focused primarily on quanta. Lorentz discussed the first topic proposed by Nernst, the application of the equipartition theory of energy to radiation. Most speakers directly addressed the established agenda: The Danish physicist Martin Knudsen discussed "The kinetic theory and the observed properties of ideal gases"; Jeans described the central problem of "the kinetic theory of specific heats according to Maxwell and Boltzmann"; and Planck spoke on "the laws of heat radiation and the hypothesis of the elementary quantum of action." In paired reports, Warburg and Rubens discussed the experimental investigation of Planck's radiation law for black-body radiation and the domain of long wavelengths. Sommerfeld and Nernst enlarged on the application of the quantum theory to physical and physical-chemical problems. Outside the spectrum of quantum theory lay the papers of J. Perrin (proofs for the true existence of molecules); H. Kamerlingh Onnes (on electric resistivity); and P. Langevin (the kinetic theory of magnetism and magnetons). Einstein presented the closing report, a synthetic evaluation of "The current status of the problem of specific heats." Einstein's characteristic style is evident in the similarity of titles chosen over the years: His 1909 paper, published before the Salzburg conference, had been entitled "On the current status of the radiation problem."

The papers of Sommerfeld, Planck, Einstein, and Nernst elicited the liveliest and longest discussions. Examination of the quantum hypothesis presented insuperable difficulties, not least among these being Planck's latest postulation of a second interpretation whereby emission of radiating energy was considered to occur in quanta, whereas adsorption was relegated to the prior "classical" continuous process. The only generally negative attitude was expressed by Poincaré, who was particularly hostile

45. Einstein to Lorentz, 15 February 1911, Lorentz Papers, roll 4. In his letter, Einstein mentions how fruitful had been his meeting with Kamerlingh Onnes at Lorentz's house.

in the discussion periods following Jeans's and Warburg's reports. The reports reflected a cautious attitude toward overly optimistic pronouncements. Jeans concluded his paper thus:

> However, I hardly believe in the prospect that the classical theory, together with some other new hypothesis regarding the mechanism of radiation that still relies on canonic equations, will ever lead to formulae that will reflect the facts as well as Planck's equations.[46]

Jeans, and to some extent Poincaré, were not impressed by the experimental data. Warburg and Rubens limited themselves to similar conclusions – namely, that although available empirical data did not "contradict Planck's formula," it did not yet provide "sufficient evidence" – in particular since in the domain of short wavelengths, Wien's formula was, in their opinion, better suited.[47] Sommerfeld was troubled by the problem of *Anschaulichkeit* and physical reality, and concluded by saying: "At the time I am inclined not to ascribe any physical reality to the emission quanta, nor to the earlier energy quanta."[48]

The most important conclusion to emerge from the First Solvay Congress in Physics was essentially that while the quantum hypothesis was an experimentally confirmed theoretical presupposition, most participants promoted the hope of *incorporating the quantum concepts into a modified classical theory*, although admitting that deep contradictions persisted at the time.[49]

Einstein has been historically portrayed as one of the earliest conceptual supporters of the quantum hypothesis of radiation and matter, particularly in light of his contributions of 1905 and 1907. In our perceptions of the acceptance of the quantum theory, we seem to rely greatly on the chronology of his scientific publications; and yet his private correspondence, coupled with his report and comments at the Solvay congress, are inconsistent with a firm and convinced attitude regarding the solidity and correctness of the quantum hypothesis – at least as it stood in 1911. The congress was twice referred to by Einstein as the "Witches' Sabbath." Thus, he wrote to Zangger upon his return: "It was highly interesting in Brussels. . . . Poincaré was simply generally negative and showed, despite his sharp intelligence, little understanding for the situation. Planck is steeped in some doubtlessly wrong preconceived opinions . . . but nobody really knows anything. The whole story would have been a delight for the diabolical Jesuit fathers."[50]

46. Eucken, 1914, p. 59.   47. Eucken, pp. 71, 75.
48. Eucken, p. 297.   49. Eucken, p. 365.
50. Einstein to Heinrich Zangger, 15 November 1911; see also Einstein to Besso, 21 November 1911; both in The Einstein Archives. In his contribution to Seelig, 1956, p. 43.

The First Solvay Congress in Physics did not achieve consensus or clarity on the issues discussed. Its primary significance resided in Lorentz's and Nernst's initiative to "talk" and to disseminate information, attract younger scientists, and coordinate certain aspects of experimental and theoretical investigations. In his opening address, Lorentz had stated:

> It is easy now to draft the program which we would follow. We first have to make explicit the incompleteness of the old theories, by establishing as precisely as possible the cause of their faults. Then we should examine the idea of the energy elements under the various forms that can be attributed to them; in the same manner *we would deal both with careful and systematic investigations and with daring hypotheses*. We should strive to separate the important from the secondary, and to reach a clear notion of the degree of probability of the individual hypotheses. The highest reward which we could expect would be to come somewhat closer to this future mechanics. What will be the result of this meeting? I do not dare to prophesy since I do not know what surprises are possibly in store for us. But it is probably wisest not to rely on surprises, and I would therefore consider it probable that we will contribute little to direct progress.[51]

The Solvay conference showed that the brightest minds present all understood the provisional character of their findings. Poincaré took Lorentz's advice literally, effecting a balanced evaluation yet leaving final pronouncements open to interpretation. Indeed, there was a measure of *operational utility* that they all ascribed to the theory, coupled with a genuine feeling of dissatisfaction. During the closing deliberations, Einstein declared that two principles would have to be adhered to above all – namely, the conservation of energy and Boltzmann's definition of entropy through probability. However, this well-known position obscures Einstein's previous remark, in which he said:

> We all agree that the so-called quantum theory of today, although a useful device, is not a theory in the usual sense of the word, in any case not a theory that can be developed coherently at present. On the other hand, it has been shown that classical mechanics ... cannot be considered a generally useful scheme for the theoretical representation of all physical phenomena.[52]

Zangger inserts in his quotation of the letter quoted here a mention of Poincaré's negativity "against the theory of relativity" – although there is no indication of that in the original.
51. Eucken, 1914, p. 6; emphasis added.   52. Ibid., p. 353.

Incidentally, the definition of the quantum theory as a "useful device," a *"Hilfsmittel,"* had been used by Nernst in 1909 and 1910; and this remark has caused historians to categorize Nernst, as opposed to quantum theoreticians, as an outsider to the quantum discourse. I would argue, however, that Einstein, too, agreed with Lorentz's and others' view as to the operational utility of the quantum. Lorentz, Planck, and Einstein certainly considered Nernst as a major participant. Thus Lorentz wrote: "The beautiful hypothesis of the elements of energy, which had first been pronounced by Mr. Planck and enlarged for many other phenomena by Mr. Einstein, Mr. Nernst et al., appeared like a wonderful ray of light. It has opened to us unexpected perspectives, and even those who view it with a certain mistrust must acknowledge its importance and fruitfulness."[53] Einstein's references in correspondence to the "Witches' Sabbath" and the "delight for the diabolical Jesuit fathers" can be explained only if one views the status of the quantum theory in 1911 as essentially provisional and unsatisfactory from the point of view of physical interpretation. Nevertheless, within several years of the meeting, substantive contributions and a major change in climate did take place.

The experimental confirmations of the quantum hypothesis in the years immediately following the Solvay conference were extensively discussed in an appendix to the German proceedings of the congress, edited by Arnold Eucken – one of Nernst's assistants and a serious contributor to his research program – and published in 1914. The Nernst-Lindemann formula for specific heats, which included two frequencies that related in the proportion 2:1, had provided surprisingly good experimental fits as compared with Einstein's initial suggestion; but systematic deviations seemed to appear for substances with very low atomic heats, as well as for diamond at certain temperatures. In 1912 and 1913, Debye, Born, and von Karman applied the idea of trains of waves to the solution of the vibrations of atoms in a solid. Their approach considered the total number of vibrations in a given interval of wavelengths, a procedure previously applied only to elastic continuous media. They thus developed a temperature function of specific heats, proportional to the third power of temperature, which in the domain of the lowest temperatures coincided much better with experimental results. This treatment differed from that of Einstein and Nernst in that it attempted to account for the whole heat movement of the body as opposed to their consideration of only individual atoms. The major change in the transition from Einstein's to Debye's theory, however, was the following: While the former localized the discontinuities on the specific atoms of the solid, Debye's treatment distributed the energy quanta over the whole macroscopic body. The localization of energy quanta is thus made impossible, and as Eucken wrote: "To imagine localized energy quanta is maybe not a necessity, but it leads us even farther away from a perceptual understanding of the quantum theory."[54]

Nernst attempted in 1913 to develop a theory according to which quanta remain distributed on the atoms or on a complex of atoms, and a model for which the synchronic vibration probabilities of n atoms were calculated. The model resulted in a function similar to Debye's, with a substantial difference in the maximal threshold of vibrations. The presupposition that a quantum effect should be observable also in the rotation of a gas molecule was first introduced by Nernst. During 1911, Eucken measured experimentally the specific heat of hydrogen and used Planck's suggestion in his second radiation theory – further elaborated by Einstein and Stern in 1913 – of a zero point energy, whereby the atom would retain an energy of half a quantum $h/2$ at the lowest temperatures. Ehrenfest tried to avoid this assumption, starting from the energy of a gas molecule as given by $nh/2$ where n is an integer, which resulted in a much improved fit with experiment.

In many ways, the tangible results of the congress materialized in the long-term success of future Solvay meetings in both physics and chemistry. Lorentz became the energetic and dedicated organizer behind the Solvay scientific institutes. He presented Solvay with a clear program of action: to encourage the study of fundamental questions regarding natural phenomena, work that could be done primarily by individuals but for which "the exchange of ideas among those who practice research is very fruitful."[55] Lorentz opposed Ostwald's efforts on behalf of a chemical Solvay institute devoted to international classification and nomenclature. Ramsay, who was contacted later, lobbied in favor of the use of Solvay funds for the publication of systematic monographs on chemical subjects. Both topics were rather unappealing to Solvay, one major reason for the delay in the convocation of the First Solvay Congress in Chemistry until 1921.

Lorentz decided that fellowships for younger researchers were the best avenue; they would encourage Belgian students to do research in estab-

54. At this time, Paul Ehrenfest wrote to Lorentz: "Again Debye has done an enormously funny thing: The finite heat conduction of crystals is thus conceived (better yet: is thus correctly calculable), that the average temperature of the stick produces incidental regions of intensification and dilution (see opalescence considerations) onto which the "acoustical" heat waves then scatter. Quantitatively this comes out very beautifully! . . . Bohr's work on the quantum theory of the Balmer formula (Phil. Mag.) has driven me to despair. If this path leads to the goal, I have to give up the preoccupation with physics." P. Ehrenfest to A. Lorentz, 25 August 1913, p. 6, Lorentz Papers, roll 4.
55. Lorentz to Solvay, 4 January 1912, Pelseneer n.d., p. 20.

lished laboratories. He desired to seek *"une certaine publicité"* by contacting some of the leading scientific societies and journals – such as the *Journal de physique,* the *Physikalische Zeitschrift,* and the British *Nature* – in order to inform the scientific community of the origin and goals of the new Solvay Institute. He wished it to become known that "the Institute disposes of a small number of subsidies for use by scholars engaged in experimental research on subjects which the Foundation encourages in particular." In order to avoid a "flood of applications," Lorentz proposed to limit the field to research "more or less directly linked to the theory of the elements of energy (radiation phenomena, specific heats, molecular theories, low temperatures)." Despite the fact that Ernest Solvay himself had donated the funds for the purpose of research in physics and physical chemistry in general, "for deepening the knowledge of natural phenomena," Lorentz considered restricting the grants to topics discussed at the 1911 congress, in order to "provide focus [and] avoid scattering."[56]

The members of the newly constituted scientific board, the *conseil,* met in September 1912 in Brussels, a gathering at which they hoped to address a number of subjects. At hand were the published proceedings of the "Quanten-Kongress"[57] – which was proceeding rather slowly in Nernst's opinion – as well as the topics and priorities of research grants and details of publicity and contacts with journals and societies. The meeting sought to set the agenda for the upcoming 1913 physics congress. In Lorentz's opinion, "We will definitely be able to say many new things about the 'Quanta,' but one could also choose another theme," such as the "The Magnetic and Magnetic-Optical Investigations."[58] At the ensuing meeting, however, the subject was reformulated by Lorentz, and the topic of "The Structure of Matter" was "unanimously accepted." In particular, the following were to be discussed: the structure of the atom (J. J. Thomson's model); the phenomena recently discovered by Max von Laue; pyro- and piezo-electricity; and the molecular theory of solids, with five prospective rapporteurs: J. J. Thomson (or Rutherford), Laue, Voigt (or Riecke), Grüneisen (or Einstein), and Brillouin.[59]

Despite the fact that no fundamentally novel theoretical or experimental

56. H. A. Lorentz to members of the Scientific Council of the Solvay Scientific Institute, 8 June 1912, *Paul Langevin Papers,* L10/76, pp. 1–3. The members were Rutherford, Brillouin, Madame Curie, E. Warburg, Nernst, and Kamerlingh Onnes, "in this order" – apparently by country, as specified by Lorentz to the secretary of the *conseil,* Martin Knudsen. 2 June 1912, *Paul Langevin Papers,* L10/76, pp. 2–3.

57. Nernst to Lorentz, 24 June 1912, *Paul Langevin Papers,* L10/233, p. 1.

58. Lorentz to Knudsen, 20 September 1912, *Paul Langevin Papers,* L10/94, p. 4.

59. Protocol of the first meeting of the *conseil* – renamed committee – distributed 26 November 1912, *Paul Langevin Papers,* L10/196, pp. 1–8.

research was presented at the congress, or in its immediate aftermath, it eventually came to be regarded as having extended participation beyond the previously confined territory of at most a dozen theoreticians. But this appreciation of its significance emerged much later, in retrospective accounts. As we have seen, many of the essential points of debate had already been discussed among some of the Solvay congress participants prior to the Brussels meeting. Setting an agenda for the Solvay meeting and deciding what needed to be explained and who would be doing the judging was part of a persuasive enterprise. The Solvay congress was an enterprise of active intervention and public display, designed by Nernst as a collaborative effort, employing a good measure of rhetoric, and convened not only to create and articulate shared attitudes and knowledge but also to expand the institutional and financial bases of organized physics and physical chemistry.

The intensified quantum discourse among the relatively small number of scientists, primarily German, had become complicated by Planck's insistence on his "second" quantum theory, one in which there was quantum emission but continuous absorption, and which led both Planck and Einstein to speculate on the existence of zero energy. Nernst continued to work on the quantum theory of specific heats, publishing numerous papers during 1912 and 1913, and inducing students and colleagues, such as Arnold Eucken, Niels Bjerrum, and Otto Sackur, to collaborate on or expand some of his suggestions. The proposals made by Nernst – that the quantization of atoms in the solid should be applied to rotating gas molecules – led not only to his own investigations but also to those of Bjerrum, Eucken, and Eva von Bahr, which were published between 1911 and 1913.

   The theory of the specific heat of solids was also soon improved by the work of Max Born and Theodore von Karman, and the most acceptable formulation was elaborated the same year (1912) by Peter Debye as the T^3 Law.[60] Incidentally, in later years, the Nernst-Lindemann formula proved to be a better description of the spectra of solid bodies than Debye's Law, which described more accurately the curves at very low temperatures. Nernst's prediction of gas degeneracy was later followed up, and is widely applied in the theory of metals and in astrophysics.

   The year 1911 was one of many contacts among theoretical and ex-

---

60. In a letter to his friend H. Zangger, Einstein wrote in the summer of 1912: "In the quantum theory Sackur has obtained an important progress (chemical constant of gases). The improvement of Born and Karman, respectively Debye, of my theory of specific heat is also a real progress." Albert Einstein to H. Zangger, Summer 1912, 1 page, A. Einstein Papers, Boston University.

perimental physicists and physical chemists. Upon his return from Brussels in October, Nernst threw himself even more vigorously into the consolidation of his research project on specific heats and, more generally, into strengthening and expanding his Berlin laboratory. He requested from the education ministry an annual increase of his 15,300-mark budget by another 5000 marks. In the previous two fiscal years, the laboratory had spent 26,000 per year. The excess of 10,000 M had so far been covered by private donations of 25,000 M collected at the inauguration of the institute extension. Even if they were to exercise restraint, Nernst expected a minimum annual expenditure of 20,000 M. For the 1912 fiscal year, this deficit of 5,000 M could "fortunately be covered from the donation of Mr. Solvay." But Nernst emphasized that Solvay was supporting only the continuations of existing thermodynamic research, and that "these funds could not be considered as a supplement to the current budget." Nernst reported that during that year, thirty-three persons were carrying out research in his institute. Among them, in addition to himself, were two professors and four or five assistants. The budget of 20,300 M was, at best, a minimum needed, in particular if his laboratory were "to keep up with the two newly founded Kaiser Wilhelm Institutes. Complicated researches will have to be funded by special, additional funds from the Academy, from the Jagor Foundation, by the director, or from private sources."[61] Upon an inquiry from the ministry, Nernst showed that already in 1909–10 expenses had exceeded the allocated government budget, and that donations had been used for the acquisition of "extremely sensitive apparatus, which will last for several years."[62]

Like many of his colleagues, Nernst continually sought to strengthen the financial basis of his laboratory and to supplant his expenses from private sources. He not only benefited from Solvay's magnanimity. He also pursued, quite successfully, his long acquaintance with Robert Goldschmidt, the young Belgian who had served as one of the secretaries of the Solvay congress.

Goldschmidt, an engineer, was heading the telegraph installations in the Belgian Congo and was apparently a confidant of the Belgian king. The German embassy in Brussels contacted Goldschmidt with regard to the development of telegraphy in Africa, and was interested in his attitude toward Germany. Apparently, to the ambassador's chagrin, Goldschmidt had "left no doubt that he is indifferent, if not even opposed to possible German intentions to install telegraphs in Africa." Goldschmidt had evidently "made it clear" that the reason for his frosty manner toward the

61. Nernst to Education Ministry, 25 March 1912, Merseburg.
62. Nernst to Education Ministry, 24 May 1912, Merseburg.

German authorities was "the denial of a decoration requested for him by Nernst." Goldschmidt felt "offended" by this refusal and was "more inclined to respond to French intentions to connect its African colonies by telegraph." The wealthy 35-year-old with an annual income of several hundred thousand francs, who "made available to the King large sums for the telegraphic installation in the colonies," was nonetheless "willing to discuss an agreement between Belgium and Germany about certain details of telegraphy in return for the award of a German decoration." He explained to the Germans that it would be easy to connect German East Africa with Cameroon and German South West Africa by wireless telegraphy over the relay system in the Congo. French cable politics made telegrams on French cable almost twice as expensive as those on the English. If an agreement were to be drawn between Germany and Belgium, and if Germany were to build a big station in Duala, it would be easy to send all Congo telegrams by wireless to Duala, whence the information would be transmitted via Monrovia over the German/South American cable. Goldschmidt also impressed the Germans with "good results with wireless telephone transmission" and even hinted that "it would be easy to detect iron deposits by wireless telegraphy."[63]

Nernst, who was involved in these confidential deliberations at the Foreign Office in Berlin, pressed his point. Goldschmidt, who had studied in Göttingen, had worked for a year in Nernst's laboratory on dissociation at very high temperature and on the heat conduction of liquids. "His main interest is technical," Nernst reported to the ministry. "He has built trucks, three dirigible airships, and recently equipped the Congo with wireless telegraphy." So that the point would not be missed, Nernst reiterated that Goldschmidt was the secretary of the foundation established by Solvay, whose 1 million francs for physics research were expected to be spent within thirty years. Moreover, Nernst made sure to mention that Goldschmidt had recently been made an officer of the French Legion d'honneur. Eventually, in early 1913, the Emperor Wilhelm conferred a minor decoration on Goldschmidt, the Red Eagle 4th Class.[64] Himself a Geheimrat, or Privy Councillor, Nernst played the political game, and battled the "expensive" and "warlike" front lines of scientific research.[65]

By 1912, strengthened by institutional recognition and ample funds from state and industry, Nernst's very active laboratory was flourishing. It included at any time fifteen to twenty doctoral candidates, and within

---

63. 15 July 1912, Flotow to Reichschancellor von Bethmann-Hollweg, from German embassy in Bruxelles, reporting on meeting with Prof. Goldschmidt. SbPK.
64. Nernst to Naumann, 9 October 1912; 17 October 1912; 7 January 1913.
65. Johnson, *The Kaiser's Chemists*, p. 101.

the span of eight years – from the enunciation of his heat theorem early in 1906 to the outbreak of the First World War – they produced some 120 papers, with a high point of activity between 1910 and 1913.[66] Nernst had rearranged his textbook of *Theoretical Chemistry* to include four new chapters that reflected recent advances in physics and chemistry on "The Molecular Theory of the Solid State," "The Atomistic Theory of Electricity," "The Metallic State," and "Radioactivity." These were based not only on existing literature but also on insights and results obtained primarily in his own laboratory. This was the period during which the quantum hypothesis was extensively tested by experimental measurements of specific heats, heats of vaporization, and spectroscopic studies, being accompanied notably by the development of new apparatus.

66. Horst Kant, "Zum Problem der Forschungsprofilierung am Beispiel der NERNSTschen Schule während ihrer Berliner Zeit von 1905 bis 1914," *NTM-Schriften. Gesch. Naturwiss., Technik, Med.*, Leipzig 11 (2):58–68 (1974).

12

Simply a Matter of Chemistry? The Nobel Prize

For the corpus of his thermodynamic researches, as well as his earlier fundamental elaboration of a theory for the working of the galvanic cell (or battery) formulated in 1889, Nernst was nominated, beginning in 1905, to the Nobel committees for prizes in physics and chemistry.

At the end of 1921, he eventually received the 1920 prize for chemistry.[1] The large number of nominations outlining the correctness, the experimental evidence, and the usefulness of the heat theorem persuaded the Swedish judges by then that a consensus had been obtained within the international scientific community.[2] However, consensus had certainly not prevailed a decade earlier. Moreover, personal, ethical, and professional ingredients played just as significant a role in Nernst's candidacy as scientific criteria or institutional guidelines.[3]

Reports and evaluations dating from 1912–13 show that the committee members had expressed specific reservations concerning the experimental validity of Nernst's account of electrolytic processes, the status of the heat theorem, and the precise prize for which Nernst would be eligible. It was argued, for example, that it had taken "somewhat longer for [Nernst's theory of electromotive forces] to be recognized in the chemical community" than the work of other physical chemists who had received a Nobel Prize. J. H. van't Hoff and Svante Arrhenius, the founders of the ionic theory, had each been recognized with a Chemistry Prize for their re-

---

1. The prize for 1920 was "reserved" and awarded a year later. During the years 1905 to 1912, prizes in chemistry were reserved for 1914, 1916, 1917, 1918. The prize for 1914 was awarded in 1915 to T. W. Richards; the one for 1918 in 1919 to F. Haber.
2. This agrees with other analyses of the prize awards. The most important sources are to be found in Bernhard, Crawford, and Sörbom, eds. (1982); Crawford, *The Beginnings* (1984); Crawford, "Arrhenius" (1984); Friedman (1981); Friedman (1989).
3. The prize itself, and its administration, are characteristic cultural symbols of the period of transition, of "uncomfortable adjustment" in the self-representation and position of scientists at the turn of the century. Heilbron, 1982. My thesis supports Shapin's suggestion that modern scientists continue to belong to a "core-set," one that interacts face-to-face, where credibility and trust rely just as heavily on individual relationships as in premodern scientific culture. Shapin, 1994.

spective theories of electrolytic dissociation and osmotic pressure. According to the statutes of the Nobel foundation, Nernst's electrolytic work dating from 1889 could have been rewarded if its relevance had been only recently confirmed. Citing the incompatibility of the theory with experimental results regarding the special case of extremely dilute solutions, the reports concluded that no substantial advance beyond an illustrious predecessor's work – that of Hermann von Helmholtz – had been achieved by Nernst.[4]

Some of the nominators, who had been sent the necessary statutes of the Nobel foundation, were probably familiar with or foresaw possible formal objections, objections which may nevertheless have weighed heavily with the committees. Thus, for instance, Carl Schall[5] of Leipzig University wrote a comprehensive recommendation for the year 1913, in which he focused on the thorny procedural issue of awarding the prize for "only the most important chemical discovery or improvement during the past year."[6] Relying on the language of the statutes, Schall argued that much of chemical research, although often yielding spectacular results of immediate benefit not only to academic chemistry but to industry and "humanity in general," had been furthered by new discoveries, as well as the "equally stressful process of elaborating and securing new theoretical conceptions, by improving and transforming older views, which often provide the first impulse to outstanding new discoveries." He emphasized the valuable contribution of theoretical work and its necessary refinements over time, rather than exclusively spectacular, palpable, and immediate discoveries. Nernst's initial formulation of the heat theorem had led to a "total transformation [*Umwälzung*] of older conceptions" of major consequences for the whole domain of chemistry, and would contribute to future "outstanding inventions and discoveries."[7]

Schall's exposition was exceptionally candid and detailed, containing a

---

4. App. 3 to NK protocols, 19 Sept. 1916, pp. 2–3. The report also noted that "conclusions drawn from the theory of osmotic pressure on the origins of the electromotive force as regards chemical reactions attributed to Nernst in fact originate in great part from Ostwald, as far as they had not been indicated previously by Helmholtz and van't Hoff." Ibid., p. 10.

5. J. F. C. Schall (1856–1939) was an associate professor of chemistry in Leipzig, who worked on molecular weight determinations, capillarity, physical properties of organic solutions, and electrochemistry.

6. The *Code of Statutes of the Nobel Foundation* stipulated that the funds should be apportioned "to the person who shall have made the most important Chemical discovery or improvement," and contributed "most materially to benefit mankind during the year immediately preceding." Crawford, *The Beginnings of the Nobel Institution*, Appendix B, pp. 221–3.

7. Ibid, pp. 1–2.

separate section on the objections raised by 1912 to the heat theorem.[8] The only public criticism, published in 1910 by Ph. Kohnstamm and L. S. Ornstein, concerned the heat theorem's admitted incompatibility with van der Waals's 1880 equation of corresponding states. Its authors were pupils of the respected J. D. van der Waals, who had submitted their paper to the Dutch Royal Academy.[9] Leonard Salomon Ornstein became chair of theoretical physics in Utrecht as a successor of P. Debye, but focused his research on experimental physics and became the director of the physics laboratory in 1925. His interests were mostly in kinetic theory and the measurements of light intensity. Philip Abraham Kohnstamm, who was born in Germany, received his Ph.D. in Amsterdam in 1901. He became van der Waals's successor as professor of physics in Amsterdam in 1908, and in 1910 had started work on the theory of reaction rates. He continued research for only seven years, becoming increasingly interested in pedagogy.

Kohnstamm and Ornstein's rather impassioned article appeared after intensive research on this problem had been carried out for more than a generation in the laboratory of Heike Kamerlingh Onnes in Leiden.[10] In 1881 he had reformulated and proved independently van der Waals's theorem. Nevertheless, reflecting upon his work more than a decade later, Kamerlingh Onnes wrote in 1894 that despite his systematic researches on the isothermals of gases at very low temperatures, refining and cor-

---

8. The practice of including known objections in letters of recommendation was not unique. See further discussions below. Other significant nominations included one from Naunyn, who proposed Einstein for the Physics Prize, for the quantum theory and the theory of relativity, and Nernst for the Chemistry Prize for his heat theorem. Le Blanc considered Nernst's work as the "most outstanding in the field of physical chemistry during the past 25 years," and referred to a number of Nernst's contributions, deeming the heat theorem as "wide ranging and leading to surprising results and relationships" in his research on the specific heats of solids at very low temperatures.

9. Meeting of the Koninklijke Akademmie van Wetenschappen in Amsterdam, Saturday, 24 December 1910.

10. The law of corresponding states, as formulated by van der Waals, says: "If we express the pressure in terms of the critical pressure, the volume in terms of the critical volume, and the absolute temperature in terms of the absolute critical temperature, the isothermal for all bodies becomes the same. . . . This result, then, no longer contains any reference to the specific properties of various bodies, the 'specific' has disappeared." J. D. van der Waals, *The Continuity of the Liquid and Gaseous States,* Physical Memoirs of the Physical Society of London, London (translation from the German of J. D. van der Waals's thesis, published originally in Dutch in 1873). Trans. R. Threlfall and F. Adair. p. 454, quoted in Kostas Gavroglu and Yorgos Goudaroulis, *Methodological Aspects of the Development of Low Temperature Physics, 1881–1956: Concepts Out of Context(s),* Science and Philosophy Series (Dordrecht/Boston/London: Kluwer Academic Publishers, 1989), p. 47.

roborating the law of corresponding states and featuring prominently the liquefaction of gases, in particular of helium, he had been frustrated:

> Various methods have been [empirically] tried . . . but without obtaining a good agreement with the observations over a whole range of equations of state. Neither was I successful in similar attempts which were repeatedly occasioned by my continued research on the corresponding states. . . . Whenever I seemed to have found an empirical form, I discovered after having tested it more closely that it appeared useful only within a limited range to complete what had been found in a purely theoretical way by van der Waals and Boltzmann.[11]

Thus, although Kamerlingh Onnes continued research on the equation of state, coupled with studies of the free-energy surfaces of monatomic and diatomic gases, it was evident by 1901, and long before the debate between Kohnstamm-Ornstein and Nernst, that the van der Waals equation was not a sufficiently established theoretical construct. In his response to his Dutch critics, Nernst similarly argued that

> it would be entirely unjustified to consider the new theorem of heat refuted on this ground; it is indeed only experiment which has to decide this question. . . . [E]xperiment has proved long ago that van der Waals's formula and even the general theory of corresponding states too are often in flagrant opposition to experiment. . . .[12]

Twice in his reply Nernst adduced support for his theorem from the "theory of indivisible units of energy," Planck's 1900 quantum theory of radiation and Einstein's 1907 quantum theory of solids.[13] While Kohnstamm

---

11. Ibid, p. 51. "Hence it appeared to me more and more desirable to combine systematically the entire experimental material on the isothermals of gases and liquids as independently as possible from theoretical considerations and to express them by series." H. Kamerlingh Onnes, "Expression of the equation of state of gases and liquids by means of series," *Comm. Phys. Lab. Univ. Leiden* 71: 3., in Gavroglu and Goudaroulis, pp. 51–2.

12. Nernst, 1911, p. 203. Nernst stressed that strongly undercooled liquids assume a rigid glassy state at low temperatures, as had been found by Tammann's investigations, and that "nobody but Messrs. Kohnstamm and Ornstein would ever think of applying van der Waals's formula to amorphous quartz and similar substances." In these cases, he argued, "there is no longer present unchecked movement of the molecules, and this is entirely in conflict with the premises from which van der Waals's formula was derived. Indeed, the new theorem of heat is intended to account for the entirely different circumstances found here; for the rest it necessarily follows from the theory of indivisible units of energy. Messrs. Kohnstamm and Ornstein therefore try to refute my theoretical considerations by evidently inaccurate, nay even inadmissible formulae." Nernst, ibid., p. 204.

13. By 1911, Nernst used the compatibility of his heat theorem with the predictions of the quantum theory as an important corroborative and confirmatory tool. Previously in his paper he had written: "For the rest it is also easy to derive from molecular theory even

and Ornstein acknowledged the temperature-dependent decrease of specific heats, and the substantial reduction of the total and free energy changes, they questioned the – thermodynamically indeed unproven – assumption that these changes would equal zero at absolute zero temperature. More fundamentally, they quarreled with Nernst's claim that his theorem would account for the behavior of matter at absolute zero.

During the summer of 1911, Nernst had sought advice from Lorentz. Nernst stressed that his theorem was applicable to chemical transformations, such as that of sulfur – an example which he introduced into his earliest papers on the heat theorem – assuming that they take place under their own vapor pressure, and that at low temperatures the contribution of the vapor pressure to the entropies of the substances is minimal. Kohnstamm and Ornstein had claimed in their paper that only if one assumes such conditions would the entropies tend toward zero at absolute zero. Nernst admitted that he had "taken for granted for many years" all these considerations, and that he should have initiated earlier a "serious discussion" of all the debated points, including the case of constant pressure and of transition points.[14]

At the meeting of the chemistry committee on 28 February 1913, it was decided that the candidacy of Nernst would be taken over by the physics committee, which was to meet on 20 March. In the Nobel decision process, nominations from a number of Swedish, Scandinavian, and foreign scientists were solicited annually by the scientific committees. These nominations, due at the beginning of each calendar year, were discussed during the spring, and a short list of candidates was decided upon; individual members of the physics or chemistry committee were then officially charged with preparing a lengthy report on the short-listed candidates. These reports, due at the end of the summer, were discussed in early fall; the committee's decision on the prize recipient would then be voted upon by the entire academy. The prizes were announced during October, and the award ceremony has ever since its inception taken place on 10 December, Nobel's birthday.

The final report of the physics committee[15] questioned the validity of the heat theorem, and as a result Nernst's name no longer figured in the joint meetings of the two committees. For the year 1914, Th. Simon of

without having recourse to the new theory of indivisible units of energy, which is of course incompatible with [van der Waals's formula, that this formula cannot possibly hold for liquids at low temperatures." Ibid., p. 203. In a footnote to his statement about "inadmissible formulae," Nernst commented: "With an analogous reasoning the said authors might also have 'refuted' Planck's formula of radiation, the whole theory of indivisible units of energy etc." Ibid., p. 204.

14. Ibid.   15. NK Report of 20 March 1913, and Appendix 3.

Göttingen recommended M. Planck as a first choice for the physics prize and Nernst as a second,[16] emphasizing the progress made in "the understanding of the physics of low temperatures." An authoritative nomination for the physics prize came in 1914 from Otto Pettersson (1848–1941), himself a member of the Nobel committee until 1912.[17] Pettersson, an influential personality on the Swedish scientific scene, held the chemistry chair at the Stockholm Högskola and had advocated physical chemistry by facilitating Svante Arrhenius's appointment to a university position.[18] In his nomination, Pettersson began with a careful recapitulation of those scientists active in both physics and chemistry who had contributed to theories of matter, physical chemistry, and radioactivity, and who were eventually rewarded by prizes in either category. His recommendation was addressed to both committees, consonant with his commitment to interdisciplinary research, such as physical chemistry, colloid chemistry, and physiological chemistry.[19]

Nernst, whose candidature in physics and chemistry was to become the longest in the prize history, received even by 1914 the highest number of nominations, but procedural issues complicated the handling of his candidacy. Controversy arose at the joint meeting of the physics and chemistry committees on 24 February 1914 over whether an evaluation of Nernst's work should be prepared by one of the two committees or by each separately. Oskar Widman (1852–1930), associate professor for organic chemistry at the University of Uppsala, suggested that an expert opinion be prepared by each committee, while O. Hammarsten (1841–1932), member of the physics committee and chair of medical and physiological chemistry at Uppsala, preferred to retain the physics committee's monopoly. Eventually, each committee prepared an independent assessment, yet both excluded Nernst: Max von Laue became the physicists' choice, whereas the chemists settled on Theodore Richards.

For the year 1915, C. Benedicks (Stockholm) proposed Nernst for the physics prize, and R. Wegscheider (Vienna), E. Beckmann (Kaiser Wilhelm

16. Th. Simon to NK, 24 February 1913, 1 page.
17. O. Pettersson to NK, 22 January 1914, 3 pages.
18. Crawford, pp. 51–2.
19. One might wonder whether the Pettersson episode could have been similar to that reported by E. Crawford and R. M. Friedman with regard to Ostwald's nomination, evaluation, and prize award in 1909. At that time, O. Widman "threw in the names of Nernst and Ostwald" during a heated debate in the chemistry committee. "Having no problem in eliminating Nernst's candidature, he then argued for an award to Ostwald. . . ." Crawford and Friedman, "The Prizes in Physics and Chemistry in the Context of Swedish Science: A Working Paper," p. 318, in Bernhard et al., eds., *Science, Technology and Society in the Time of Alfred Nobel*, pp. 311–31.

Institute of Chemistry, Berlin,) and K. A. Vesterberg (Stockholm)[20] nominated him for the chemistry prize. Wegscheider wrote:

> Through his hypothesis that in condensed systems the changes of internal energy and of work become identical at absolute zero temperature, Nernst has made possible the solution of the problem [to calculate chemical equilibria from thermal data]. The hypothesis has been proven correct by experiment, and has found excellent confirmation through the recent work on specific heats at low temperatures. Nernst himself participated in the research of specific heats at low temperatures and in particular refined the most compelling measurement methods.[21]

Ernst Beckmann's letter intimated that for the year 1915, two prizes for chemistry were envisaged – apparently due to the withholding of the prize for 1914 – and Nernst, Richard Willstätter, and A. C. Harries were proposed.[22] But the chemistry committee did not discuss Nernst any further, and the prize award was made on 25 May to Willstätter.[23]

In addition to the issues regarding the theoretical and experimental proof of the heat theorem, the six "short list" (or "top candidate") evaluations prepared on Nernst between 1911 and 1921 had to address the difficult question of whether the heat theorem, in any of its versions, was indeed a physical law. The 1914 report, for instance, stated that

> [Hendrik A.] Lorentz has shown that not only the specific heat (at constant volume) tends to zero when approaching $T = $ Zero . . . but also that dv/dt, that is the expansion coefficient, approaches zero.[24] This corresponds well to the requirements of quantum theory, but is obviously not included in Nernst's original heat theorem. . . . [I]n any case one can see that Nernst's heat theorem is not identical with the proposition: "It is impossible to imagine an arrangement which enables the attainment of absolute zero," as Nernst now wishes to for-

20. Carl Benedicks to NK, 31 January 1915, 3 pages; K. A. Vesterberg to NK, 31 January 1915, 1 page; Ernst Beckmann, 28 January 1915, 1 page. R. Pschorr, Director of the Organic Chemistry Laboratory at the Berlin Technical University, proposed Nernst as a second choice for the chemistry prize, after R. Willstätter, 18 January 1915, 2 pages.
21. Rud. Wegscheider (Vienna) to NK, 22 January 1915, 1 page.
22. Willstätter's contributions had been primarily in the field of organic synthesis of natural compounds, while Harries had been active in the analytic organic chemistry of synthetic and artificial rubber and its improvement.
23. NK protocols, 25 May 1915, p. 2.
24. A. Lorentz, "On Nernst's Heat Theorem," *Chemisch Weekblad* (1913): 621, in A. Lorentz, *Collected Papers*, vol. 6 (The Hague: Martinus Nijhoff, 1938), pp. 318–24.

mulate his theorem, thereby seeking agreement with the two classical laws.[25]

Lorentz's article was, in fact, a strong declaration of support for Nernst's theorem. As mentioned above, it was the result of an apparently substantial correspondence between Lorentz and Nernst with regard to both the Solvay congress deliberations and the attack in Dutch journals by Kohnstamm and Ornstein. Lorentz wrote in his article entitled "On Nernst's heat theorem" that

> ... though it is of primary importance to compare the results, to which this theorem leads, with observation, the question arises whether, starting from more or less plausible premises, one can deduce the theorem theoretically. Nernst has done this by taking as his leading postulate the impossibility to reach zero with experiments in which only finite changes occur. The object of the following considerations is to enlarge to some extent upon the arguments used in this deduction.

Lorentz's paper was a demonstration of the principle that, if Nernst's theorem of the unattainability of absolute zero is correct, then the expansion coefficient dv/dt must also become zero at absolute zero. The report pointed out that it was the "constant change in the content of Nernst's heat theorem which contributes to the fact that it escapes criticism."[26] Nernst's revisions and improvements over a period of more than a decade were deemed confusing, and ultimately detrimental to a consensus, and indicate that the academy steadfastly insisted on unequivocal theoretical positions in addition to convincing experimental confirmation.

Nernst's candidacy as top contender for the prize in 1916 was rejected by the physics committee, warranting the chemistry committee's justification "that Nernst's work has not been considered important enough for the physics committee in order to be eligible for the prize." Unable to ignore nominations by two influential Swedish scientists, The Svedberg and Wilhelm Palmaer,[27] the chemistry committee puzzled "to what science

25. NK Protocols, 26 May 1914, p. 6.
26. The report proceeded to undermine the existing experimental proofs concerning the integration constant I = ni, discussing the criticism of H. R. Kruyt on the calculations of the transformation point of monoclinic and rhombic sulfur, an example which until then was considered as one of the most beautiful proofs for the correctness of the heat theorem.
27. In 1912, The Svedberg was appointed to the first physical chemistry chair in Sweden. His strong atomistic views were substantially influenced by his reading of Nernst's *Theoretische Chemie* of 1903, and guided his research into colloid chemistry in the years 1905–15. See Anders Lundgren, "The ideological use of instrumentation," in Svante Lindqvist, ed., *Center on the Periphery: Historical Aspects of 20th Century Swedish Physics*, Science History Publications/USA, 1993, pp. 327–46, and Boelie Elzen, "The failure of a successful artifact," ibid., pp. 347–77. Palmaer had briefly been a Nernst student.

should Nernst's contributions be primarily attributed, a question in and of itself not insignificant but secondary to the main question: to what extent do these researches fulfill all the real and formal requirements in order to be considered for the prize award in any of the possible prize categories?"[28] While conceding that the academy had already awarded several prizes for physical-chemical research on their recommendation, the chemistry committee nonetheless seemed reluctant to institutionalize such disciplinary demarcations, and decided against a prize award to Nernst.

As in the case of Max Planck's quantum theory and Albert Einstein's theory of relativity, some nominators – albeit highly enthusiastic – cautiously avoided definitive cognitive claims on behalf of Nernst's heat theorem. For instance, in 1915, C. Benedicks wrote that "Nernst's theorem is an unconfirmed but calculation-facilitating hypothesis . . . which certainly opens new possibilities of quantitative research . . . but which has not yet proved its validity." Wilhelm Palmaer and The Svedberg acknowledged in 1916 that the experimental confirmation "includes various uncertainties, in particular due to the uncertainty in the calculation of the specific heats at the lowest temperatures, which leads to the most criticism." Therefore, the negative 1916 "candidate" report also quoted Bror Holmberg's nominating letter, which mentioned that "the theoretical value of this theorem is not totally undisputed." Such remarks prompted a well-justified delay in awarding a prize, particularly since the committees were charged with recognizing a *specific scientific achievement* of undisputed merit.

Nevertheless, by 1921, Nernst garnered ninety nominations in physics and chemistry. The tenor of doubt seemed to fade away from the nominations received after 1916. The successful synthesis of artificial ammonia was considered by many to be a robust verification of the heat theorem (Haber and Le Rossignol, 1908). The refined theories of specific heats that had been developed by Albert Einstein, Max Born, and Peter Debye just before the war, coupled with the stabilization of experimental methods, led to a growing number of nominations and eventually a final positive evaluation by the academy (Einstein 1907; Born 1912; Born and v. Karman 1914; Debye 1912). The predictions of the heat theorem had not been contested for several years. On the contrary, the successes of quantum theory relied substantially on the experimental work performed primarily in Nernst's laboratory. While Heike Kamerlingh Onnes achieved remarkable results in his low-temperature experiments at the Leiden laboratory, Nernst and his collaborators built a different, rather successful small-scale liquefier and calorimeter and demonstrated that specific heats

28. App. E. to NK protocols, 19 September 1916, p. 8.

of solids decrease substantially with temperature (Nernst 1910, 1911; Nernst and Lindemann, 1911).

By 1921, Nernst had reached the pinnacle of professional accomplishment and honors. He had become director of the Chemical Institute in Berlin, rector of the university, and also the director of the foremost German national laboratory of standards and measurements, the Physikalisch-Technische Reichsanstalt. He had traveled extensively in Europe and America, had guided scores of productive students, and had expanded the institutional bases for physical chemistry. Nevertheless, the documents in the Nobel Archives indicate that the academy, although voting almost unanimously to award the prize to Nernst, was treading carefully: The draft of the prize award citation was modified at the last minute to indicate that Nernst would be rewarded for his thermochemical work, deleting the remark "with special emphasis on [his] heat theorem."[29]

One could therefore argue that since experimental data had accumulated slowly, primarily during the years 1909 to 1914, and because during the war no prizes had been awarded owing to the political situation, Nernst's turn to receive a prize was not unduly delayed. One could reasonably view Nernst's prize as a final step in the legitimation of the discipline of physical chemistry. J. H. van't Hoff, Wilhelm Ostwald, and Svante Arrhenius had all received a Nobel Prize. With this last award, to the youngest and to many the most accomplished member of the founding group, the academy recognized a change that Nernst had brought to chemistry, infusing it with theoretical, physicalist conceptions.

All six critical candidate reports to the Nobel committees cited earlier, however, had been penned by Svante Arrhenius. To be sure, Arrhenius was among the first to telegraph Nernst and congratulate him on the prize. He was hoping that, upon their arrival in Stockholm for the Nobel ceremonies, Nernst and his wife could attend a private reception in their honor at his house. He advised Nernst to keep his prize money in a Swedish bank, as had Max Planck and Fritz Haber, "who must not have regretted this decision," since runaway inflation and poverty were crippling Germany that year.[30] But the tender of hospitality masked a long, complex history between the two men that profoundly influenced the Nobel selection process.

Walther Nernst and Svante Arrhenius had known each other for thirty-five years by the time of the Nobel celebration at Christmas 1921. As mentioned earlier, they had first worked together as students in the Würzburg

29. NK protocols, 10 November 1921, Apps. D and E.
30. In Arrhenius to Riesenfeld, 11 November 1921, 2 pages, Riesenfeld Papers, KVA.

University physics laboratory and went on from there to study together with Boltzmann, becoming close friends.

Arrhenius, whose 1887 theory of electrolytic dissociation gained him respect among European colleagues, had anxiously waited for appropriate recognition in Sweden in the 1890s. Nernst's insistent inquiries as to his progress and career produced feelings of inadequacy and resentment in Arrhenius. In 1891, after having been appointed associate professor in Göttingen, Nernst wrote:

> In the meantime you must have become accustomed, as I have, to the new position [as lecturer], but one has to acknowledge that the professor title allows one to feel very well, especially here in Göttingen, where the leap from Privatdocent to Professor is quite considerable. How is your laboratory doing and what are you doing in it? Are you a sugar man or an ultraviolet?[31] Do you have good help?

In the same letter, Nernst quite bluntly criticized aspects of Arrhenius's original work on the heats of dissociation, describing some of the errors as "fatal" and intimating that substantial corrections would be required, corrections which he himself intended to undertake.[32] At the time, such wrangling on matters scientific were taken in their stride by the two friends. They challenged each other and formulated their disagreements in the shape of bets, bets which would have to materialize in "an equivalent dinner or supper with wine."[33]

When in 1895 Ludwig Boltzmann resigned his professorship in Munich, Nernst was proposed for the physics chair, and lest he forsake Prussia, its powerful secretary of culture, W. Althoff, matched the offer. Nernst remained in Göttingen and was rewarded with the first physical-chemical institute built for this purpose in Europe, which opened with much fanfare in the summer of 1896. Soon thereafter, Nernst visited Stockholm. "Der liebe kleine Nernst" stayed in Arrhenius's bachelor den, and together the friends relived "the dear good old times." But Arrhenius recounted that something had changed. Nernst was engrossed in "the great prob-

31. This refers to Arrhenius's original work on sugar solutions which had led him to the dissociation theory. The ultraviolet refers to some investigations which Arrhenius had begun in Boltzmann's laboratory in March of 1890, where he did research on the conduction of gases. He sought to prove that, according to van't Hoff's theory, "gases should also dissociate electrically." He continued this work later in the year in Leipzig, before returning to Sweden at the end of the year, working with Hammersten on the kinetic aspects of osmotic pressure. Arrhenius to J. H. van't Hoff, 28 October 1890, Archief 200f, Boerhaave Museum.
32. Nernst to Arrhenius, 29 October 1891, Arrhenius Papers, KVA.
33. Nernst to Arrhenius, 4 November 1891, Arrhenius Papers, KVA.

lems of electrochemistry," and Arrhenius acknowledged that he could only with "difficulty follow Nernst's train of thought when he spoke about scientific problems. But the meeting was extremely enjoyable and informative."[34]

A patronizing tone regarding the progress of Arrhenius's research began to seep into Nernst's letters following this visit. Both friends realized that Arrhenius was not quite as active and involved in recent developments. After 1895, when he was appointed professor of physics at Stockholm, Arrhenius "largely lost interest in experimental research." He was concerned with "applying theoretical chemical laws, especially his own theory of ionic dissociation, to a wide range of phenomena within meteorology, geo- and astrophysics, cosmology and biology."[35] The modest facilities available to Arrhenius as professor of physics at the Högskola disappointed Nernst.[36] Late in 1896, when Arrhenius was appointed rector, an insouciant Nernst strongly urged that the "position as rector will remain a secondary matter, that is, that it will not cost you too much time and attention." Nernst feared that administrative tasks might diminish the "complete time and full concentration" necessary for work in physical chemistry. "At least this is what happens to me. Perhaps you can entrust your nerves with much more than I can," he wrote.[37]

At the time, Nernst himself was working on several projects: polarization measurements with the "telephone," residual rays (*Reststrahlen*),[38] gas chains, "beautiful" work with his assistant Dolezalek on a "new electrometer," as well as dielectrical measurements with fast vibrations. In addition, he reported on des Coudres's "beautiful things with cathode rays, [but he] cannot however bring himself to publish." Nernst was pleased that through some consulting work he had obtained 500 marks, which he earmarked for a vacation with Arrhenius on the Italian Riviera.

It was during the following year, however, that a fateful gathering, intended to revive "the good old times," produced rather more friction than intimacy between them. Arrhenius, Nernst, and Tammann met in Sweden

34. Arrhenius to Tammann, 14 September 1896, 2 pages, and Nernst to Arrhenius, 9 October 1896, Arrhenius Papers, KVA.

35. Franz Luttenberger, "Arrhenius vs. Ehrlich on Immunochemistry: Decisions About Scientific Progress in the Context of the Nobel Prize," *Theoretical Medicine* 13 (1992): 150.

36. The relatively poor facilities are described by Crawford, *The Beginnings. . . .*, p. 50, especially as compared to Nernst's brand new institute.

37. Nernst to Arrhenius, 31 December 1896, Arrhenius Papers, KVA.

38. Kangro states that "residual rays" were mentioned for the first time by Rubens and his assistant, Emil Aschkinass, in late 1897. These rays were defined as that part of the total radiation of a substance that persist after many multiple reflections on the surfaces of the substance. It is clear that the term was in use previously. Kangro, p. 165.

in the summer of 1897. During his sojourn, Nernst gave a demonstration of his famous electrolytic lamp. Upon attempting to operate the lamp in a Stockholm hotel, Nernst blew the fuses of the entire electrical circuit. Arrhenius seems to have laughed "a little bit too hard," and that seems to have been "the beginning of the end" of the friendship between Arrhenius and Nernst.[39]

The ultimate breach occurred three years later. At the end of July 1900, Arrhenius and Nernst met during the electrochemical congress in Zürich, where it appears that Nernst told Arrhenius "face-to-face that the conditions in Stockholm were very unfavorable for scientific work," the "best example" being Arrhenius himself. According to Arrhenius, Nernst had previously made similar remarks to other colleagues, a rumor to which Arrhenius had not given credence until then. In Zürich, Arrhenius "turned his back" on Nernst, "firmly deciding to have as little to do with him in the future as possible." When Arrhenius and Nernst met again in mid-October at the opening of the Hoffmann House of Chemistry in Berlin, Nernst repeatedly attempted to approach Arrhenius and alleviate the tension. But Arrhenius interpreted these as "political . . . flattery," rejecting any reconciliation with a "diplomatic friend."[40] From then on, their interactions became overtly confrontational.

Arrhenius's debate – or "polemic," as he called it – with Nernst was initiated with the publication of an article by the Berlin chemist H. Jahn in 1900. Hans Max Jahn (1853–1906), student of Bunsen and Hoffmann and a close friend of Planck, was initially trained as an organic chemist, worked in Graz until 1889, and then moved to Berlin. He remained a Privatdozent for a long time, lecturing on electrochemistry and focusing in his research on the thermodynamics of electrochemical and conductivity studies of dissociation phenomena. Nernst attempted to obtain Jahn's promotion to associate professor shortly before the latter's sudden death. Jahn had found that the mobility of ions increased by approximately 8.4 percent when the concentration of electrolyte is doubled, rather than a deviation in the opposite direction, as had been predicted by Arrhenius in

39. I am grateful to Elizabeth Crawford for this personal communication (9 December 1989), in which she refers to Euler's recollections in *Svensk kemiskt tidskrift* 71 (1959): 288–96. Crawford and Friedman have pointed out that Arrhenius "showed an excessive preoccupation with the regard in which his work was held by the Uppsala physicists," and have stressed the major significance of Arrhenius's membership on the Physics Committee and the "unofficial membership" on the Chemistry Committee. Crawford and Friedman, pp. 312ff. Tammann also recalled the event: "How Nernst works became clear to me one morning at the exhibition in Stockholm." Tammann to Arrhenius, 23 September 1898, Arrhenius Papers, KVA.
40. Arrhenius to Tammann, 29 October 1900, 2 pages, Arrhenius Papers, KVA.

1892. In essence, Jahn questioned the validity of Arrhenius's formula $\alpha =$ μ/μ, whereby the degree of dissociation is given by the ratio of electrical conductivities measured at a given concentration and at very high dilution. Jahn attacked what was to remain for a long time the weakest spot in Arrhenius's theory of dissociation: the case of the strong electrolytes. Jahn expressed serious misgivings about the applicability of the conductivity measurements method to the calculation of the precise degree of dissociation of strongly dissociating electrolytes in solution. Instead, Jahn applied the method of measuring electromotive forces according to Nernst's theory of 1889, and found serious deviations from Arrhenius's theory.

In his response to Jahn, Arrhenius, claimed that Nernst's method for the calculation of the degree of dissociation through the measurement of electromotive forces was nothing more than a simple derivation of Helmholtz's work. Arrhenius, therefore, denied the value of what had been until then considered Nernst's most important original scientific work. In addition, in his article he included criticism of Planck's thermodynamic treatment of the dissociation theory, which became the cause of an altogether unpleasant exchange between Arrhenius and Nernst during 1901 in the pages of the *Zeitschrift für physikalische Chemie.*[41]

While preparing his manuscript, Arrhenius anxiously consulted Ostwald, the journal's editor and publisher. If anything were to be wrong in his calculations, he feared that the three interested parties, Jahn, Nernst, and Planck, would "fall upon [him] and thrash [him] downright." Although confident, Arrhenius was still eager to hear Ostwald's judgment regarding the article's formal and substantive aspects, being aware of the "great responsibility" in sending the paper to press. One month later, apparently disturbed by the absence of a response, Arrhenius repeated his insistent inquiries in a New Year's letter, reporting to Ostwald that he had lately rewritten the criticism of Jahn-Nernst-Planck, since "special care was needed: If any mistake occurs, the three will perform a Red-Indian triumphal dance around me."[42]

41. S. Arrhenius, "Zur Berechnungsweise des Dissociationsgrades starker Electrolyte," *Z. phys. Chem.* 36 (1901): 28–40; "Zur Berechnungsweise des Dissociationsgrades starker Electrolyte II," *Z. phys. Chem.* 37 (1901): 315–22; W. Nernst, "Erwiderung auf einige Bemerkungen der Herren Arrhenius, Kohnstamm, Cohen, und Noyes," *Z. phys. Chem.* 36 (1901): 596–604; "Zur Theorie der Lösungen," *Z. phys. Chem.* 38 (1901): 487–506; H. Jahn, "Über die Nernstschen Formeln zur Berechnung der elektromotorischen Kraft von Konzentrationselementen. Eine Erwiderung an Herrn Arrhenius." *Z. phys. Chem.* 36 (1901): 453–60.

42. Arrhenius to Ostwald, 19 November 1900, Körber, vol. 2., p. 160. This refers to Arrhenius's first paper. S. Arrhenius, "Zur Berechnungsweise des Dissociationsgrades starker Elektrolyte," *Z. physik. Chem.* 36 (1901): 28–40. Arrhenius to Ostwald, 22 December 1900, Körber, vol. 2, pp. 160–1.

Ostwald offered neither encouragement nor criticism but simply published the paper at the beginning of 1901. In March, Jahn countered, pointing out that Arrhenius had not appropriately understood Nernst's formula and its dependence on the dilution of solutions. Arrhenius privately acknowledged this oversight, being well aware of spurning Nernst: "And now Nernst's turn will soon come. It seems as though I won't escape responding again."

Ostwald alerted Arrhenius of the forthcoming "unpleasant" rebuttal by Nernst, a matter which had occupied him for a considerable time. Eventually, Ostwald accepted Nernst's article, albeit in a milder form: "Nernst's mood seems to me somewhat pathological. As to the matter itself, it is probably just as well that it was you who gave the stimulus for the discussion." While composing a second article, it appears that Arrhenius genuinely wished "the thing to be completely finished and done with," and that a quick publication would "make further disputes superfluous." He gave serious consideration to the possible consequences of this episode, rewriting his replies three times "in order not to leave points open to attack." In his view, Nernst had "tempted" Jahn "out onto the ice," encouraging him to be aggressive.

By now Ostwald was deeply enmeshed in "unfortunate mischief" among two of his most successful disciples. Nernst had contacted him immediately upon the publication of Arrhenius's first paper:

> I would not have expected A.[rrhenius] capable of such terrible nonsense; he seems to have forgotten the most elementary principles of the solution theory. I regret this in his own interest; but it wouldn't annoy me any further. What has indeed angered me is his procedure to assign the theory of electromotive forces from now on to Helmholtz. In this I see an act of most explicit enmity and consequently I am enclosing my reply.[43]

Five days later Nernst continued:

> Now A. has attacked me without mercy on several topics in the field in which I have mainly worked; . . . it would be foolish on my side to further restrain myself.
>
> As far as I know, the cause of A[rrhenius]'s animosity against me – besides subjective instances on his part, which I do not want to elaborate upon – is the following: *It is my strong belief, founded on many proofs, that a very mean fraud is being perpetrated in Sweden with regard to the Nobel estate, by which the monies will be used for com-*

43. Nernst to Ostwald, 19 February 1901, Ostwald Papers, Sign. 2126, AAW.

pletely different purposes (institutes, etc.) than those for which they were designated, and the prizes have been withheld now for 5–6 years without cause.

A[rrhenius] knows my view on the affair very well; but he himself is unfortunately interested in a most fatal way in the misappropriation of the Nobel millions, because – ipsissima verba! – he hopes to become the director of the institutes to be built with the stolen money!...[44]

In private, Nernst opposed the construction of Nobel institutes, a stance which was known to Arrhenius, who speculated that "since [Nernst] is so nervous," he must be concerned that "as a consequence of [their scientific] debate his prospects have diminished." It seems, therefore, that the improbability of Nernst's winning a Nobel Prize had dawned upon Arrhenius even prior to the award of the very first Nobel Prize on 10 December 1901.[45]

As the dispute on electrolytes was being set aside, a new topic regarding the applicability of physical-chemical concepts to immunochemistry was becoming the focus of debate. Arrhenius campaigned energetically in favor of a physical-chemical physiological movement. In 1901, he sent for publication to Ostwald's journal a paper by Th. Madsen,[46] in whose Copenhagen institute for serum therapy he had worked during the summer of 1900.[47] In their joint work, Madsen and Arrhenius showed that the binding of toxin and antitoxin proceeds according to the mass-action law. Arrhenius intended to work in Frankfurt with P. Ehrlich and his assistant Morgenroth during the summer of 1903 in order to investigate the nature of amboceptors by physical-chemical methods. Madsen was convinced that "physiological chemistry would develop more and more as a practical application of physical chemistry," and Arrhenius hoped that this "development would bring the best fruits for both sciences."[48] But the association between Arrhenius and Ehrlich soured. During their prolonged controversy, according to Arrhenius, Ehrlich "ran to Nernst, who

44. Nernst to Ostwald, 24 February 1901, Ostwald Papers, Sign. 2126, AAW (emphasis added).
45. After elaborating on his dispute with Nernst, and Nernst's "nervousness," Arrhenius immediately and openly referred to the first Nobel Prize award. Arrhenius to Tammann, 8 June 1901, 2 pages, Arrhenius Papers, KVA.
46. Arrhenius to Ostwald, 21 January 1901. Körber, vol. 2, p. 163. Th. Madsen, "Versuche über die Abhängigkeit der Hydrolyse von der Temperatur," *Z. phys. Chem.* 36 (1901): 290–304.
47. Arrhenius to Ostwald, 4 January 1902, Körber, vol. 2, p. 168.
48. Arrhenius to Ostwald, 14 September 1902, and 25 December 1902, Körber, vol. 2, pp. 171–2.

had anyway spoken badly about me," and the two "decided to run me down. The great battle of Bonn was arranged. There, at the Bunsen Congress, the two called all their supporters from the whole world and really treated me in a very unfriendly way. At that time only Ostwald stood by my side."[49]

The construction of a Nobel physical chemical institute for Arrhenius was a slow and disappointing process, which prompted Arrhenius to employ his connections outside of Sweden in order to enhance his position. In November 1904, he wrote to van't Hoff:

> The academy of sciences has decided to build here an institute for physical chemistry for me. The King has given his approval. This all happened after Althoff had offered me a position at the [Prussian] Academy with conditions similar to yours.

On 8 October, Arrhenius had privately spoken in Berlin with Althoff, the Prussian under-secretary of education. However, Arrhenius seemed to need two letters confirming that an offer had indeed been made. He intended to write to E. Fischer and Behn on this matter, alluding that van't Hoff should do the same.[50] He was again requesting help to counter opposition at home, complaining that Hasselberg and Mittag-Leffler were "collecting letters against" him, letters denying the stories of a call to the Prussian Academy of Sciences.[51] Van't Hoff tried to clarify the apparent confusion, and was told outright that the possibility of a position "had been discussed very privately by two members of the [Prussian] Academy with Arrhenius" who had been too hasty in publicizing the matter.[52]

Nernst had actually helped Arrhenius with his first professorship in 1895, when Lord Kelvin had communicated a disapproving evaluation of

49. Quoted in letter of E. Riesenfeld, Arrhenius's brother-in-law, to his mother, 1 September 1910, in Körber, vol. 2, pp. 177–8, n 1. See W. Biltz, "Zur Immunochemie. Eine Erinnerung an die Hauptversammlung der Deutschen Bunsengesellschaft 1904," *Z. angew. Chem.* 7: 169–70. At the meeting of the Bunsengesellschaft that took place in Bonn 12 to 14 May 1904, Arrhenius presented a paper entitled "Serum Therapy from the Physical-Chemical Viewpoint." S. Arrhenius, "Die Serumtherapie vom physikalisch-chemischen Gesichtspunkte," *Z. f. Elektrochem.* (1904): 661–8. W. Biltz, Privatdozent in Göttingen, Prof. H. Zangger from Zürich, Ehrlich, and Nernst, together with Ostwald, participated in the long discussion that followed after the presentation, published in great detail. "Diskussion zu dem Vortrage von Arrhenius über Serumtherapie," *Z. f. Elektrochem.* (1904): 668–79. For more details, see Luttenberger, 1992, and Crawford, 1996.
50. Arrhenius to van't Hoff, November 1904, Archief 200f, Mus. Boerhaave.
51. Arrhenius to van't Hoff, 31 January and 5 February 1905, Archief 200f, Mus. Boerhaave. Auwers and H. C. Vogel from the Potsdam Observatory had written to Sweden that Arrhenius would in no way be accepted into the Berlin Academy. Arrhenius asked for letters from van't Hoff and Kohlrausch to help him with the nomination.
52. H. C. Vogel to van't Hoff, 11 February 1905, Archief 200f. Mus. Boerhaave.

Arrhenius to the faculty in Stockholm. Upon hearing the news, Nernst had made three concrete proposals to Arrhenius. He urged him to write to Boltzmann and "tell him the whole story. At your request, he would certainly whip the old Kelvin on the head." Secondly, Nernst offered to take the matter up with H. Landolt and E. Fischer, at the time Berlin's two chemistry chair holders. Inasmuch as he was on the board of the German Chemical Society, Nernst himself "would immediately put through your nomination as an honorary member, if you agree to that. So send me the material. . . . I will only use it very discreetly." And finally, Nernst encouraged Arrhenius to "write to Ostwald, so that he can make you an honorary member of the Electrochemical Society.[53] Otherwise, in my opinion, you can only gain in the future if you now play the martyr for a short time. I would approach the whole matter with humor since a brilliant rehabilitation will soon come by."[54]

An official offer to join the Prussian Academy in 1904 was never made, but Arrhenius obtained his Nobel institute, although proper laboratory installation and scientific research were delayed for some time. His isolation from Germany increased and his animosity toward Berlin and Nernst persisted.[55] The memory of their recent encounters lingered. Arrhenius avoided international scientific meetings on the grounds that he was either too preoccupied with his research or that he could not request permission for business trips. He also genuinely feared that the pleasures of a foreign excursion "might be spoiled, as had been [his] painful experience in Bonn and Zürich."[56]

When Arrhenius was finally given his own Nobel research institute in 1905, albeit without a permanent building and with only modest funds and apparatus, he planned to continue his physiological work on hemolysis "which requires small means." Arrhenius seemed to think that the debates with Ehrlich were of advantage to him: He wrote to Ostwald that since Ehrlich was "almighty in Germany" and received support from Nernst, all those who wished to advance – and "in Germany almost only Semites work in this field" – had to "distance themselves from us"; therefore "we" can work with a kind of monopoly in this "splendid little field, unobstructed by small pushers." He hoped that in a few years they would "have put all the old views out of circulation."[57]

53. Which indeed was taken care of shortly thereafter.
54. Nernst to Arrhenius, 29 May 1895, Arrhenius Papers, KVA.
55. Arrhenius to Tammann, 2 April 1904, 2 pages, Arrhenius Papers, KVA.
56. Arrhenius to Tammann, 28 April 1906 and 18 September 1907, Arrhenius Papers, KVA. Arrhenius at the time hoped that he would be able to move into the new institute in the fall of 1908.
57. Arrhenius to Ostwald, 24 August 1905, Körber, vol. 2, pp. 189–90.

News of Paul Drude's suicide in 1906 gave Arrhenius cause to criticize Berlin and its scientific establishment severely. He was distressed that a man as strong as Drude had fallen victim to the Berliners' zeal, to "this notion that the scientists necessarily have to be the most noble in the world," an attitude which was straining people beyond their capacities. Arrhenius believed that Emil Fischer, the renowned organic chemist, could survive because he had "excellent nerves." The foremost physicists of Berlin, A. Kundt and Fr. Kohlrausch, suffered from the febrile metropolitan life, and Arrhenius wondered how Emil Warburg could stand it. Helmholtz was able to manage only by neglecting his teaching duties, something a lesser person could not dare to do; but

> then there is the other alternative: one obtains the finest instruments in the world and takes the most accurate measurements with them, and therefore imagines that science consists of precision measurement and scolds all other projects. . . . But this alternative is increasingly vanishing, because the Americans have much more money and less imagination, and therefore more patience. Consequently, the Berliners will soon either become the greatest geniuses in the world or will have no justification for existing. This crazy system cannot continue for ever.

Arrhenius was content to have been unsuited for Berlin and had instead remained in Stockholm, in "modest conditions," where nobody demanded the impossible. The "senseless" urge for scientific supremacy seemed manifest to him in the pages of the new journal *Biochemisches Zentralblatt*, which he thought of as being directed against Madsen and himself, "since the Berliners want to be the top in this field."[58]

Arrhenius's complete unwillingness to grant Nernst the recognition of a Nobel Prize was justified in his eyes not solely by his distaste for the "ossified Geheimrat's" precision measurement program, nor by his purely personal antagonism. Rather, he explained to his friends that the deepest reasons for denying Nernst a prize were *ethical and moral failures,* of which the Swedish scientists were allegedly cognizant. In an impassioned letter, he listed obstacles and allegations to Tammann, who had proposed Nernst for the Nobel Prize in 1910: Nernst's heat theorem "will hardly be tenable," while his "older things" were spread over a great field and dated back quite a while. "But everything might work out . . . since the competition diminishes all the time," Arrhenius conceded,

> if only there were not another hitch. Nernst has earned, or more correctly, has made, a lot of money with the lamp. He has sought to use

---

58. Arrhenius to van't Hoff, 11 July 1906, Archief 200f, Mus. Boerhaave.

this money as profitably as possible, and found that night cafés bear solid interest. When the Berlin students want to go to the Café National, they say: One has to help Geheimrat Nernst, doesn't one? People here believe that there are enough night cafés without support from the Nobel funds. Besides, N. has had two very unpleasant libel suits. I was very sad to hear this about a former friend and the leading man in my field in Germany. *Therefore, N. would have good chances [for the prize] if one could prove that this information is excessive, but I fear one cannot. And as long as this stain remains unwashed, his chances are minimal. Unfortunately, morals and science do not always coexist, but I have known few examples as grave as this one.*[59]

Tammann's vigorous denial of these charges and of any unethical behavior did not sway Arrhenius, who undeniably breached the rules of confidentiality regarding prize awards. In 1901, on the occasion of the first prize decisions, he promptly informed van't Hoff on the names and numbers of nominating individuals and universities. Although the secret academy reports, of which only ten copies existed, were "at the binder," Arrhenius reported that van't Hoff had obtained an "absolute majority" vote. Van't Hoff received the first Nobel Prize in chemistry for 1901, of which Arrhenius informed him before the decision was officially made public.[60]

The unfortunate animosity between Nernst and Arrhenius had quite dramatic effects. What Nernst could not have known, officially at least, was that for sixteen years, the only referee charged with evaluating his work had been his "good old friend" Svante Arrhenius. As a member of the Nobel physics committee, Arrhenius succeeded repeatedly in planting misgivings concerning the merits of Nernst's work, doubts that effectively blocked the award of either the physics or the chemistry prize. Although initially Arrhenius had been appointed a member of the Swedish Academy of Sciences in 1901 "over strong opposition,"[61] he nonetheless became the most influential physical chemist in Sweden and the director of a Nobel institute. Therefore, to request his expertise on Nernst's research as a referee seemed eminently warranted.

His most extensive report to the physics committee, that of 1914, was attached to all subsequent discussions and decisions. Each year, Arrhenius

59. Arrhenius to Tammann, 22 December 1910, 2 pages, Arrhenius Papers, KVA. In part also quoted by Crawford, *The Beginnings . . .* , p. 257, n. 51.
60. Arrhenius to van't Hoff, 2 January 1901 and 13 November 1901, Archief 200f, Museum Boerhaave.
61. H. A. M. Snelders, "Svante Arrhenius," *Dictionary of Scientific Biography*, vol. 1, p. 297.

amended his previous evaluations, but they all centered on several major objections: The electrochemical work was too old to be considered for a prize; the heat theorem had insufficient or controversial experimental backing; its status as a natural law was debatable. Arrhenius rigorously surveyed the literature on specific heats, phase transitions, and entropy calculations. He found a few – albeit minor – published critiques of Nernst's work.[62] He argued, for example, that atomic heats at high temperatures rise over Nernst's theoretical value of 5.95. Quoting experimental results by Theodore Richards, he stressed that this behavior could be explained neither by Einstein's 1907 version of the quantum theory of specific heats, nor by Nernst and F. A. Lindemann's reformulation of 1910–11:

> Nernst often says that it is irrelevant whether we say that $dQ/dt = 0$ or is very small. This is in principle correct, if we deal with equilibrium conditions at normal temperatures. . . . But theoretically Nernst's claim is incorrect. His heat theorem and his deductions then collapse.[63]

Objections that Nernst supported the heat theorem by selectively interpreting the data were voiced several times in Arrhenius's reports: "Nobody doubts that whatever results appear, Nernst's apprentices will always succeed in developing interpolation formulas which will save his fundamental equation,"[64] he wrote, adding that one obtains "a clear picture of what an arsenal of explanatory possibilities and calculation methods are available to the Nernst school when it comes to strengthening the three laws [of thermodynamics]."[65]

In 1917, Albert von Ettingshausen, Nernst's old mentor and doctoral supervisor, and Franz Streintz, chair of physics at the Technical University in Graz, recommended Nernst for the physics prize.[66] Ettingshausen's presentation of Nernst's work addressed almost exclusively the heat theorem as a contribution to the development of thermodynamics, recapitulating the main theses: its relationship to the two laws of thermodynamics, the molecular-kinetic content of its predictions, as well as the

---

62. H. R. Kruyt, "Die dynamische Allotropie des Schwefels," *Z. phys. Chem.* 81 (1913): 726–48 [with experiments performed by H. S. van Klooster, from Groningen and M. J. Smits, from Utrecht]; section on Nernst theorem, pp. 738–40. Interestingly, however, Kruyt showed that the critique of Kohnstamm and Ornstein against Nernst was not valid, since regarding the transformation of sulfur, they had relied on untenable experimental data published by Smits and Leeuw, *Versl. Jon. Ak. Wet. Amsterdam* 20 (1911): 400. Nernst's reply in "Das Gleichgewichtsdiagramm der beiden Schwefelmodifikationen," *Z. phys. Chem.* 83 (1913): 546–50.
63. Arrhenius, ibid., p. 15.   64. Arrhenius, ibid., pp. 23–4.   65. Arrhenius, ibid., p. 25.
66. Ettingshausen to NK, 2 January 1917, 4 pages; Streintz to NK, 3 January 1917, 2 pages.

experimental proofs brought since 1906, by both Nernst and his students.[67] But in the chemistry committee, Nernst's candidacy was not discussed any further, while the physics committee decided not to award any prize for 1917.

Nernst received only one nomination for 1918,[68] and none for 1919. No prize was awarded for 1918. The committee decided a year later that the chemistry prize for 1918 would go to Fritz Haber, the physics prize for 1918 to Max Planck, and the physics prize for 1919 to Johannes Stark. But for the 1920 chemistry prize, Nernst received eleven nominations, primarily from German scientists.[69] In characteristically elegant prose, Fritz Haber, in his proposal, recounted that through his intimate association with problems related to chemical equilibria, the magnitude of Nernst's achievement during the previous fifteen years was particularly manifest to him. The ability to calculate chemical affinity from purely thermal data, of which one could earlier only say that it was "an uncertain possibility, whose pursuit was bound to bring many surprises, difficulties, and contradictions," became a "simple and experimentally secure concept of very general importance. Thermodynamics and its application to chemistry therefore made fundamental progress." Haber broached the delicate issue of the priority claims that Th. Richards had voiced with regard to the formulation of the heat theorem.[70] It probably referred to a comment found

---

67. In chemistry, nominations were sent in by Emil Fischer, Fritz Ephraim (Bern), G. Tammann (Göttingen), and Robert Luther and Fritz Foerster (both Dresden). Luther to NK, 27 January 1917, 2 pages; Ephraim to NK, 4 January 1917, 2 pages; Tammann to NK, 21 January 1917; Foerster to NK, 18 January 1917, 3 pages; Fischer to NK, 6 November 1916, 1 page. Ephraim stressed that although at the time,

> the law had been formulated only as a hypothesis, it needed justification . . . which was brought through a rich observation material over the past decade. However, it still retains the character of a hypothesis, since an exact proof cannot be brought forward, just as in the case of the two laws of mechanical heat theory.

Ephraim, pp. 1–2.

68. R. Wegscheider (Vienna,) to NK, 25 January 1918, 1 page.

69. M. Le Blanc to NK, 26 December 1919, 1 page; A. Hantzsch to NK, 31 December 1919, 1 page; F. Haber to NK, 2 January 1920, 2 pages; Waldeyer Hartz to NK, 10 January 1920, 1 page; Wilh. Palmaer to NK, 31 January 1920, 1 page; F. Weigert to NK, 14 January 1920, 3 pages; Naunyn to NK, 26 January 1920, 1 page; C. Drucker to NK, 20 January 1920, 2 pages. Schall again proposed Nernst, this time to share the prize with Fr. Fichter, of Basel, for his electrochemical work. Schall to NK, 16 February 1920, 8 pages. Late nominations came from Sieverts, prof. of physical chemistry at Greifswald, 27 March 1920, 1 page, and 29 April 1920, 1 page; and B. Rassow (Leipzig), 6 February 1920, 1 page.

70. For a more detailed analysis of the criticism leveled by Richards and Kruyt, see Diana

in a 1909 paper by Richards, in whose conclusions regarding the electrochemistry of amalgams we read:

> The temperature coefficient of the electromotive force of a cell made
> from liquid amalgams is as a matter of fact approximately equal to the
> ideal potential of the cell (calculated from the relative concentrations
> of the amalgams) divided by the absolute temperature. . . . [H]ence ac-
> cording to the theorem recently advanced by one of the present au-
> thors, the free energy output of the chemical part of the change may
> be expected to be equal to the total energy output, and both would be
> expected to remain invariable with the temperature. . . ."[71]

Here Richards quoted his own paper of 1902. He wrote in the footnote:
"This theorem has been elaborated by Nernst in a very interesting way,"
referring to Nernst's *Silliman Lectures*.[72] Richards's priority claims relied
upon his pronouncements in the 1902 paper regarding the equality of
$dA/dT = dU/dT$ at absolute zero temperature. Nernst never fully acknowledged a debt to Richards's postulation, formulated as follows:

> The change of the available or free energy of a reaction with the tem-
> perature must have some fundamental connection with the change of
> the total energy with the temperature. This fundamental connection
> becomes manifest on comparing the actual values of dU/dT with
> those of dA/dT.[73]

Richards tabulated the data for dU/dT and dA/dT for ten chemical reactions and found a proportionality relationship of $dA/dT = -M\, dU/dT$ "in
which the value of M averages about 2." On extrapolating graphically U
and A, Richards found that "it is evident that each pair tends to converge
at a point not far from absolute zero. There is no reason for surprise at
this fact; indeed such a result is a necessary consequence of the equation
of Helmholtz, in which $A - U = 0$ when $T^0 = 0$. The interest centers about
the fact that U always increases when A decreases with the temperature,

Kormos Barkan, "Walther Nernst and the Transition to Modern Physical Chemistry,"
Ph.D. diss., Harvard University, 1990.

71. T. W. Richards and R. N. Garrod-Thomas, "Electrochemical Investigation of Liquid
Amalgams of Zinc, Cadmium, Lead, Copper, and Lithium," *Carnegie Institution of
Washington Publication* No. 118; p. 60.

72. T. W. Richards, "The Significance of the Change of Atomic Volume," *Proc. Am. Acad.*
36 (1902): 300, also published in *Zeitschr. phys. Chem.* 42 (1902): 138. Richards referred the reader to W. Nernst, "Thermodynamics and Chemistry," *Silliman Lectures*,
p. 56.

73. Ibid., p. 300.

and vice versa." And he continued: "It will be noticed that in order to converge at the absolute zero these lines must not be exactly straight, but slightly curved. . . ."[74]

It is, however, significant that the insight into the tangential shape of the free-energy and internal-energy curves constituted a major change in Nernst's perception versus Richards's pronouncements. In a draft for a book review of Nernst's *Die theoretischen Grundlagen des neueren Wärmesatzes* in 1919, Arnold Sommerfeld judged the heat theorem to be justly called

> the third law of thermodynamics. In the same sense and to the same extent as the first and second laws, Nernst's heat theorem is also an empirical law [*Erfahrungssatz*]: It not only reproduces immediate experience, but refines and condenses crude experience to an idealized law and to a degree of exactitude which reaches beyond immediate experience. The raw data which Nernst has refined in his heat theorem, especially the Thomsen-Berthelot rule (heat content as approximate measure of affinity in chemical reactions) and Thomson's equation (heat content of a galvanic cell as approximate measure of the electromotive force of elements). With rare boldness and acuity Nernst formulated therefrom his heat theorem: The curves of energy and free energy have at absolute zero not only the same coordinate but also the same (horizontal) tangent.[75]

The magnitude of this progress was not diminished, in Haber's eyes, by the fact that "Mr. Richards and Mr. van't Hoff and myself were close to finding a starting point. . . . The great accomplishment does not reside in the intuition of the correct context, but in the fact that [Nernst] has grasped it with such certainty and has pursued it in all directions with such power and persistence that the result has been lifted from the sphere of interesting possibilities to the height of lasting possessions of science. This we owe to Herr Nernst."[76]

74. Ibid., p. 301.
75. Arnold Sommerfeld Papers, DM., Bestand Bopp. Letters to AS. Editorial board of *Stahl und Eisen*, Verein Deutscher Eisenhüttenleute, Düsseldorf, to A. Sommerfeld, 6 November 1919; MS draft of review on back of letter. A. Sommerfeld in letter to the board, 9 August 1919, had promised the book review.
76. Haber, pp. 1–2. Haber stressed that "the importance of the heat theorem has been confirmed not only in isolated cases but in all cases. . . . Maybe doubts would be raised whether a reward by a physics prize should not replace the chemistry prize. But the same doubt could be voiced in the case of van't Hoff, Arrhenius and Ostwald. . . . These examples therefore constitute a precedent for rewarding Nernst with the Chemistry prize." Ibid., p. 3.

Evidently, Haber was well informed about the obstacles to Nernst's candidacy, which he carefully addressed. It is safe to assume that through personal and public exchanges, the absence of a prize for Nernst had become somewhat of a *cause célèbre*. The timing of this nomination was significant. Haber had just been awarded the Nobel Prize himself, and while professional rivalry had marked his interactions with Nernst prior to World War I, the post-1918 years bound them closely together: Both had worked in classified military research during the war, and were at the time the personalities most identified with German chemical warfare. By 1918, both were listed as enemy civilians by the Allies, fearing not only ostracism but internment, which induced both to travel to Switzerland in order to avoid extradition to a possible war tribunal.[77] Haber wished his fellow scientist to be rewarded at a time when both he, and German science in general, were in dire straits. Past enmities and frictions were softened or set aside at this time when many in the scientific community showed tolerance and compassion toward Nernst, who had lost both sons in the war.

Wilhelm Palmaer, whose previous efforts on behalf of Nernst had been thwarted, reiterated his recommendation in 1920, quoting an important passage from the newest edition of Max Planck's classic and influential *Vorlesungen über Thermodynamik* (1917):

> For the fifth edition I have reworked the whole content of the book, in particular the section on *Nernst*'s heat theorem, in its most far-reaching version, which has in the meantime found complete confirmation and has therefore passed into the secure domain of theory.

In April, the chemistry committee again commissioned an evaluation from Arrhenius. On 10 May 1920, he submitted a complicated document.[78] Arrhenius agreed that the equivalence of internal and free energies at absolute zero temperature was probably *correct*:

$$dA/dt = dU/dt = 0 \text{ at } T = 0 \qquad [1.]$$

> The further development of Planck's quantum theory by Einstein, and in particular by Debye, has in the meantime led to the result that the

---

77. The archives of the [then] Prussian Academy of Sciences contain a letter addressed to the scientific academies of neutral countries, protesting against the intention "of the allied and associated powers to bring its members, Professors Walther Nernst and Fritz Haber, in front of a military tribunal, being charged according to article 228 of the peace treaty, to have acted against the laws and customs of war." However, it seems that between "February 1919 and August 1920, the mail log of the Academy did not register any letter with corresponding contents having been sent." On 30 August 1920, the letter was deposited to the archives. AAW, Personalia Sign. II-III.38, pp. 76–7.

78. Arrhenius report, 10 May 1920, in the protocols of the session of 18 May 1920 of the NK for Chemistry, 7 pages in MSS. Also typed as Appendix 11 to NK protocols, 16 September 1920, 10 pages.

specific heats for all solids in the vicinity of absolute zero are proportional to T^3 according to which it follows from [1.] that dU/dt is there zero. Furthermore, one can prove with classical mechanics . . . that it is impossible to imagine in a finite space a process by which bodies can be cooled down to absolute zero temperature. From this proposition one can also derive Nernst's heat theorem, that is, dA/dt is also zero at $T = 0$.[79]

Yet, if one "follows this theorem, difficulties appear in calculations." Arrhenius selectively quoted the American chemist Hugh S. Taylor, whose research had provided a confirmation of the theorem, but who had suggested in 1916 that "a considerable increase of experimental evidence relative to the Nernst theorem [was] eminently desirable." However, in his article, Taylor had concluded explicitly that

> [t]he data have been used in an investigation of the assumptions of the Nernst Heat Theorem and of the methods of testing the same. *So far as the present experimental evidence relative to specific heats may be employed, results favorable to the theorem have been obtained* [emphasis added].[80]

Taylor made full use of the quantum theoretical apparatus put forward by Einstein, Nernst, Lindemann, and Debye for the theoretical calculation of specific heats. This is of note since before 1916, few American physicists or chemists had as yet turned their attention to quantum theoretical developments. Born in St. Helens, Lancashire, Taylor had worked on doctoral research with Arrhenius on acid-base catalysis during 1912 and 1913, and spent the following two years in Max Bodenstein's laboratory in Hannover. After obtaining his degree in Liverpool in 1914, Taylor moved to Princeton, where he became chairman of the chemistry department.

Arrhenius conceded that Nernst's "principle is probably not wrong, but its application is prevented by the very complicated calculations and the time consuming determinations of specific heats needed for these calculations. . . . For this reason, this principle has not been as "useful" for chemistry as stipulated by the testament of Nobel." Despite a substantial change of opinion, Arrhenius concluded that Nernst should not be awarded a prize for work done in Germany's greatest physical-chemical laboratory in the years 1906–17; credit should not automatically be given to the leader of a large research group.[81]

79. Ibid., p. 3.
80. Ibid., p. 5. Hugh S. Taylor, "The Thermodynamic Properties of Silver and Lead Iodide," *JACS* 38 (1916): 2309.
81. Arrhenius, ibid., p. 9.

In its meeting on 3 June 1920, the chemistry committee decided not to propose Nernst for the prize. In the broader academy meeting on 30 October, when Palmaer submitted a thirteen-page memorandum on behalf of Nernst, reiterating that Nernst had received the highest number of nominations to date (68 nominations by 42 different scientists, 58 of them in chemistry and 10 in physics,) two chemistry committee members voiced reservations.[82] Despite the evidently strong rally in favor of an award, no prizes were recommended.[83]

For the year 1921, Nernst was nominated in chemistry by seven German scientists. Several nominators were associated with the University of Breslau, where A. Eucken, and later F. Simon, established a physical-chemical research school that had emerged directly under the influence of Nernst, and that was to shape to a great extent the field now known as solid-state physics and quantum chemistry. But most significantly, a notable Scandinavian group coalesced in order to make a definitive statement on behalf of Nernst. Fourteen Swedish and Norwegian scientists cosigned two letters of nomination: Collenberg, Lindemann, Riiber, and Schmidt-Nielsen (all from Trondheim), and Palmaer, Kullgren, Oden, Vesterberg, Strömholm, Svedberg, Barthel, Ramberg, Petren, and Soden (from Stockholm and Uppsala).

At the session of the chemistry committee of 23 February 1921, a major event was initiated when for the first time, an outsider was charged with the drafting of the "candidate report." Hans K. A. S. von Euler-Chelpin (1873–1964) was a German-born scientist who, among many other subjects, had studied physics and chemistry, and had worked in Nernst's Göttingen institute for two years before becoming Arrhenius's assistant in Stockholm in 1897.[84] On 16 March, following the academy's approval, the chemistry committee asked Euler to write a report and participate in the debates.[85] Euler's detailed exposition confronted Arrhenius's six pre-

82. App. 13 to protocols of NK Chemistry, 6 September 1920, 2 pages.
83. Palmaer report, 30 October 1920. Palmaer's position was supported by Söderbaum and by von Euler-Chelpin. The speaker of the section suggested that the final wording of the nomination should be left open, enabling members who supported Nernst to reach a common formulation. In the vote for the prize in chemistry for 1919, Hammarsten and Widman agreed with the majority to move the prize to the special fund. Klarson proposed Le Chatelier, while the other five chemistry committee members voted for Nernst (Söderbaum, Hedin, Ekstrand, von Euler-Chelpin, and Palmaer). The 1919 prize was moved and the one for 1920 was reserved for the following year.
84. NK protocols, 23 Feb. 1921, p. 6, par. 3–4. In the same session, an application by Euler to use interest from the special fund for research on enzymes in the amount of 4000 Kroner was approved. Euler's application letter is dated 23 September 1920, and appears as App. b., 2 pages. Euler received a Nobel Prize in 1929 for his studies on fermentation.
85. NK protocols, 16 March 1921, p. 1., par. 1.

vious memoranda, and centered on the two major topics: electrochemistry and thermochemistry.[86] He argued that the correspondence of theory and experiment had been facilitated by quantum theory. He pointed out that the precision work of Nernst and his collaborators on atomic heats, specific heats, and chemical equilibria and the creation of new instrumental methods were in themselves similar to work previously honored with a prize, regardless of the content of the heat theorem. Euler considered the determination of chemical affinity from thermal data, that is, the calculation of chemical constants, of the greatest value for chemistry. In his view, the formulation of a new thermodynamic law belonged to the inventions and discoveries which have "brought humanity the greatest service," from the standpoint of both chemistry as well as physics. He thereby insisted that Nernst's work fulfilled not one but all of Nobel's stipulations.

In a rather crafty rhetorical move, Euler quoted from Arrhenius's 1900 *Textbook of Theoretical Electrochemistry,* in which Arrhenius had asserted that Nernst's 1889 theory of electromotive forces had been applied much more frequently to electrochemical and chemical problems than that of Helmholtz. This he contrasted with Arrhenius's evaluation report of 1911. Without directly referring to the change in relationship between Nernst and Arrhenius after 1900, Euler thereby openly reminded the academy of personal disagreements that must have colored Arrhenius's evaluations. That the strained relationship between Nernst and Arrhenius had become a matter of open record is also confirmed by a very early biography of Emil Fischer. Fischer had expressed his admiration for Nernst in the nomination for a Nobel Prize, and in a letter to Richard Willstätter had written that the mathematician Ciamician and Nernst were the only ones recommended by him who did not receive the prize – "Nernst primarily due to persoanl disagreements."[87]

In the second part of his report, Euler discussed the heat theorem and related experiments, including Nernst's own results and his revision of older data from other authors. He listed specific examples, such as the transformation of rhombic-monoclinic sulfur, the free-energy and internal-energy curves for the Clark element,[88] and others, and concluded that

86. App. 1 to NK protocols, 19 May 1921, received 30 April 1921. Part A on electrochemistry, 22 pages; part B on Nernst Heat Theorem, 21 pages. Euler also submitted an abstract to be read to the academy, but made no direct reference to Arrhenius's memoranda. App. B. to NK protocols, 8 September 1921.

87. Hoesch, Kurt. *Emil Fischer. Sein Leben und sein Werk.* (Berlin: Verlag Chemie, 1921).

88. The Clark element is an example of the application of Nernst's theorem on the calculation of electromotive forces from thermal data. The common example is the reaction of zinc with sulfuric acid in water, resulting in the hydrated zinc sulfate and mercury deposit. The Clark element at temperatures below $-7°C$ was considered a condensed

"the heat theorem has been confirmed (with few exceptions such as the melting of water)[89] for condensed systems." He also underscored the importance of the Nernst chemical constants. Finally, he expanded on the significance of the work for chemistry:

> Professor W. Nernst has developed two major programs of extreme importance for chemistry, each worthy of a Nobel. However, since one group of researches was performed more than thirty years ago, it seems to me most appropriate to reward him with one of this year's available prizes for his thermochemical work, especially his new heat theorem, since his thermochemical works are much more general and of greater significance than his contribution to electrochemistry.[90]

The chemistry committee decided on 19 May to propose Nernst for the 1920 prize "for his thermochemical work."[91]

However, Arrhenius prepared one last report addressed to the physics committee, dated 14 July 1921,[92] in which he tried to refute all of Euler's arguments. He took up Nernst's concept of solution pressure in osmotic theory and conductivity studies, and declared that its value had been greatly exaggerated by Nernst's admirers, and that he himself considered the old electrolytic methods satisfactory without introducing the concept of solution pressure, which had often been used but which carried little practical relevance. Arrhenius thus harked back to issues that had divided him and Nernst twenty years earlier. His position indicates to what extent Nernst's criticism regarding the limited applicability of Arrhenius's dissociation theory to strong electrolytes must have wounded Arrhenius's pride. On the heat theorem, Arrhenius reiterated his view that it could not be applied exactly, that its usefulness for chemistry was limited, due to the

---

substance, and the theorem was applied. The presence of ice, however, complicated the calculation of the free- and internal-energy curves, which did not correspond to the observed data and for which correction factors had to be introduced. See F. Pollitzer, *Die Berechnung chemischer Affinitäten nach dem Nernst'schen Wärmetheorem* (Stuttgart: F. Enke, 1912).

89. Water is one of the few examples for which discrepancies of the order of a few calories per degree Kelvin per mole between thermochemical and statistical thermodynamic entropies have been found. In all cases, the thermochemical entropy was found to be too low. These are examples in which the freezing produces a glass and not a crystalline solid. See A. W. Adamson, *A Textbook Physical Chemistry* (New York: Academic Press, 1973), pp. 211–12. The problem of water was a legitimate topic for disagreement, even in the 1920s. The crystalline structure that exposed the complex nature of ice has only recently been analyzed, thanks to neutron diffraction studies and more precise knowledge of interatomic bonds.

90. Euler, cf. n. 86, p. 21.   91. NK protocols, 19 May 1921, p. 4, par. 5.

92. App. A. to NK, 8 September 1921, 11 pages. The document is dated 14 July 1921.

difficult measurements of specific heats. He again quoted H. R. Kruyt's criticism, and – tendentiously out of context – more recent work by Taylor.[93] He wrote that Nernst's theorem was rather a mixture between Einstein's formula and earlier researches. He ascribed recent technological successes of chemical industry to van't Hoff's Law of Equilibrium Displacement, rather than to Nernst's theorem.[94]

Two months later, Euler refuted Arrhenius's criticism that Nernst had based his revision of the law of Dulong-Petit on measurements performed earlier by Weber, Behn, and Trowbridge. Euler wrote that "Nernst's precision measurements of the atomic-heat curves in the most interesting temperature domain, near absolute zero, have been secured experimentally, and it was he who essentially investigated the deviations from the Dulong-Petit law."[95] He admonished Arrhenius for basing his objections on Kruyt, "a minor authority." Arrhenius had also denied the practical importance of the theorem by attributing the innovations in Haber's process of ammonia production to the invention of appropriate catalysts and the design of resistant reaction vessels. Euler emphatically replied that "no factory would have dared approach the task of overcoming extraordinary difficulties" without a secure "theoretical evaluation of the production yield that could be achieved."[96] He dismissed Arrhenius's charges that Nernst's theorem required tedious calculations, since time consuming procedures were a matter of varying judgment. Finally, he

---

93. H. S. Taylor and G. St. John Perrott, "The Thermochemical Data of Cadmium Chloride and Iodide," *JACS* 43 (1921): 484–93. Arrhenius referred to p. 493, where the authors recommended that "further specific heat data at low temperatures are required." The article, published in March 1921, quoted on p. 8.

94. Arrhenius, ibid., pp. 10–11.

95. App. B, 26 August 1921, to NK report, 8 September 1921, , p. 5.

96. Ibid., p. 7. The utilitarian argument in favor of the heat theorem has been subsequently used. In their textbook on low-temperature physics, M. and B. Ruhemann wrote in the mid-1930s:

> We shall not repeat here all the bulk of material that has been brought forward in the course of time to corroborate the theorem. *The most convincing proof of the applicability of Nernst's theorem can be seen in the increasing demand on the part of chemical industry of all countries for measurements of specific heats at low temperatures.* The material published every day in the scientific and technical journals, especially of the U.S.A., is enormous and even now it could well fill a volume. . . . *This popularity is to be explained by the applicability of the theorem to reactions with gaseous components.* These provide the majority of industrial processes, and although the theorem contains an explicit statement on condensed systems only, it nevertheless enables us to draw important conclusions concerning the chemical behavior of gases. [Emphasis added.] M. and B. Ruhemann, *Low Temperature Physics* (Cambridge University Press), 1937, pp. 184–5.

quoted Einstein, Planck, and Haber, as well as Max Born, who had written in 1920:

> The task of totally eliminating chemical measurements and to base
> the calculation of chemical affinities exclusively on physical quanti-
> ties has only been solved by Nernst, who thus opened a third period
> in the development of thermodynamics. The practical importance of
> the Nernst heat theorem and its relation to the fundamental princi-
> ples of nature (quantum theory) justify the designation of the theo-
> rem as the Third Law of Thermodynamics.[97]

On 8 September 1921, the chemistry committee unanimously recommended Nernst for the prize, "for his thermochemical work." On 10 November 1921, the academy decided to award Nernst the Nobel Prize in
chemistry for the year 1920, and to reserve the prize for 1921.[98]

Epilogue

It was ironic that the two most unlikely candidates, who had been nominated for many years, should eventually receive the prize simultaneously:
Anatole France's award had been expected yet delayed for a long time
since, as the German papers liked to report, the committee in Christiania
"does not like revolutionaries."[99] Walther Nernst, on the other hand, was
widely identified with German gas warfare, a detail which did not go un-
noticed in the non-German press.[100] At the time, and for several years to
come, German scientists and intellectuals were excluded from international meetings; scientific cooperation ceased, as had visits, correspondence, and student exchanges. For a time, Nernst feared the wrath of an

97. Euler, ibid., p. 9. "This, indeed, is a great achievement – the determination of chemical
    processes by purely physical measurements. It is a fusion of chemistry with physics,
    which apparently leaves nothing to be desired." Born, 1923, p. 51.
98. NK protocols, App. D., p. 9. In this first version, the phrase "for his heat theorem" was
    deleted and "for his thermochemical work" added, all in MS. In App. E., typewritten,
    only the latter phrase was included. The protocol was signed by all members of the NK
    for chemistry: O. Hammarsten, P. Klason, O. Widman, A. G. Ekstrand, H. G. Söder-
    baum, and also H. v. Euler.
99. *Berliner Tageblatt* excerpt, n.d., Archiv der Bunsengesellschaft, Darmstadt.
100. The *New York Times*, 13 November 1921, announced:

> NOBEL PRIZE WINNER MADE GERMAN GAS – Professor Nernst, How-
> ever, Was Rewarded for His Work in Electro-Chemistry. Professor Walter
> Nernst of the University of Berlin, who received the Nobel prize of 1920 for
> chemistry, was the head of a scientific staff which produced poison gas for Ger-
> many during the war.

international tribunal. He had signed the infamous declaration of intellectuals supporting Germany's war aims, while Anatole France belonged to the isolated French intellectuals who had not slipped into chauvinism.

The president of the Nobel Foundation could not ignore these contradictions. The tenth of December 1921 was the twenty-fifth anniversary of Alfred Nobel's death. A quarter of a century demanded a certain examination of collective conscience: Had the Nobel Foundation done justice to Alfred Nobel's wishes? The answer could only be discouraging:

> For the realization of Nobel's wish of establishing peace, our efforts weigh less than a feather, and one would have to delve deep into the human past to find life conditions as desolate as those which reign at the present hour. Alfred Nobel's initial intentions had been that the Nobel Foundation would seek out some dreamer, some unknown and obscure scientist or poet, and provide him with the economic means necessary for a fruitful development of his discoveries for the benefit of humanity or to continue a literary career endangered by misery or daily worries. But Nobel's intentions came to naught because obscure geniuses cannot be empirically discovered.

Therefore, the president conceded, the prizes were often awarded to already recognized celebrities in the world of science and arts.

However, the secretaries of the Royal Swedish Academy of Sciences and of the Swedish Academy of Arts presented the year's awards as an emblem of cooperation between the two former continental enemy nations. The official publications of the Nobel Foundation, therefore, recorded that at the end of the ceremonies, a moving episode took place: The frail octogenarian Anatole France, after having been assisted back to his seat, advanced toward Nernst and warmly took his hand, holding it firmly and insistently, a sign of national reconciliation to which the attending public responded with warm applause.

On 10 December 1921 at 5 o'clock in the afternoon, the King of Sweden and his family were greeted with the royal anthem upon entering the grand hall of the Swedish Academy of Music, sumptuously decorated with baroque flags and fresh flowers. In attendance were members of the diplomatic corps, high military and civilian dignitaries, university professors and students, and representatives of the academies, of the arts, sciences, and the press. A goodly number of ladies had been invited as well. On the seven-part afternoon program were the performance of four musical pieces by Swedish composers, an opening speech by the president of the administrative council of the Nobel Foundation, and the award of two Nobel Prizes: the first, in chemistry, to Professor Walther Nernst of Berlin; the second, in literature, to Anatole France. The permanent secretary of the Academy of Sciences concluded the prize presentation as follows:

> Herr Geheimrat Nernst!
> The discovery of fire, which even during classical antiquity had been ascribed to Prometheus, is one of the oldest and most important of all discoveries. For a long time, chemists have diligently sought after the suspected relationship between the liberation of heat and chemical affinity in the combustion of coal and other transformations. Due to your work, this relationship has been found, for which you have exploited with outstanding clarity your masterful experimental researches on specific heats and chemical equilibria. With the aid of the heat theorem discovered by you, it has become possible to calculate from heats of reaction and specific heats, on the one hand, chemical affinities, on the other the equilibrium of unknown reactions. The academy of science has decided to award you, in recognition of your thermochemical researches, the Nobel Prize in chemistry.

Nernst's Nobel Prize odyssey exemplified the growing appreciation of his work by both chemists and physicists. The force of Nernst's argument had gradually emerged, and his insights had been substantially worked out over the decades that had elapsed since 1906. Even those among his cohorts who had deplored his sometimes abrasive nature had come to recognize the corpus of his contributions to both physics and chemistry and the central role that he had played in the German academic environment. The prize eventually came as a validation of the new directions toward which Nernst had impelled many of his students, and of the untiring ingenuity with which he had found fundamental scientific topics on which to work.

Conclusion

In 1921, Nernst became more and more convinced that he had found a solution to the dreaded "heat death" consequence of the Second Law of Thermodynamics. In response to the challenge he had set himself upon reading Boltzmann several decades earlier, he stated emphatically:

> [E]very scientific theory of the cosmos will assume that, in total contradiction to the [heat death] consequence of thermodynamics, the universe is in a stationary state, that is, that on average as many stars disappear due to extinction as new ones light up.[1]

Foreshadowing the steady-state cosmology that was to emerge in the mid-1950s, Nernst presented his astrophysical theory in a popular lecture series at the Prussian Academy of Sciences on 19 February 1921 and again, in an expanded version, to the Vienna Engineers' Association (Wiener Ingenieurverein) and the Prague Urania society. These lectures gave him a welcome opportunity to discuss his ideas with various astronomers and to "diminish the danger of gross mistakes which generally occur when a scientist ventures into fields removed from his own specialty." He felt that physical chemists were not the worst prepared scientists to form a judgment on physical and chemical aspects of astronomical questions. With his characteristic love for diminutives, Nernst referred to this essay as a *"kleines Schriftchen"* – indeed a small volume of sixty-three pages in comparison to his textbook of chemistry which, by 1921, had reached nine hundred pages.[2]

1. "Denn ein Zweifel daran, dass obige Konsequenz des zweiten Wärmesatzes von höchster Unwahrscheinlichkeit ist, kann wohl kaum ernstlich gehegt werden; vielmehr wird jede naturwissenschaftliche Theorie des Kosmos davon ausgehen müssen, dass ganz im Gegenteil zu der erwähnten Konsequenz der Thermodynamik das Weltall sich in einem *stationären* Zustande befindet, dass also im Mittel ebensoviel Sterne im Kosmos durch Erlöschen ausscheiden, wie neue erglühen." Nernst indicated in his book that he had initially put forth the hypothesis of a steady-state universe in 1912, in a lecture to the Naturforscherversammlung in Münster, entitled "Zur neueren Entwicklung der Thermodynamik," which he reproduced in the text. Nernst, *Das Weltgebäude im Lichte der neueren Forschung* (Berlin: J. Springer, 1921), p. 1.
2. W. Nernst, *Theoretische Chemie*, 18th ed. (Stuttgart: Ferdinand Enke, 1921).

Nernst vividly remembered the impact that Boltzmann's resigned scepticism to one of the most pressing and disturbing issues in late-nineteenth-century cosmology and philosophy had made on him as a 20-year-old student. It is quite probable that Nernst's reminiscences accurately reflect the motivation for his consistent interest in fundamental thermodynamic problems. Boltzmann's influence looms large in Nernst's career, not only in his atomism but in the tolerant and antidogmatic philosophy and history of science transmitted to many of those who descended from the Nernst school. Nernst's versatility in physics and chemistry, and his investigations of the structure of matter, the nature of electrical conductivity, chemical kinetics, and thermochemistry – coupled with keen interest in instrumentation and the application of experimental results to practical purposes – represented the model of a scientist less interested in fundamental unifying theories than in establishing analogies and relationships among sometimes remote fields of inquiry.

The early 1920s marked Nernst's swan song to science. His circle of colleagues and friends was steadily disintegrating. And so was Germany.

As rector of the university between 1921 and 1922, Nernst vehemently protested the brutal murder of his friend and associate, Walther Rathenau, and preached moderation and tolerance to his students. He also instituted rather harsh regulations against the Communist student organizations. Shortly after having been awarded the Nobel Prize, he became for a brief time the president of the Physikalisch-Technische Reichsanstalt. His years in office were difficult. Rampant postwar inflation and a straitlaced new government seriously impaired the conduct of science. Nernst enjoyed the position of president, but he resigned after only two years, in 1924, and returned for a decade to the university, where he assumed the directorship of the prestigious Physical Institute as the successor to Heinrich Rubens. He there turned to yet other new and exploratory research fields, such as photochemistry and astrophysics, although he devoted much attention to developments in electrochemistry. The low-temperature experimentation was carried on by Francis Simon, first in Breslau and later at the Clarendon Laboratories at Oxford. In the years after the war, additional instrumental and financial resources led to a shift of the low temperature work from the university laboratories to the PTR.

Nernst was briefly courted by certain political circles of the new Weimar Republic. After the signing of the "Peace of Berlin" between the United States and Germany on 25 August 1921, the search for a suitable ambassador in the new political climate was headed by the "unusually influential personnel director, minister Schlüter, who on such occasions would preferentially look among the outsiders." For the Washington position, Schlüter suggested to Edgar von Haniel, then secretary of state in the foreign office, the name of Nernst, whom he appreciated for his industrial inventions and knowledge of economic issues. It seems that both the foreign minister, Rosen, and the Reich chancellor, Wirth, agreed to offer Nernst the position. "The great scholar of small stature" was extremely surprised at this proposition. Although evidently pleased, Nernst immediately and regretfully refused the offer. He felt that he could not fulfill one of the most important presuppositions for successful negotiations in Washington: the English language, with which he was unfamiliar. Acquiescing to the secretary of state's energetic reassurances that this deficiency had no bearing in light of Nernst's international reputation and personality, and that professional diplomats would stand at his side, Nernst was persuaded to give the offer a second thought. At the next meeting, however, Nernst returned only to decline the honor for the same reasons. "What almost moving modesty reveals itself in this attitude," gushed an eyewitness. "In retrospect it must evoke sad admiration, at a time when almost all international contacts took place through interpreters."[3]

Nernst lavishly entertained colleagues, in the company of political leaders of the Weimar Republic, such as Gustav Stresemann, then foreign minister, and C. H. Becker, then minister of culture. He also enjoyed an extended cruise around the Mediterranean, alighting in Baalbeck, Jerusalem, Cairo, Athens, and Constantinople.[4]

Nernst continued to dabble. In the early 1930s, he invented a new kind of piano that lacked a sounding board and was appointed with an electrical amplifier. In cooperation with the respected piano manufacturers Bechstein and with the Siemens firm, Nernst proceeded to develop this "radical invention." The "thoroughly tried instrument for universal use, not an experimental curiosity" was placed on the market in 1931. Nernst referred to the piano as "both a musical instrument and a scientific instrument of precision, because its tone production is completely controlled."

The Nernst-Bechstein-Siemens, as it was called, looked like a baby grand piano. It was regarded as the "first fundamental structural departure since the piano came into existence." The sounding board was replaced by an electrical amplifying device, whose strings were arranged radially in sets of five and passed under magnets. Smaller-than-usual hammers produced the strings' vibrations, which in turn induced an electric current that ran through the amplifier and emerged from a loudspeaker. "Electrical control allows tones to be held indefinitely and makes them swell or die away

3. *Frankfurter Allgemeine Zeitung*, 19 July 1964. Letter from Dr. Georg Ahrens, a retired diplomat, from Wiesbaden, wrote to the paper after having read their "beautiful commemoration" of Nernst.
4. Nernst to Quesada, 15 April 1929, SBPK, Handscriftenabteillung Autogr. 1601/1.2.

gradually," Nernst explained to the press, adding that unlike the bass "of the ordinary piano, that sounds overtones almost exclusively," the new instrument brings out fundamental tones. Speaking of the upper registers of the ordinary piano, Nernst told the journalists: "My friend Einstein, who, you know, is very musical, says they sound like porcelain getting smashed." The electrical piano was purported to bring even "greater improvements in the upper registers, which for many years baffled the efforts of piano manufacturers to make them mellow."

Nernst himself "disclaimed musical competence," although he had trained his ear by listening to good music at Bayreuth and elsewhere. Despite being married to the daughter of an extremely gifted musician, Nernst, unlike Einstein, thought of himself as essentially unmusical: "I approached the problem altogether from the viewpoint of physics. If I have been musically influenced the afflatus must have come from my residence [as president of the PTR] in what formerly was the home of the great Helmholtz, father of musical acoustics." Nernst had essentially invented the forerunner of the modern electronic synthesizer: It was expected to function "as spinet, harmonium, phonograph and radio receiver. The last two features permit of unusual combinations, such as switching in on an orchestral concert while playing the piano. Control mechanisms permit adjusting the volume of tone to any size room and acoustics and varying the quality of tones according to one's desire." However, the price of the new instrument was $650, a small fortune in the depression year 1931.[5]

In the fall of 1933 at age 69, Nernst retired from academic duties. He spent the rest of his life on his estate in Eastern Prussia, and gradually gave up his residence in Berlin. The sumptuous villa Am Karlsbad was abandoned in favor of a series of small apartments. He continued to attend the meetings of the Prussian Academy of Sciences but was increasingly isolated as the result of political developments. He rejected Nazism, and his retirement was in many ways a political statement against the unjust treatment that many of his colleagues, both at the university and at the Kaiser Wilhelm institutes, were made to suffer.

Two of his three daughters, Angela and Hilda, married to Jewish husbands, left Germany. After the dramatic loss of his two sons in the war, the separation from his daughters and grandchildren was an almost unbearable pain. His health declined precipitously. The first signs of serious disease appeared in April 1939 when he suffered a stroke.

Nernst and his wife Emma visited their daughter Hilda for the last time on an airplane trip to England in the summer of 1938. The other daugh-

5. *New York Times*, 30 August 1931. "New German Piano Lacks Sounding Board. Electrical Amplifier Takes Its Place – Instrument Is Said to Have Superior Tone Quality."

ter, Angela, settled in Sao Paolo in July 1939. During the last years, Nernst solicited the good services of his old friend Wilhelm Palmaer, who would serve as a go-between mailbox for the separated Nernsts, unable to correspond with one another directly after the beginning of hostilities in 1939. Thus he wrote to Palmaer:

> Dear Friend,
>   One of our daughters lives with her children and her husband, of course, in London. Since it is probable that from now on letters from us here will not be forwarded to her, we would like to ask you for friendly permission and only in rare cases, to forward letters addressed to my daughter. If I do not hear any negative answer from you, I will proceed thus and thank you for your trouble . . . [6]

Palmaer generously forwarded mail, while other "unknown friends" gave the Nernsts news about the family "through other channels."[7]

Nernst spent the last year of his life on his estate, where he celebrated his seventy-fifth birthday in a small family circle, surrounded by a few devoted friends, such as Max von Laue. A press release by the Pressestelle des Reichs und Preussische Akademie announced that congratulations on behalf of the academy had been conveyed to Nernst by Peter Debye. It listed "his electrolytic work in youth, the heat theorem in manhood, his recent ideas on the development of the cosmos which have had almost prophetic character, and [his] newly confirmed ideas on the transformation of radiation into matter and vice versa." But Nernst's title was given as that of an emeritus "Professor at the Technische Hochschule Berlin," rather than the University.[8] He had already been abandoned to the vagaries of faulty history, while he himself did not feel "strong enough to write his memoirs during his last days. . . ."[9]

Nernst died on 18 November 1941, at age 77. He was cremated a week later, and his ashes were interred in Berlin. After the war, his remains were moved to Göttingen, the city of his greatest scientific achievement. About his years in that city, Nernst had written to Arnold Sommerfeld three years earlier:

> When I think back to Göttingen, then Felix Klein in particular appears in front of my eyes, and you certainly think of him with the same devotion and gratitude; Felix Klein, Harnack, and Emil Fischer

6. Nernst to Palmaer, 3 September 1939, Palmaer Papers, KVA.
7. Nernst to Palmaer, 18 November 1939, Palmaer papers, KVA.
8. Pressenotiz, 24 Juni 1939, Zentrales Archiv Historische Abt. II Akten der Preuss. Akademie d. Wiss., Personalia der Mitglieder Signatur II-III.85, AAW.
9. Emma Nernst to E. Riesenfeld, 2 April 1942, 2 pages, Riesenfeld Papers, KVA.

Nernst lectures in the late 1920s

are for me the ideal figures of the bygone German professorate, while among the higher administrators Althoff and your father-in-law, each in his own way, have a unique place in my memory; all of them columns of a vanished splendour![10]

Nernst died at the opening of the "physicists' war." He had belonged to a different generation, one with its own momentous struggles. Enthusiastically at first, a generation of chemists had aided in the construction of weapons of unprecedented destructive force, one that was unimaginable before 1914, and which no one in 1941 thought would be so soon surpassed. At the end of World War I, Arrhenius had written to Ostwald, regarding the Nobel Prizes, that "the chemists have been somewhat pushed into the shadows, since they have almost all been busy during the war either with explosives or with deadly gases and incendiary bombs. In the future this will not exactly be one of chemistry's honors."[11]

The physicists' story in the second half of our century often obscures

10. Nernst to Sommerfeld, on Sommerfeld's 70th birthday, from Zibelle, 3 December 1938. Arnold Sommerfeld Nachlass, Mappe 70th, 75th and 80th birthday, DM.
11. Arrhenius to Ostwald, 17 December 1918, 2 pages, Arrhenius Papers, KVA.

the significance of chemists and chemistry in the quantum revolution, in the practice of science, and in the industrial and technological concerns of the preceding decades. Nernst and his fellow chemists had been severely criticized for their work on chemical weapons. The physicists' version came to dominate our understanding of the history of physical science. The impact that physics had already made upon the twentieth century was expressed in the *New York Times* in one of many editorials and obituaries published in the United States upon Nernst's death:

> To scientists Nernst stands beside the great in physics – beside such men as Boltzmann, Planck and the rest of the bright band that gave physics, and especially atomic physics, new purpose and direction. For Nernst goes down as the formulator of what is now called the third law of thermodynamics and as a founder of physical chemistry.[12]

Even Einstein helped promulgate, in his typically terse and simple prose, this version of Nernst's life and work. In 1941, Einstein described Nernst as an ascendant from Arrhenius, Ostwald, and van't Hoff, "as the last of a dynasty which based their investigations on thermodynamics, osmotic pressure and ionic theory." Einstein saw Nernst "essentially restricted to that range of ideas" until 1905, and referred playfully to Nernst's "somewhat elementary . . . theoretical equipment." He was amused by Nernst's invention of "the witty null-method of determining the dielectric constant of electrically conducting bodies by means of Wheatstone's Bridge (alternating current, telephone as indicator, compensating capacity in comparison-bridge branches)." But Einstein considered Nernst's work in electrochemistry as merely an elaboration of "principles which had already been known before Nernst." Without any further detail Einstein stated laconically:

> This work led [Nernst] gradually to a general problem which is characterized by the question: Is it possible to compute from the known energy of the conditions of a system, the useful work which is to be gained by its transition from one state into another? Nernst realized that a theoretical determination of the transition work A from the energy-difference U by means of equations of thermodynamics alone is not possible. There could be inferred from thermodynamics that, at absolute zero, the temperature of the quantities A and U must be equal. But one could not derive A from U for any arbitrary temperatures, even if the energy-values or differences in U were known for all conditions. This computation was not possible until there was introduced, with regard to the reaction of these quantities under low temperatures, an

12. *New York Times*, 20 November 1941, op-ed page.

assumption which appeared obvious because of its simplicity. This assumption is simply that *A* becomes temperature-independent under low temperatures. The introduction of this assumption as a hypothesis (third main principle of the theory of heat) is Nernst's greatest contribution to theoretical science. Planck found later a solution which is theoretically more satisfactory; namely, the entropy disappears at absolute zero temperature. From the standpoint of the older ideas on heat, this third main principle required very strange reactions of bodies under low temperatures. To pass upon the correctness of this principle, the methods of calorimetry under low temperatures had to be greatly improved. The calorimetry of high temperatures also owes to Nernst considerable progress.[13]

Although the English version of this obituary did a great disservice to both the author and subject of the above paragraph, nevertheless, the problems that this book has attempted to elucidate are all implicitly raised by Einstein's essay. And since biographies of Nernst are rare, historians have relied heavily on the above, rather flawed, chronological and explanatory schema. How did Nernst move from well-trod territory in solution theory to issues of chemical reactivity and work in solid and gas reactions? Did high-temperature calorimetry indeed follow only as an afterthought?

In retrospect, one may well assent to the notion that the injection of a certain physicalist thought into thermochemistry, electrochemistry, and theoretical chemistry, broadly defined, prevailed long before mid-century, and that the physicists' era dawned long before the Manhattan Project of the 1940s. Whatever identity crises the generations of chemists experienced, these were resolved by the end of World War I. By 1913, physical-chemical textbooks contained what is now the accepted chronology for physical chemistry, one that rendered obsolete the historical postulations of Ostwald, van't Hoff, and Arrhenius, but one that also cast its own tradition, carved out its own space in the expanding world of science.

By the time a new generation of scientists was entering an essentially changed academic environment, the shift in emphasis from chemistry to physics had been complete. Lavoisier, electrochemistry, and heat theories retreated into the distant past. No longer was physics a remote ancestor, but rather a direct and immediate influence at all stages on the history of physical chemistry. Time was measured by major theoretical landmarks, rather than by experimental discovery or synthetic ingenuity, punctuated

13. Albert Einstein, *Out of My Later Years,* rev. repr. ed. (Westport, Conn.: Greenwood Press, 1975). Originally published 1950 by the Philosophical Library, New York.

by moments in history at which various "groups of ideas" entered "chemistry from physics."[14]

The historical, genealogical, tradition-seeking temperament can be found in many, much more recent, new scientific fields. The introductory remarks in many inaugural volumes of scientific journals attest to the tension inherent in the process of scientific specialization. The aspiration to be innovative, to join high-powered research at the frontiers of knowledge, counterbalanced by the need to justify a new scientific enterprise in terms of a long established ancestry, compels scientists to produce programmatic statements of both rhetorical and practical value. Specialization, one of the most salient features of modern scientific culture, is thus fraught with perils for the identity of the specialist.

The years 1890 to 1914, however, constitute one of the most dynamic periods in the history of modern physical sciences. The rapid succession of intense experimental and observational discoveries took place simultaneously with fundamental changes in the theoretical conceptions regarding the structure of matter and radiation, in particular the internal constitution of molecules and atoms. These changes did not come about as the result of any single or particular experimental observation, but were part of a gradual process during which internal inconsistencies, as well as discrepancies with observational data, led to the formulation of the quantum hypothesis. The lengthy process of refinement and incorporation of quantum concepts was accompanied by substantial transformations in the practice of not only theoretical and experimental physics but also that of related disciplines, such as physical and theoretical chemistry. A receptive audience for the quantum hypothesis was created by physical chemists, in particular Nernst and his collaborators, who extended thermodynamical and chemical investigations to the extremes of the temperature range. The fact that low-temperature physics is the domain primarily governed by quantum phenomena was recognized, albeit tentatively, by Nernst as early as 1908, mainly due to insights gained in the study of the behavior of solids at low temperatures. The refinements of quantum treatments provided by Einstein, Nernst, Lindemann, Debye, Born, and v. Kármán in the years 1907 to 1912, reinforced by electrical conductivity studies at low temperatures in the Cryogenic Laboratory at Leiden under the direction of Kamerlingh Onnes, provided the backdrop for significant advances in quantum physics after the hiatus brought about by World War I. These particularly "discovery intense" years preceded the postulation of

---

14. See, for example, the monograph of Karl Jellinek, *Physical Chemistry of Homogeneous and Heterogeneous Gas Reactions with Special Consideration of the Theory of Radiation and Quanta and Nernst's Theorem*, preface, n.p., 1913.

Bohr's atomic model, and the vitality of the discourse among various scientists indicates that the sources for the elaboration of the Bohr model may be more varied than previously thought.

The debate over the validity of the heat theorem as it related to the award of the Nobel Prize constitutes a study in the scientific, political, and personal dimensions of the scientific reward system.[15] The variety and coherence of certain methodological and instrumental characteristics of Nernst's work underscore the continuity in his approach toward seemingly disparate problems in electrochemistry, thermodynamics, and molecular and atomic physics and chemistry. The persistence of certain instrumental setups and apparatus that Nernst adapted for different purposes facilitated the transition from high-temperature investigations to low-temperature research. Poised at the interface of disciplinary domains, a central figure of interlacing networks, Nernst was in many ways ideally located, both intellectually and professionally, to promote actively the quantum hypothesis. His elaborations upon Einstein's formula and the experimental confirmations of the quantum theory of matter were a major catalyst in the process of the transformation of classical physical sciences.

This book has focused on two specific points: innovation and motivation. It aimed to illuminate how "neglected possibilities in the available conventions" were exploited. The late nineteenth century constituted one of the boldest experiments with many novel and simultaneous forms of perspective. It encouraged the proliferation of points of view, points of attack, new handles on well established subjects, "a ferment in which virtually every option . . . had a new chance at life." This was the case within physics, chemistry, and other related and blossoming scientific fields. Just as in art, inventions, projects, instruments, and ideas made for one purpose were adapted to new purposes.[16]

Electrical conductivity, black-body radiation, spectroscopy, and the study of chemical decompositions at the extremes of the temperature range were intimately related in the work of the Nernst group, since knowledge about the temperatures of gases at dissociation temperatures was accessible only

15. Controversies regarding the status of the heat theorem have not abated. See, for example, Martin J. Klein and Solomon J. Glass, "Gruneisen's Law and the Third Law of Thermodynamics," *Philosophical Magazine* 3 (1958): 538–9; Solomon J. Glass and Martin J. Klein, "Sublimation and the Third Law of Thermodynamics," *Physica* 25 (1959): 277–80; more recently, Irwin Oppenheim "Comment on 'An equivalent theorem of the Nernst theorem,'" *J. Phys. A: Math.Gen.* 22 (1989):143–4: "In a letter entitled 'An equivalent theorem of the Nernst theorem,' Yan and Chen purport to derive the third law of thermodynamics on the basis of the second law and the conclusion that heat capacities tend to zero as the temperature approaches absolute zero. The proof is faulty and the third law is an additional postulate in thermodynamics."
16. Varnedoe, pp. 49ff.

through estimates of heat radiation. Nernst's low-temperature work was initially meant to elucidate the shape of the temperature-dependency curve for specific heats. His concern was to explain why and at what temperatures certain substances begin to require ever greater amounts of energy in order to increase their radiation capabilities; and also to understand gas reactions in the high-temperature range of technical and industrial combustion processes. It is, of course, an important feature of such research that all properties of interest are interconnected. Therefore, a better understanding of the behavior of specific heats provided vital information about molecular constitution and chemical equilibrium. Conversely, establishing when a reaction has reached equilibrium could furnish information about the composition of a system.

Nernst wished to find an electrical filament that would withstand high temperatures and fluctuations in electric current, yet provide a steady, bright, light. He needed to find the best conducting materials, which would reach incandescence without decomposing too fast in the alternate or direct currents of varying frequency employed at the time. He needed to establish a relationship between "heat and energy," as he explicitly stated, for these solid electrolytic conductors, a classical problem that involved knowledge of the composition of the alloys he was testing, including the purity of the materials used. Therefore, the origins of Nernst's heat theorem do not reside in the openly definable program of predicting the direction and rate of chemical reactions, although these aspects were intimately connected. Nor did they simply or naturally fit into the oft-cited physical-chemical research tradition of Wilhelm Ostwald. Instead, Nernst's outstanding experimental ingenuity and theoretical boldness allowed him to extend knowledge, methods, skills, and modes of asking and answering questions acquired early in his career into electrical conductivity studies, electrochemistry, and the mechanical power of electromotive forces, molecular-weights determinations, gas reactions, thermometry, the solid state – all subjects that at first seem unrelated but are seen to flow gradually and in a complicated fashion from one into the next.

Moreover, what we have come to see as a rather tightly interconnected *practice* applies equally well to important *theoretical* developments. The work on electrolytic conduction carried out by Nernst and his collaborators during the 1890s shows that Weberian particulate theories of electric conduction, elaborated upon by Wilhelm Giese in Berlin and his student Eduard Riecke in Göttingen, were substantially influenced by the "ionic" theory of the physical chemists. In the transition from macroscopic Maxwellian electromagnetism to a "true" microphysics, Nernst provided theoretical physicists a concrete examination of the similarities and discrepancies among electrical conductivity in metals, solid electrolytes, and liquid electrolytes. Therefore, certain aspects of the transition to microphysics were accomplished precisely through Nernst's experimental interest in his "lamp."[17]

The study of modern science leads to the recognition that technology, the rise of the university system, standardization, communication, the radio, telephone, railroad, and automobile made innovation possible. The post-1895 revolution in the physical sciences happened individually, collectively, and in a variety of locales. It did not occur only because industry, the state, and the rising middle classes impinged and "led" to this transformation; or because science and scientists "contributed" to these social, political changes. Rather, it happened because particular experimental, theoretical, technological agendas were developed by specific people, research groups, countries, in unique ways because of distinctive, special circumstances.

17. Jed Z. Buchwald, *From Maxwell to Microphysics* (Chicago: University of Chicago Press, 1985).

Bibliography

Archival Sources

England

Bodleian Library, Oxford University.
  Kurt Mendelssohn Papers.

France

Bibliothèque Nationale, Paris.
  Manuscripts Collection.
Institut de France, Bibliothèque et Archives, Paris.
École Supérieure de Physique et de Chimie Industrielles de la Ville de Paris.
  Paul Langevin Papers.
  Marie Curie Papers.

Germany

AAW: Archiv der Akademie der Wissenschaften, Berlin.
  Wilhelm Ostwald Nachlass, Sign. 2126.
  Hermann von Helmholtz Nachlass.
  Zentrales Archiv Historische Abteilung. II. Akten der Preussischen Akademie der Wissenschaften.
  Personalia der Mitglieder Signatur II–III.85.
  Die Schenkung vom Hr. Boettinger: Mesothorium.
MPG: Bibliothek und Archiv zur Geschichte der Max-Planck Gesellschaft, Berlin.
  Generalverwaltung, Akten.
  Hans Hartmann Nachlass.
  Sammlung Kangro.
  Otto Hahn Nachlass.
  Karl F. Bonhoeffer Nachlass.
  Fritz Haber Nachlass.
  F. A. Paneth Papers.
  KWI für Chemie.

SbPK: Staatsbibliothek Preussischer Kulturbesitz, Berlin.
  Handschriftenabteilung
  Autographensammlung
SPK: Geheimes Staatsarchiv Preussischer Kulturbesitz, Berlin and Meresburg.
PTB: Physikalisch-Technische Bundesanstalt, Berlin.
  Physikalisch-technische-Reichsanstalt. Unterlagen.
NSuUB: Archiv der Niedersächsischen Staats- und Universitätsbibliothek, Göttingen.
AUL: Archiv der Karl-Marx-Universität Leipzig.
  Walther Nernst, Personalakte. Sign. PA 773.
  Wilhelm Ostwald, Personalakte. Sign. PA 787.
Archiv der Bunsengesellschaft, Institut für Physikalische Chemie der Technischen Hochschule, Darmstadt.
Sondersammlung und Bibliothek des Deutschen Museums, München.
  Arnold Sommerfeld Nachlass. Bestand Bopp.

Holland

Archive and Library of the Museum Boerhaave, Leiden.
  H. Kamerlingh Onnes Papers.
  Manuscript Collection.
Archive of the University of Amsterdam.
  J. D. van der Waals Papers.
Kamerlingh Onnes Laboratory, University of Leiden.
  Kamerlingh Onnes Reprint Collection.

Sweden

KVA: Royal Swedish Academy of Sciences, Stockholm.
  Svante Arrhenius Papers.
  Wilhelm Palmaer Papers.
  Ernst Riesenfeld Papers.
NK: Protocols of the Nobel Komites for Chemistry and Physics.

United States of America

AIP: American Institute of Physics, New York.
  H. A. Lorentz Papers. Microfilm.
  AHQP Oral Interviews.
  Boston University.
  The Collected Papers of Albert Einstein.
  Library of Congress, Manuscript Division, Washington, D.C. Irving Langmuir Papers.

Printed Sources

Achinstein, Peter, and Owen Hannaway, eds. *Observation, Experiment, and Hypothesis in Modern Physical Science.* Cambridge, Mass.: MIT Press, 1985.

Adamson, A. W. *A Textbook of Physical Chemistry.* New York: Academic Press, 1973.

Albert Einstein – Hedwig und Max Born Briefwechsel: 1916–1955. Introduction by Bertrand Russell. Preface by Werner Heisenberg. München: Nymphenburger Verlagshandlung, 1969.

Albert Einstein–Michele Besso Correspondence. Edited by P. Speziali. Paris: Hermann, 1972.

Aris, Rutherford, H. Ted Davis, and Roger H. Stuewer. *Springs of Scientific Creativity: Essays on Founders of Modern Science.* Minneapolis: University of Minnesota Press, 1983.

Arrhenius, Svante. "Les oscillations seculaires de la Temperature." *Revue generale des Sciences: 1899; Philosophical Magazine* (1896): 267; *Naturw. Rundschau* 11 (1896): 328.

——— "Zur Berechnungsweise des Dissociationsgrades starker Electrolyte." *Zeitschr. f. phys. Chem.* 36 (1901): 28–40.

——— "Zur Berechnungsweise des Dissociationsgrades starker Electrolyte II." *Zeitschr. f. phys. Chem.* 37 (1901): 315–22.

——— "Die Serumtherapie vom physikalisch-chemischen Gesichtspunkte," *Zeitschr. f. Elektrochem.* (1904): 661–8.

Bachelard, Gaston. *Le nouvel esprit scientifique.* Paris: PUF, 1983.

Badash, Lawrence. *Kapitza, Rutherford, and the Kremlin.* New Haven and London: Yale University Press, 1985.

Bartel, Hans-Georg. *Walther Nernst.* Biographien hervorragender Naturwissenschaftler, Techniker und Mediziner. Vol. 90. Leipzig: Teubner, 1989.

Barus, Carl. "The Progress of Physics in the Nineteenth Century." *Congress of Arts and Science, Universal Exposition, St. Louis, 1904.* Boston, 1906.

Bennewitz, K. "Der Nernst'sche Wärmesatz." *Handbuch der Physik* 9 (1926): 141.

Bennewitz, K., and F. Simon. "Zur Frage der Nullpunktsenergie." *Zeitschr. Phys.* 16 (1923): 183–99.

Bensaude-Vincent, Bernadette. *Langevin, 1872–1946: science et vigilance.* Paris: Belin, 1987.

Bernal, J. D. *The Extension of Man: A History of Physics Before the Quantum.* Cambridge, Mass.: MIT Press, 1972.

Bernhard, C. G., E. Crawford, and Per Sörbom. *Science, Technology and Society in the Time of Alfred Nobel.* Oxford: Pergamon, 1982.

Biltz, W. "Zur Immunochemie. Eine Erinnerung an die Hauptversammlung der Deutschen Bunsengesellschaft 1904." *Zeitschr. f. angew. Chem.* 7: 169–70.

Bjerrum, Niels. "Über die spezifische Wärme der Gase." *Zeitschr. f. Elektrochem.* 17 (1911): 731–4.

Bodenstein, Max. "Gedächtnisrede auf Walther Nernst von Hrn. Bodenstein." *Jahrbuch d. Preuss. Akad. d. Wiss.* 1942 (1943): 140–2.

Boltzmann, Ludwig. *Wied. Ann.* 57 (1896).
   Festschrift Ludwig Boltzmann. Leipzig: J. A. Barth, 1904.
   Populäre Schriften. Leipzig: Amb. Barth, 1905.
   Vorlesungen über Gastheorie. 2 vols. Leipzig, 1896–8. Translated as *Lectures on Gas Theory* by S. G. Brush. Berkeley: University of California Press, 1964.
   Populäre Schriften. Edited with a Foreword by Engelbert Broda. Braunschweig: Fr. Vieweg, 1979.
   Leben und Briefe. Edited by Watler Höflechner. Graz: Akademische Druck-u. Verlagsanstalt, 1994.
Born, Max. *The Constitution of Matter. Modern Atomic and Electron Theories.* Translated by E. W. Blair and T. S. Wheeler. London: Methuen & Co., 1923. (From the 2nd rev. German ed., 1921. First ed. 1920.)
   Ausgewählte Abhandlungen. 2 vols. Göttingen: Vandenhoek und Ruprecht, 1963.
   My Life. Recollections of a Nobel Laureate. New York: Charles Scribner's Sons, 1978.
Born, Max, and Th. v. Kármán. "Über Schwingungen in Raumgittern." *Physik. Zeitschr.* 13 (1912): 297–309.
   "Zur Theorie der spezifischen Wärme." *Physik. Zeitschr.* 14 (1913): 15–19.
Bredig, Georg. *Über die Chemie der extremen Temperaturen.* Habilitationsvorlesung, 9 February 1901. Also published as *Physik. Zeitschr.* Sonderdruck II. Leipzig: Verlag Hirzel, 1901.
Broglie, Louis de. *La physique nouvelle et les quanta.* Paris: Flammarion, 1937.
Broglie, Maurice de. *Les Premiers Congrès de Physique Solvay, et l'orientation de la physique depuis 1911.* Paris: Editions Albin Michel, 1951. Cahiers de la collection Sciences d'Aujourd'hui, dirigés par André George.
Brush, Stephen. G. *The Kind of Motion We Call Heat: A History of the Kinetic Theory of Gases in the 19th Century.* 2 vols. Vol. 6 of *Studies in Statistical Mechanics.* Amsterdam/New York: North Holland, 1976.
   Statistical Physics and the Atomic Theory of Matter, from Boyle and Newton to Landau and Onsager. Princeton, N.J. : Princeton University Press, 1983.
Brush, Stephen G., and Lanfranco Belloni, eds. *The History of Modern Physics: An International Bibliography.* New York: Garland, 1983.
Buchwald, Jed Z. *From Maxwell to Microphysics: Aspects of Electromagnetic Theory in the Last Quarter of the Nineteenth Century.* Chicago: University of Chicago Press, 1985.
   The Creation of Scientific Effects: Heinrich Hertz and Electric Waves. Chicago: Chicago Unversity Press, 1994.
Bunge, Mario, and William R. Shea. *Rutherford and Physics at the Turn of the Century.* New York: Dawson and Science History Publications, 1979.
Cahan, David. *The Physikalisch-technische Reichsanstalt.* Ph.D. diss. Johns Hopkins University, 1980.
   "The institutional Revolution in German Physics, 1865–1914." *HSPS* 15 (1985): 1–65.
   An Institute for an Empire: The Physikalisch-Technische Reichsanstalt, 1871–1918. New York: Cambridge University Press, 1989.

Caneva, Kenneth L. "From Galvanism to Electrodynamics: The Transformation of German Physics and Its Social Context." *HSPS* 9 (1978): 63–159.

Cantor, G. N., and M. J. S. Hodge, eds. *Conceptions of Ether: Studies in the History of Ether Theories, 1740–1900.* New York: Cambridge University Press, 1981.

Casimir, Hendrik. "Superconductivity." *Proceedings of the International School of Physics 'Enrico Fermi':* Course LVII, *'History of Twentieth Century Physics.'* London, 1977.

Haphazard Reality: Half a Century of Science. New York: Harper & Row, 1983.

Chamberlin, T. C. "The method of multiple working hypotheses." *Science* 15 (1890): 92–6. Reprinted in Albritton, C., ed. *Philosophy of Geohistory.* Stroudsburg, PA: Dowden, Hutchinson and Ross, 1975, pages 126–31.

Chronik der Königlichen Friedrich-Wilhelms-Universität zu Berlin. Berlin: Berlin University, n.d.

Clay, J. "On the change with temperature of the electrical resistance of alloys at very low temperatures." *CPL* 107d (1908): 31–51.

Cohen, Ernst. *Jacobus Henricus van't Hoff. Sein Leben und Wirken.* Leipzig: Akademische Verlagsgesellschaft, 1912. Series *Grosse Männer. Studien zur Biologie des Genies.* Vol. 3. Edited by Wilhelm Ostwald.

Cohen, I. Bernard. *Revolution in Science.* Cambridge, Mass./ London, England: The Belknap Press of Harvard University Press, 1985.

Crawford, Elisabeth. "Arrhenius, the Atomic Hypothesis, and the 1908 Nobel Prizes in Physics and Chemistry." *ISIS* 75 (1984): 503–522.

The Beginnings of the Nobel Institution: The Science Prizes, 1901–1915. Cambridge: Cambridge University Press, 1984.

Arrhenius: From Ionic Theory to the Greenhouse Effect. Canton, Mass.: Science History Publications, 1996.

Crawford, Elisabeth, and R. M. Friedman. "The Prizes in Physics and Chemistry in the Context of Swedish Science: A Working Paper." In Bernhard et al., eds. *Science, Technology and Society in the Time of Alfred Nobel,* pages 311–31.

Crawford, Elisabeth, J. L. Heilbron, and Rebecca Ulrich. *The Nobel Population 1901–1937: A Census of the Nominators and Nominees for the Prize in Physics and Chemistry.* University of California, Berkeley: Office for History of Science and Technology; Uppsala University, Uppsala: Office for History of Science, 1987.

Curd, Martin V. *Ludwig Boltzmann's Philosophy of Science.* Ph.D. diss. University of Pittsburgh, 1973.

Curie, Eve. *Madame Curie.* Gallimard: Paris, 1938.

Dahl, Per F. "Kamerlingh Onnes and the discovery of superconductivity: The Leyden Years, 1911–1914." *HSPS* 15 (1984): 1–37.

"Superconductivity after World War I and circumstances surrounding the discovery of a state B = 0." *HSPS* 16 (1986): 1–58.

Debye, Peter. "Zur Theorie der spezifischen Wärme," *Ann. d. Phys.* 39 (1912): 789–839. In *The Collected Papers of Peter J. W. Debye,* New York: Interscience, 1954, pages 650–96.

de Haas, W. J., and E. C. Wiersma. *Physica* 1 (1934): 779–80; 2 (1935): 81–6, 335–40, 438.

Dewar, James. "The nadir of temperatures, and allied problems." *Proceedings Royal Society London,* 73 (1904): 244–51.

Dictionary of Scientific Biography. Edited by Charles C. Gillispie. 15 vols. New York: Scribner's 1970–8 (DSB).

Die physikalischen Institute der Universität Göttingen. Leipzig/Berlin: B. G. Teubner, 1906.

Dolby, R. G. A. "Debates over the theory of solution: A study of dissent in physical chemistry in the English-speaking world in the late nineteenth and early twentieth centuries." *HSPS* 7 (1976): 297–404.

"The Case of Physical Chemistry." In G. Lemaine et al. *Perspectives on the Emergence of Scientific Disciplines.* Mouton/Aldine, 1976, pages 63–73.

"Thermochemistry versus Thermodynamics: The Nineteenth Century Controversy." *Hist. Sci.* 22 (1984): 375–400.

Drennan, O. *Electrolytic Solution Theory: Foundations of Modern Thermodynamical Considerations.* Ph.D. diss. University of Wisconsin, 1961.

Drude, Paul. "Zur Elektronentheorie der Metalle." *Ann. Phys.* 1 (1900): 566–613.

Dubpernell, George, and J. H. Westbrook, eds. *Selected Topics in the History of Electrochemistry. Proceedings of the Electrochemical Society.* Vol. 78. Princeton, N.J.: 1978.

Duhem, Pierre. "Une science nouvelle. La chimie physique." *Revue Philomatique de Bordeaux et du Sud-Ouest* (1899): 205–19, 260–80.

The Aim and Structure of Physical Theory. Translated by Phillip P. Weiner. Princeton: Princeton University Press, 1954.

Einstein, A., and O. Stern. "Einige Argumente für die Annahme einer molekularen Agitation beim absoluten Nullpunkt." *Ann. d. Phys.* 40 (1913): 551–60. Submitted 5 January 1913.

Einstein, Albert. *The Collected Papers of Albert Einstein.* Vol. 1, *The Early Years: 1879–1902.* Vol. 2, *The Swiss Years: Writings, 1900–1909.* Edited by John Stachel. Princeton, N.J.: Princeton University Press, 1987–9.

"Die Plancksche Theorie der Strahlung und die Theorie der spezifischen Wärme." *Ann. d. Phys.* 22 (1907): 180–90.

"Zum gegenwärtigen Stand des Strahlungsproblems." *Phys. Zeitschr.* 10 (1909): 185–93.

"Eine Beziehung zwischen dem elastischen Verhalten und der spezifischen Wärme bei festen Körpern mit einatomigem Molekül." *Ann. d. Phys.* 34 (1911): 170–4.

"Zum gegenwärtigen Stande des Problems der spezifischen Wärme." In A. Eucken, ed. *Die Theorie der Strahlung und der Quanten. Verhandlungen auf einer von E. Solvay einberufenen Zusammenkunft* (30. Oktober bis 3. November 1911). Halle a.S.: Wilhelm Knapp, 1914, pages 330–52.

"Quantentheorie des einatomigen idealen Gases." *Berl. Ber.* (1925): 3–16.

Out of My Later Years. Revised reprint edition. Westport, Conn.: Greenwood Press, 1975. Originally published 1950 by the Philosophical Library, New York.

Elkana, Yehuda. "Two-Tier Thinking: Philosophical Realism and Historical Relativism." *Social Studies of Science* 8 (1978): 309–26.

Encyclopädie der mathematischen Wissenschaften. Vol. 4, *Mechanik.* Edited by F. Klein and C. Müller. Leipzig: Teubner, 1901–8.

Eucken, Arnold. "Die Molekularwärme des Wasserstoffs bei tiefen Temperaturen." *Berl. Ber.* (1 February 1912): 141–50.

——— ed. *Die Theorie der Strahlung und der Quanten. Verhandlungen auf einer von E. Solvay einberufenen Zusammenkunft* (30. Oktober bis 3. November 1911). Halle a.S.: Wilhelm Knapp, 1914.

——— *Lehrbuch der chemischen Physik.* Leipzig: Akademische Verlagsgesellschaft, 1930. [3rd ed. of Eucken, *Grundriss der physikalischen Chemie.*]

Eucken, Arnold, and F. Fried. "Über die Nullpunktsentropie kondensierter Gase." *Zeitschr. f. Phys.* 29 (1924): 36.

Eucken, Arnold, E. Karwat, and F. Fried. "Die Konstante I der thermodynamischen Dampfdruckgleichung bei mehratomigen Molekülen." *Zeitschr. f. Phys.* 29 (1924): 1.

Eve, A. S. *Rutherford: Being the Life and Letters of the Rt Hon. Lord Rutherford, O.M.* With a Foreword by Earl Baldwin of Bewdley. New York: Macmillan, 1939.

Falkenhagen, H. "Die Elektrolytarbeiten von Max Planck und ihre weitere Entwicklung." In *Max-Planck-Festschrift 1958.* Edited by B. Kockel, W. Macke, and A. Papapetrou. Berlin: VEB Deutscher Verlag der Wissenschaften, 1959, pages 11–34.

Farber, Eduard. *The Evolution of Chemistry: A History of Its Ideas, Methods, and Materials.* New York: Ronald Press Co., 1952.

Ferber, Christian v. *Die Entwicklung des Lehrerkörpers der deutschen Universitäten und Hochschulen 1864–1954.* Göttingen: Vandenhoek und Ruprecht, 1956.

Feuer, Lewis. S. *Einstein and the Generations of Science.* New York: Basic Books, 1974.

Feynman, Richard P. "Application of Quantum Mechanics to Liquid Helium." In *Progress in Low Temperature Physics, I.* Edited by C. J. Gorter. Amsterdam: 1955, pages 17–53.

Fischer, Emil. *Festmahl zu Ehren der Herren E. Solvay und H. von Böttinger am 1. Juli 1909 im Kaiserlichen Automobilklub zu Berlin.* Pamphlet.

——— *Aus meinem Leben.* Edited with a Foreword by M. Bergman. Berlin: Julius Springer, 1922.

Forman, Paul. "The Environment and Practice of Atomic Physics in Weimar Germany: A Study in the History of Science." Ph.D. diss. University of California, Berkeley, 1967.

——— "Weimar Culture, Causality, and Quantum Theory, 1918–1927: Adaptation by German Physicists and Mathematicians to a Hostile Intellectual Environment." *HSPS* 3 (1971): 1–115.

Forschen und Wirken: Festschrift zur 150-Jahr-Feier der Humboldt-Universität zu Berlin 1810–1960. Vol. I. Berlin: VEB Deutscher Verlag der Wissenschaften, 1960.

Forman, Paul, John L. Heilbron, and Spencer Weart. "Physics circa 1900: Personnel, Funding, and Productivity of the Academic Establishments." *HSPS* 5 (1975): 1–185.

Fowler, R. H., and T. E. Sterne. "Statistical Mechanics with Particular Reference to the Vapour Pressures and Entropies of Crystals." *Review of Modern Physics* 4 (1932): 635–708 (plus appendices).

Frank, Philipp. *Einstein: His Life and Times.* A. Knopf: New York, 1947. Translated from a German manuscript by George Rosen. Edited and revised by Shuichi Kusaka.

Freund, Ida. *The Study of Chemical Composition; an Account of Its Method and Historical Development.* With a Foreword by Laurence E. Strong and a biographical essay by O. Theodor Benfey. New York: Dover Publications, 1968.

Galison, Peter. *How Experiments End.* Chicago and London: The University of Chicago Press, 1987.

Image & Logic: A Material Culture of Microphysics. Chicago: University of Chicago Press, 1997.

Gavroglu, Kostas. "The Reaction of the British Physicists and Chemists to van der Waals' Early Work and to the Law of Corresponding States." *HSPS* 20 (1990): 199–237.

Gavroglu, Kostas, and Yorgos Goudaroulis. "Some Methodological and Historical Considerations in Low Temperature Physics: The Case of Superconductivity 1911–57." *Ann. Sci.* 41 (1984): 135–49.

"Some Methodological and Historical Considerations in Low Temperature Physics II: The Case of Superfluidity." *Ann. Sci.* 43 (1986): 137–46.

Methodological Aspects of the Development of Low Temperature Physics, 1881–1956: Concepts Out of Context(s). Science and Philosophy Series. Dordrecht/Boston/London: Kluwer Academic Publishers, 1989.

Geertz, Clifford. *Local Knowledge.* New York: Basic Books, 1983.

Giauque, W. F., and D. P. MacDougall. *Phys. Rev.* 43 (1933): 768; *J. Am. Chem. Soc.* 57 (1935): 1175–85.

Girnus, Wolfgang. *Grundzüge der Herausbildung der physikalischen Chemie als Wissenschaftsdiziplin. Eine wissenschaftshistorische Fallstudie zur Disziplinengenese in der Wissenschaft.* Ph.D. diss. Berlin: Akademie der Wissenschaften der DDR, 1982.

Graham, Lauren, Wolf Lepenies, and Peter Weingart, eds. *Functions and Uses of Disciplinary Histories.* Volume VII. Dordrecht /Boston/ Lancaster: D. Reidel, 1983.

Grüneisen, E. "Das Verhältnis der thermischen Ausdehnung zur spezifischen Wärme fester Elemente." *Zeitschr. f. Elektrochem.* 17 (1911): 737–9.

Guéron, J., and M. Magat, "A History of Physical Chemistry in France." *Ann. Rev. Phys. Chem.* 22 (1971): 1–23.

Günther, Paul. "Zu Walther Nernst's 75. Geburtstag." *Zeitschr. f. Elektrochem.* 45 (1939): 433–5.

Guye, C. E. *Physico-Chemical Evolution.* Translated by J. R. Clarke. New York, 1925.

Haar, D. ter, ed. *The Old Quantum Theory.* New York: Basic Books, 1966.

Haber, F., and R. le Rossignol. "Bestimmung des Ammoniakgleichgewichtes unter Druck." *Zeitschr. f. Elektrochem.* (1908): 181–96.

Haberditzl, Werner. "Walther Nernst und die Traditionen der physikalischen Chemie an der Berliner Universität." In *Forschen und Wirken. Festschrift zur 150-Jahr-Feier der Humboldt-Universität zu Berlin 1810–1960*. Vol. 1. Berlin: VEB Deutscher Verlag der Wissenschaften, 1960, pages 401–16.

Hacking, Ian. *Representing and Intervening*. Cambridge: Cambridge University Press, 1983.

Hahn, Otto. *My Life: The Autobiography of a Scientist*. Translated by E. Kaiser and E. Wilkins. New York: Herder and Herder, 1970.

Harnack, Adolf, ed. *Geschichte der Königlich preussischen Akademie der Wissenschaften zu Berlin*. 3 vols. Berlin: Reichsdruckerei, 1900.

Hartmann, Max. *Max Planck als Mensch und Denker*. Basel, Thun, Düsseldorf: Ott Verlag, 1953.

Hasenöhrl, Friedrich. "Über die Grundlagen der mechanischen Theorie der Wärme." *Physik. Zeitschr.* 12 (1911): 931–5.

Heilbron, John L. *The Dilemmas of an Upright Man: Max Planck as Spokesman for German Science*. Berkeley: University of California Press, 1986.

Heilbron, John L., and T. S. Kuhn. "The Genesis of the Bohr Atom." *HSPS* 1 (1969): 211–90.

Heilbron, J. L., and Bruce Wheaton. *Literature on the History of Physics in the 20th Century*. Berkeley: Office for the History of Science and Technology, U. C. Berkeley, 1981.

Helm, Georg. *Die Energetik nach ihrer geschichtlichen Entwicklung*. Leipzig, 1898.

Helmholtz, Hermann v. "Die Thermodynamik der chemischen Vorgänge." *Berl. Ber.* 2nd. ser. 1 (1882): 22–39.

Hermann, Armin. *The Genesis of the Quantum Theory*. Translated by Claude W. Nash. Cambridge, Mass.: MIT Press, 1971.

Hiebert, Erwin. N. *The Conception of Thermodynamics in the Scientific Thought of Mach and Planck*. Ernst-Mach-Institut, Bericht Nr. 5. Freiburg i.Br., 1968.

"The Energetics Controversy and the New Thermodynamics." In *Perspectives in the History of Science and Technology*. Edited by Duane H. D. Roller. University of Oklahoma Press, 1971, pages 67–86.

"Nernst, Walther Hermann." *DSB* Supplement (1978): 432–53.

"Nernst and Electrochemistry." In *Selected Topics in the History of Electrochemistry*. Edited by George Dubpernell and J. H. Westbrook. Proceedings of the Electrochemical Society, vol. 78. Princeton, N.J.: 1978, pages 180–200.

"Boltzmann's Conception of Theory Construction: The Promotion of Pluraism, Provisionalism, and Pragmatic Realism." In *Pisa Conference Proceedings*. Edited by J. Hintikka, D. Gruender, and E. Agazzi. Boston: D. Reidel, 1980, vol. 2: 175–98.

"Developments in Physical Chemistry at the Turn of the Century." In *Science, Technology and Society in the Time of Alfred Nobel*. Edited by C. G. Bernhard, E. Crawford, and Per Sörbom. Oxford: Pergamon, 1982, pages 97–115.

"Walther Nernst and the Application of Physics to Chemistry." In *Springs of*

Scientific Creativity: Essays on Founders of Modern Science. Eited by Rutherford Aris, H. Ted Davis, and Roger H. Stuewer. Minneapolis: University of
Minnesota Press, 1983.

Hiebert, E. N., and H.-G. Körber. "Ostwald, Wilhelm." *DSB* Supplement (1978):
455–69.

Hildebrand, Joel. "Das Königsche Spektralphotometer in neuer Anordung und
seine Verwendung zur Bestimmung chemischer Gleichgewichte." *Zeitschr. f.
Elektrochem.* (1908): 349–353.

Hobsbawm, Eric, and Terence Ranger, eds. *The Invention of Tradition.* Cambridge:
Cambridge University Press, 1983.

Holton, Gerald. *Thematic Origins of Scientific Thought.* Cambridge, Mass./
London, England: Harvard University Press, 1973.

"The Roots of Complementarity." *Daedalus* 99 (1970): 1015–55.

Hufbauer, Karl. *The Formation of the German Chemical Community, 1720–
1795.* Berkeley : University of California Press, ca. 1982.

Ihde, Aaron J. *The Development of Modern Chemistry.* New York: Dover Publications, 1982.

Instituts Solvay. *La structure de la matière. Rapports et discussions du Conseil de
Physique tenu à Bruxelles du 27 au 31 Octobre 1913.* Paris: Gauthier-Villars,
1921.

Jahn, Hans. "Über die Nernstschen Formeln zur Berechnung der elektromotorischen Kraft von Konzentrationselementen. Eine Erwiderung an Herrn Arrhenius." *Zeitschr. f. phys. Chem.* 36 (1901): 453–60.

Jarausch, Konrad H. *Deutsche Studenten 1800–1970.* Frankfurt a. Main: Suhrkamp, 1984.

Jeans, James. *Report on Radiation and the Quantum-Theory.* London: The Elec-
trician, 1914; 2nd ed. London: Fleetway, 1924.

Jost, W. "The First 45 Years of Physical Chemistry in Germany." *Ann. Rev. Phys.
Chem.* 17 (1966): 1–14.

Jungfleisch, E. "Notice sur la vie et les travaux de Marcellin Berthelot." *Bulletin
de la Société Chimique de France* (1913).

Kamerlingh Onnes, Heike. "On the Cryogenic Laboratory at Leiden and on the
Production of Very Low Temperatures." *Comm. Phys. Lab. Univ. Leiden.* 14:
1–30.

"Remarks on the Liquefaction of Hydrogen, on Thermodynamical Similarity,
and on the Use of Vacuum Vessels." *Comm. Phys. Lab. Univ. Leiden* 23: 1–23.

"Expression of the equation of state of gases and liquids by means of series."
Comm. Phys. Lab. Univ. Leiden 71: 1–25.

Kangro, Hans. *Vorgeschichte des Planckschen Strahlungsgesetzes. Messungen
und Theorien der spektralen Energieverteilung bis zur Begründung der Quantenhypothese.* Wiesbaden: Franz Steiner Verlag, 1970.

Kant, Horst. "Zum Problem der Forschungsprofilierung am Beispiel der NERNST-
schen Schule während ihrer Berliner Zeit von 1905 bis 1914." *NTM-Schriftenr.
Gesch. Naturwiss., Technik, Med.,* Leipzig 11(2): 58–68 (1974).

Kargon, Robert A. *The Rise of Robert Millikan.* Ithaca: Cornell University Press,
1982.

Karlik, Berta, and Erich Schmid. *Franz Serafin Exner und sein Kreis. Ein Beitrag zur Geschichte der Physik in Österreich.* Wien: Verlag der Österreichischen Akademie der Wissenschaften, 1982.

Kim, Mi Gyung. "Labor and Mirage: Writing the History of Chemistry." *Studies in History and Philosophy of Science* 26 (1995): 155–66.

King, Christine M., and Keith J. Laidler. "Chemical Kinetics and the Radiation Hypothesis." *Arch. Hist. Exact Sci.* 30 (1984): 45–86.

Kirsten, Christa, and H. G. Körber, eds. *Physiker über Physiker.* Berlin: Akademie-Verlag, 1975.

Kirsten, Christa, and H. J. Treder, eds. *Albert Einstein in Berlin 1913–1933.* Berlin: Akademie-Verlag, 1979.

Klein, Martin. "Einstein, Specific Heats, and the Early Quantum Theory." *Science* 148 (1965): 173–80.

Paul Ehrenfest. Vol. 1, *The Making of a Theoretical Physicist.* Amsterdam and London: North Holland, 1970.

"Thermodynamics and Quanta in Planck's Work." *Physics Today* 19 (11): 23–32.

Koenigsberger, Johannes. "Über die Atomwärmen der Elemente." *Zeitschr. f. Elektrochem.* 8 (1911): 289–93.

Koenigsberger, J., and O. Reichenheim. "Über ein Temperaturgesetz der elektrischen Leitfähigkeit fester einheitlicher Substanzen und einige Folgerungen daraus." *Phys. Zeitschr.* 7 (1906): 507–78.

Koenigsberger, J., and K. Schilling. "Über Elektrizitätsleitung in festen Elementen und Verbindungen." *Ann. Phys.* 32 (1910): 179–230.

Koenigsberger, Leo. *Hermann von Helmholtz.* 3 vols. Braunschweig: Vieweg, 1908.

Kohler, Robert E. "The Lewis-Langmuir Theory of Valence and the Chemical Community, 1920–1922." *HSPS* 6 (1975): 431–68.

Kohnstamm, Ph., and L. S. Ornstein. "On Nernst's Theorem of Heat." *Proceedings Royal Academy Amsterdam* 13 (1910): 700–15.

Kopperl, Sheldon J. "Richards, Theodore William." *DSB* Vol. 11:416–18.

The Scientific Work of Theodore William Richards. Ph.D. diss. University of Wisconsin, Madison, 1970.

Körber, Hans-Günther, ed. *Aus dem wissenschaftlichen Briefwechsel Wilhelm Ostwalds.* Berlin: Akademie Verlag: 1969.

Kossel, W. "Über Molekülbildung als Frage des Atombaus." *Ann. Phys.* 49 (1916): 229–362.

Kox, A. J. "Einstein and Lorentz: More than Just Good Colleagues." *Science in Context* 6 (1993): 43–58.

Kraul, Margaret. *Das deutsche Gymnasium 1780–1980.* Frankfurt a. Main: Suhrkamp, 1984.

Krüger, Lorenz, et al., eds. *The Probabilistic Revolution.* Vol. 1., *Ideas in History.* Cambridge, Mass.: MIT Press, 1987.

Kruyt, H. R. "Die dynamische Allotropie des Schwefels." *Zeitschr. f. physik. Chem.* 81 (1913): 726–48.

Kuhn, Thomas S. *The Structure of Scientific Revolutions.* 2nd ed. Chicago: Chicago University Press, 1970.

Black-Body Theory and the Quantum Discontinuity 1894–1912. Oxford/New York: Oxford University Press, 1978.

Kürti, Nicholas, P. Laine, B. V. Rollin, and F. Simon. *Compt. Rend.* 202 (1935): 1576–8.

Kürti, Nicholas, and F. Simon, *Proc. Roy. Soc. (London)* (A) 149 (1935): 152–76; 151 (1935): 61–623.

Ladenburg, Albert. *Vorträge über die Entwicklungsgeschichte der Chemie in den letzten hundert Jahren.* Braunschweig: Vieweg, 1869.

Laidler, Keith J. "Chemical Kinetics and the Origins of Physical Chemistry." *Arch. Hist. Exact Sci.* 32 (1985): 43–75.

Lange, F. "Untersuchungen über die spezifische Wärme bei tiefen Temperaturen." *Zeitschr. f. phys. Chem.* 110 (1924): 343.

Langevin, Paul. "The Relations of Physics of Electrons to Other Branches of Science." *Congress of Arts and Science, Universal Exposition, St. Louis, 1904.* Vol. 4. Boston: Houghton Mifflin, 1905.

Langevin, Paul, and Maurice de Broglie, eds. *La théorie du rayonnement et les quanta. Rapports et discussions de la reunion tenue à Bruxelles, du 30 octobre au 3 novembre 1911.* Paris: Gauthier, 1912.

Laudan, Rachel. "The Method of Multiple Working Hypotheses and the Development of Plate Tectonic Theory." In *Scientific Discovery: Case Studies.* Edited by Thomas Nickles. Dordrecht, Holland; Boston, USA; London, England: D. Reidel, 1980.

Leicester, Henry M. *A Source Book in Chemistry, 1900–1950.* Cambridge, Mass.: Harvard University Press, 1968.

Lepsius, Richard. "Zur hundersten Wiederkehr des Geburtstages von Walther Nernst." *Chemiker Ztg.* 88 (1964): 603–6.

Lewis, G. N. "The Development and Application of a General Equation for Free Energy and Physico-chemical Equilibrium." *Proceedings American Academy of Arts and Sciences* 35 (1899): 3–38.

"The Atom and the Molecule." *JACS* 38 (1916): 762–85.

Lewis, G. N., and M. Randall. *Thermodynamics and Free Energy of Chemical Substances.* New York: McGraw-Hill, 1923.

Lindauer, Maurice W. "The Evolution of the Concept of Chemical Equilibrium from 1775–1923." *J. Chem. Ed.* 39 (1962): 384–90.

Lindemann, Frederick A. "Untersuchungen über die spezifische Wärme bei tiefen Temperaturen. IV." *Berl. Ber.* (1910): 316–21.

The Physical Significance of the Quantum Theory. Oxford: Clarendon Press, 1932.

Lindemann, F. A., and F. Simon. "Walther Hermann Nernst (1864–1941)." *Obituary Notices of Fellows of the Royal Society* 4 (1942): 101–12.

Livingston, Burton E. *Role of Diffusion and Osmotic Pressure in Plants.* Chicago: University of Chicago Press, 1903.

Lodge, Oliver. "On Electrolysis." *Report of the British Association for the Advancement of Science* 55 (1885): 723–72.

Loeb, Jacques. *The Physiological Problems of Today.* Ithaca: American Society of Naturalists, 1897.

Lorentz, Hendrik A. "On Nernst's Heat Theorem." In H. A. Lorentz. *Collected Papers*. Vol. 6 of 9 vols. The Hague: Martinus Nijhoff, 1938, pages 318–24. The article had originally appeared in Dutch in *Chemisch Weekblad* 10 (1913): 621.

Mach, Ernst. *Erkenntnis und Irrtum*. Leipzig, 1917.

Principles of the Theory of Heat. Edited by Brian McGuiness. Introduction by Martin Klein. Boston: Reidel, 1986.

Madsen, Theodore. "Versuche über die Abhängigkeit der Hydrolyse von der Temperatur." *Zeitschr. f. phys. Chem.* 36 (1901): 290–304.

Magnus, A., and F. A. Lindemann. "Über die Abhängigkeit der spezifischen Wärme fester Körper von der Temperatur." *Zeitschr. f. Elektrochemie* 8 (1910): 269–72.

Maier, Clifford. *The Role of Spectroscopy in the Acceptance of the Internally Structured Atom*. New York: Arno Press, 1981.

Manegold, Karl-Heinz. *Universität, technische Hochschule- und Industrie*. Vol. 16, *Schriften zur Wirtschafts- und Sozialgeschichte*. Edited by W. Fischer. Berlin: Duncker und Humblot, 1970.

Max-Planck-Gesellschaft. *50 Jahre Kaiser-Wilhelm-Gesellschaft und Max-Planck-gesellschaft zur Förderung der Wissenschaften 1911–1961*. Göttingen: Max-Planck-Gesellschaft, 1961.

McClelland, Charles E. *State, Society, and University in Germany, 1700–1914*. Cambridge and New York: Cambridge University Press, 1980.

McClintock, P. V. E. "Cryogenics." In Robert A. Myers, ed. *Encyclopedia of Physical Science and Technology*, vol. 3. Orlando: Academic Press, 1987.

McCormmach, Russell. "Henri Poincaré and the Quantum Theory." *Isis* 58 (1967): 37–55.

"On the Growth of the Physics Discipline in the Nineteenth Century." Address published as Editor's Foreword in *HSPS* 3 (1971): ix–xxiv.

"On Academic Scientists in Wilhelminian Germany." *Daedalus* (summer 1974): 147–71.

McCormmach, Russell, and Christa Jungnickel. *Intellectual Mastery of Nature*. 2 vols. Chicago and London: University of Chicago Press, 1986.

McKie, Douglas, and Niels H. de V. Heathcote. *The Discovery of Specific and Latent Heats*. London: Edward Arnold & Co., 1935.

Mehra, Jagdish. *The Solvay Conferences on Physics. Aspects of the Development of Physics since 1911*. Foreword by W. Heisenberg. Dordrecht and Boston: D. Reidel Publishing Company, 1975.

Mendelssohn, Kurt. *Cryophysics*. New York: Interscience Publishers, 1960.

"Prewar Work on Superconductivity as Seen from Oxford." *Rev. Mod. Phys.* 36 (1964): 7–12.

The World of Walther Nernst. The Rise and Fall of German Science, 1864–1941. London: Macmillan, 1973.

Millikan, Robert A. "Eine experimentelle Prüfung der Clausius-Mossotti-schen Formel." *Ann. Phys.* 60 (1897): 376–80.

The Autobiography of Robert A. Millikan. New York: Prentice-Hall, 1950.

Milne, E. A. *Sir James Jeans. A Biography*. With a memoir by S. C. Roberts. Cambridge: Cambridge University Press, 1952.

Moody, H. W. "A Determination of the Ratio of the Specific Heats and the Specific Heat at Constant Pressure of Air and Carbon Dioxide." *Phys. Rev.* 34 (1912): 275–95.

Morrison, G. S. "Wilhelm Ostwald's 1896 History of Electrochemistry: Failure or Neglected Paragon." In *Selected Topics in the History of Electrochemistry.* Edited by G. Dubpernell and J. H. Westbrook. Proceedings of the Electrochemical Society, vol. 78. Princeton, N.J.: The Electrochemical Society, 1978, pages 213–25.

Nagel, Bengt. "The Discussion Concerning the Nobel Prize for Max Planck." In Bernhard et al., eds. *Science, Technology and Society in the Time of Alfred Nobel.* pages 352–76.

Nernst, Walther Hermann. "Über die elektromotorischen Kräfte, welche durch den Magnetismus in von einem Wärmestrome durchflossenen Metallplatten geweckt wird." *Wied. Ann.* 31 (1887): 760.

"Bildungswärme der Quecksilberverbindungen." *Zeitschr. f. physik. Chem.* 2 (1888): 23.

"Über das thermische und galvanische Verhalten einiger Wismuth-Zinn-Legierungen im magnetischen Felde." *Wied. Ann.* 33 (1888): 474.

"Zur Kinetik der in Lösung befindlichen Körper. I. Theorie der Diffusion." *Zeitschr. f. physik. Chem.* 2 (1888): 613–37.

"Zur Kinetik der in Lösung befindlichen Körper. II. Überführungszahlen und Leitvermögen einiger Silbersalze." *Z. f. physik. Chem.* 2 (1888): 948–63; reprinted as "Rates of Transference and the Conducting Power of Certain Silver Salts." *Am. Chem. J.* 11 (1888): 106.

"Die elektromotorische Wirksamkeit der Ionen." *Zeitschr. f. physik. Chem.* 4 (1889): 129–81.

"Über gegenseitige Beeinflussung der Löslichkeit von Salzen." *Zeitschr. f. physik. Chem.* 4 (1889): 372–83.

"Ein osmotischer Versuch." *Zeitschr. f. physik. Chem.* 6 (1890): 37–40.

"Über ein neues Prinzip der Molekulargewichtsbestimmung." *Zeitschr. f. physik. Chem.* 6 (1890): 16–36.

"Über eine neue Anwendung des Gefrierapparates zur Molekulargewichtsbestimmung." *Zeitschr. f. physik. Chem.* 6 (1890): 573–7.

"Über das Henry'sche Prinzip." *Nachr. Ges. Wiss. Göttingen* (1891): 202–12.

"Verteilung eines Stoffes zwischen zwei Lösungsmitteln und zwischen Lösungsmitteln und Dampfraum." *Zeitschr. f. physik. Chem.* 8 (1891): 110–39, pages 111, 139.

"Über die mit der Vermischung konzentrierter Lösungen verbundene Änderung der freien Energie." *Nachr. Ges. Wiss. Göttingen* (1892): 428–38.

"Über die Löslichkeit von Mischkristallen." *Zeitschr. f. physik. Chem.* 9 (1892): 137–42.

"Über die Potentialdifferenz verdünnter Lösungen." *Wied. Ann.* 45 (1892): 360–9.

"Osmotischer Druck in Gemischen zweier Lösungsmittel." *Zeitschr. f. phys. Chem.* 11 (1893): 1–6.

Theoretische Chemie vom Standpunkte der Avogadroschen Regel und der

Thermodynamik. Stuttgart: Ferdinand Enke, 1893. 7th ed., 1913; 18th ed., 1921.

"Über die Beteiligung eines Lösungsmittels an chemischen Reaktionen." *Zeitschr. f. phys. Chem.* 11 (1893): 345–59.

"Über Flüssigkeitsketten." *Zeitschr. f. Elektrochem.* 1 (1894): 153–5.

"Über die Auflösung von Metallen in galvanischen Elementen." *Zeitschr. f. Elektrochem.* 1 (1894): 243–6.

"Elektrostriktion durch freie Ionen." *Zeitschr. f. phys. Chem.* 15 (1894): 79–85.

Die Ziele der physikalischen Chemie. Festrede gehalten am 2. Juni 1896 zur Einweihung des Instituts für physikalische Chemie und Elektrochemie der Georgia Augusta zu Göttingen. Göttingen: Vandenhoek & Ruprecht, 1896.

"Die elktrolytische Zersetzung wässriger Lösungen." *Ber. Dtsch. Chem. Ges.* 30 (2): 1547–63 (1897).

"Über das chemische Gleichgewicht, elektromotorische Wirksamkeit und elektrolytische Abscheidung von Metallgemischen." *Zeitschr. f. phys. Chem.* 22 (1897): 539–42.

"Über die elektrolytische Leitung fester Körper bei sehr hohen Temperaturen." *Zeitschr. f. Elektrochem.* (1899): 41–3.

"Ein elektrischer Platinofen." *Zeitschr. f. Elektrochem.* (1900): 253.

"Einiges über das Verhalten elektrolytischer Glühkörper." *Zeitschr. f. Elektrochem.* (1900): 373–6.

"Erwiderung auf einige Bemerkungen der Herren Arrhenius, Kohnstamm, Cohen, und Noyes." *Zeitschr. f. phys. Chem.* 36 (1901): 596–604.

"Zur Theorie der Lösungen." *Zeitschr. f. phys. Chem.* 38 (1901): 487–506.

"Über Molekulargewichts-Bestimmungen bei sehr hohen Temperaturen," *Zeitschr. f. Elektrochem.* 32 (1903): 622–7. Also in abstract in *Gött. Nachr.* 1903.

"Chemisches Gleichgewicht und Temperaturgefälle." In *Festschrift Ludwig Boltzmann.* Leipzig: J. A. Barth, 1904, pages 904–15.

"Über die Berechnung chemischer Gleichgewichte aus thermischen Messungen." *Nachr. Gött. Ges.* 1(1906): 1–39. Submitted to the meeting of 23 December 1905.

"Gleichgewicht und Reaktionsgeschwindigkeit beim Stickoxyd." *Zeitschr. f. Elektrochem.* (1906): 527–9.

"Über das Ammoniakgleichgewicht." *Zeitschr. f. Elektrochem.* (1907): 521–2.

Silliman Lectures: Applications of Thermodynamics to Chemistry. New York: Scribner's, 1907.

"Development of General and Physical Chemistry During the Last Forty Years." *Annual Report of the Smithsonian Institution* (1908): 245–53.

"Untersuchungen über die spezifische Wärme bei tiefen Temperaturen. II." *Berl. Ber.* (1910): 262–82.

"Untersuchungen über die spezifische Wärme bei tiefen Temperaturen. III.," *Berl. Ber.* (1911): 306–15. Presented to the academy on 23 February 1911.

"On the inconsistency of my heat theorem and VAN DER WAALS' equation at very low temperatures." *Proceedings Royal Academy Amsterdam* 14 (1911): 201–4.

"Über ein allgemeines Gesetz, das Verhalten fester Stoffe bei sehr tiefen Temperaturen betreffend." *Physik. Zeitschr.* 12 (1911): 976–9.

"Über einen Apparat zur Verflüssigung von Wasserstoff." *Zeitschr. f. Elektrochem.* 17 (1911): 735–7.

"Über neuere Probleme der Wärmetheorie." Öffentliche Sitzung zur Feier des Geburtsfestes Sr. Majestät des Kaisers und Königs und des Jahrestages König Friedrich's II. 26 Jan. 1911. *Berl. Ber.* 4 (1911): 65–90.

"Sur la determination de la l'affinité chimique a partir de données thermiques." *Journal de chimie physique* 8 (1911): 228–67.

"Thermodynamik und spezifische Wärme." *Berl. Ber.* (1 February 1912): 134–40.

"Untersuchungen über die spezifische Wärme. VII." *Berl. Ber.* (12 December 1912): 1172–6.

"Das Gleichgewichtsdiagramm der beiden Schwefelmodifikationen." *Zeitschr. f. physik. Chem.* 83 (1913): 546–50.

The Theory of the Solid State. Based on Four Lectures Delivered at University College, London, in March 1913. London: University of London Press, 1914.

"Kinetische Theorie fester Körper." In M. Planck, P. Debye, W. Nernst, M. v. Smoluchowski, A. Sommerfeld, and H. A. Lorentz, *Vorträge über die kinetische Theorie der Materie und der Elektrizität. Gehalten in Göttingen auf Einladung der Wolfskehlstiftung.* [April 1913.] With contributions by H. Kamerlingh Onnes and W. H. Keesom. Preface by D. Hilbert. Leipzig/Berlin: B. G. Teubner, 1914, pages 61–82.

Die theoretischen und experimentellen Grundlagen des neuren Wärmesatzes. Halle: Knapp, 1918.

Das Weltgebäude im Lichte der neuren Forschung. Berlin: J. Springer, 1921.

"Physico-Chemical Considerations in Astrophysics." *Journal of the Franklin Institute* 206 (1928): 135–42.

"Zum 50. Geburtstage der elektrolytischen Dissoziationstheorie von Arrhenius." *Zeitschr. f. Elektrochem.* 43 (1937): 146–8.

Nernst, Walther, and R. Abbegg. "Über den Gefrierpunkt verdünnter Lösungen." *Zeitschr. f. phys. Chem.* 15 (1894): 681–93.

Nernst, Walther, and E. Bose. "Ein experimenteller Beitrag zur osmotischen Theorie." *Zeitschr. f. Elektrochem.* 5 (1898): 233–5.

Nernst, Walther, and A. v. Ettingshausen. "Über das Auftreten elektromotorischer Kräfte in Metallplatten, welche von einem Wärmestrome durchflossen werden und sich im magnetischen Felde befinden." *Wied. Ann.* 29 (1886): 343.

"Über das Hall'sche Phänomen." *Ber. Wien. Akad.* (1887): 560–610.

Nernst, Walther, F. Koref, and F. A. Lindemann. "Untersuchungen über die spezifische Wärme bei tiefen Temperaturen. I." *Berl. Ber.* (1910): 247–61.

Nernst, Walther, and F. A. Lindemann. "Untersuchungen über die spezifische Wärme bei tiefen Temperaturen. V." *Berl. Ber.* (1910): 494–501.

"Untersuchungen über die spezifische Wärme. VI." *Berl. Ber.* (1912): 1160–71.

Nernst, Walther, and R. Pauli. "Weiteres zur electromotorischen Wirksamkeit der Ionen." *Ann. Phys.* 45 (1892): 352–9.

Nernst, Walther, and E. H. Riesenfeld. "Über elektrolytische Erscheinungen an der

Grenzfläche zweier Lösungsmittel." *Nachr. Ges. Wiss. Göttingen* (1901): 54–61, and *Ann. Phys.* 8(4) 600–8: (1902).

Nye, Mary Jo. "N-rays: An Episode in the History and Psychology of Science." *HSPS* 11 (1980): 125–56.

"Berthelot's Anti-atomism: A Matter of Taste?" *Ann. Sci.* 38 (1981): 585–90.

The Question of the Atom. From the Karlsruhe Congress to the First Solvay Conference, 1860–1911. A Compilation of Primary Sources. Los Angeles: Tomash, 1984.

Science in the Provinces. Scientific Communities and Provincial Leadership in France, 1860–1930. Berkeley: University of California Press, 1986.

"Chemical Explanation and Physical Dynamics: Two Research Schools at the First Solvay Chemistry Conferences, 1922–1928." *Ann. Sci.* 46 (1989): 461–80.

From Chemical Philosophy to Theoretical Chemistry: Dynamics of Matter and Dynamics of Disciplines, 1800–1950. Berkeley: University of California Press, 1994.

Ostwald, Wilhelm. "Grundlinien der allgemeinen Energetik." *Bericht über die Verhandlungen der mathematisch-physikalischen Klasse der Sächsischen Gesellschaft der Wissenschaften zu Leipzig* 44 (1892): 211–37.

Überwindung des wissenschaftlichen Materialismus. Lecture held at the third general session of the Association of German Scientists and Physicians in Lübeck, 20. Sept. 1895. Leipzig: Verlag von Veit & Co., 1895.

Abhandlungen und Vorträge. Leipzig, 1904.

Lebenslinien. 3 vols. Berlin, 1926–7.

Electrochemistry: History and Theory. 2 vols. Translated by N. P. Date. Washington: Smithsonian Institution and the National Science Foundation, 1980.

Ostwald, Wilhelm, and Walther Nernst. "Über freie Ionen." *Zeitschr. f. physik. Chem.* 3 (1889): 120–30.

Pais, Abraham. *'Subtle Is the Lord . . .' The Science and the Life of Albert Einstein*. Oxford: Oxford University Press, 1982.

Palmaer, Wilhelm. "Über die kapillarischen Erscheinungen." *Zeitschr. f. phys. Chem.* 36 (1901): 664–80.

Palmer, W. G. *A History of the Concept of Valency to 1930*. Cambridge: Cambridge University Press, 1965.

Parkes, G. S. "Some Notes on the History of Thermochemistry." *J. chem. ed.* 28 (1949): 262–6.

Partington, J. R. *History of Chemistry*. Vol. 4. London: Macmillan & Co., 1964.

Partington, J. R., and W. G. Shilling. *The Specific Heats of Gases*. London: Ernest Benn, 1924.

Paul, Harry W. *The Edge of Contingency: French Catholic Reaction to Scientific Change from Darwin to Duhem*. Gainesville: University of Florida Press, 1979.

Paul, Martin A. *Principles of Chemical Thermodynamics*. New York: McGraw-Hill Book Company, 1951.

Paulsen, Friedrich. *Die Deutschen Universitäten und das Universitätsstudium*. Berlin: A. Asher, 1902.

Peierls, Rudolf. *Bird of Passage. Recollections of a Physicist.* Princeton, N.J.: Princeton Univerity Press, 1985.

Pelseneer, Jean. *Historique des Instituts Internationaux de Physique et de Chimie Solvay.* Microfilm at Archives for the History of Quantum Physics, AIP, New York.

Perrin, Jean. "Radiation and Chemistry." Translated by H. Borns. *Transactions of the Faraday Society* 17 (1921–2): 546–72. Almost identical to "Matière et lumière," *Annales de Physique* 11 (1919): 1–108.

Pestre, Dominique. *Physique et physiciens en France 1918–1940.* Paris: Editions des archives contemporaines, 1984.

Planck, Max. "Ueber das Prinzip der Vermehrung der Entropie, III." *Wied. Ann.* 32 (1887): 462–503. Also in M. Planck. *PA.* Vol. 1, pages 232–73.

"Über die molekulare Konstitution verdünnter Lösungen." *Zeitschr. f. phys. Chem.* 1 (1887): 577–82. In Planck. *PA.* Vol. 1, pages 274–9.

"Über das Prinzip der Vermehrung der Entropie. Dritte Abhandlung. Gesetze des Eintritts beliebiger thermodynamischer und chemischer Reactionen." *Wied. Ann.* 32 (1887): 462–503. In Planck. *PA.* Vol. 1, pages 232–73.

"Über die Potentialdifferenz zwischen zwei verdünnten Lösungen." 40 (1890): 561–76.

"Über die Potentialdifferenz zwischen zwei verdünnten Lösungen binärer Electrolyte." *Wied. Ann.* 40 (1890): 561–76. In *PA.* Vol. 1, pages 356–71.

"Über die Potentialdifferenz zwischen zwei verdünnten Lösungen binärer Electrolyte." *Wied. Ann.* 40 (1890): 561–76. Also presented in abstract, with an introduction to the theory of solutions, in the *Verhandlungen der Physikalischen Gesellschaft zu Berlin.* Meeting of 18 April 1890. Both in Planck. *PA.* Vol. 1, pages 330–71.

"Über das Princip der Vermehrung der Entropie. Vierte Abhandlung. Gesetze des elektrochemischen Gleichgewichts." *Wied. Ann.* 44 (1891): 647–56. Also in Planck. *PA.* Vol. 1, pages 382–425.

"Bemerkungen über das Carnot-Clausiussche Princip." *Wied. Ann.* 46 (1892): 162–6, in Planck. *PA.* Vol. 1, pages 426–30.

Grundriss der allgemeinen Thermochemie. Mit einem Anhang: Der Kern des zweiten Hauptsatzes der Wärmetheorie. Breslau: E. Trewendt, 1893.

Vorlesungen über Thermodynamik. Leipzig: von Veit, 1897.

Uber neuere thermodynamische Theorien (Nernstsches Warmetheorem und Quantenhypothese), Leipzig: Akademische Verlagsgesellschaft, 1912. A lecture delivered on 16 December 1911 to the German Chemical Society in Berlin.

"Neue Bahnen der physikalischen Erkenntnis: Rede des antretenden Rektors Dr. Max Planck." In *Rektorwechsel and der Friedrich-Wilhelm-Universität zu Berlin am 15. Oktober 1913.* Berlin: Norddeutsche Buchdruckerei, 1913.

Die Entstehung und bisherige Entwicklung der Quantentheorie. Leipzig: J. A. Barth, 1920.

Erinnerungen. Edited by W. Keiper. Berlin: Keiperverlag, 1948.

Physikalische Abhandlungen und Vorträge. 3 vols. Braunschweig: Friedr. Vieweg & Sohn, 1958. (*PA*)

A *Bibliography of His Non-technical Writings*. Berkeley Papers in History of Science, Office for History of Science and Technology, University of California, Berkeley, 1977.

Planck, Max, P. Debye, W. Nernst, M. v. Smoluchowski, A. Sommerfeld, and H. A. Lorentz. *Vorträge über die kinetische Theorie der Materie und der Elektrizität. Gehalten in Göttingen auf Einladung der Wolfskehlstiftung*. [April 1913.] With contributions by H. Kamerlingh Onnes and W. H. Keesom. Preface by D. Hilbert. Leipzig/Berlin: B. G. Teubner, 1914.

Poggendorff, J. C. *Biographisch-Literarisches Handwörterbuch der exakten Naturwissenschaften*. Edited by R. Zaunick and H. Salié. Berlin: Akademie Verlag, 1959.

Pohl, R. W. *Gedächtnis-Kolloquium am 29*. November 1976. Frankfurt/Main: Musterschmidt, 1978.

Poincaré, Henri. "Sur la Theorie des Quanta." *Journal de Physique* 2 (January 1912): 37.

Pollitzer, F. *Die Berechnung chemischer Affinitäten nach dem Nernst'schen Wärmetheorem*. Sondersausgabe aus der Sammlung chemischer und chemisch-technischer Vorträge, edited by W. Herz, vol. 17. With a Foreword by W. Nernst. Stuttgart: F. Enke, 1912.

Popper, Karl R. *Quantum Theory and the Schism in Physics*. From *Postscript to the Logic of Scientific Discovery*. Edited by W. W. Bartley, III. Totowa, N.J.: Rowman and Littlefield, 1982.

Raman, V. V. "The Permeation of Thermodynamics into Nineteenth Century Chemistry." *Indian Journal of History of Science* 10 (1975): 16–37.

Reiche, Fritz. *Die Quantentheorie. Ihr Ursprung und ihre Entwicklung*. Berlin: J. Springer, 1921.

Richards, Theodore W. "The Significance of the Change of Atomic Volume." *Proc. Am. Acad.* 36: 293–317; also published in *Zeitschr. phys. Chem.* 42 (1902).

"The Significance of Changing Atomic Volume. III." *Proceedings American Academy Arts and Sciences*. 38 (1902): 291.

Richards, T. W., and R. N. Garrod-Thomas. "Electrochemical Investigation of Liquid Amalgams of Zinc, Cadmium, Lead, Copper, and Lithium." With contributions from the Chemical Laboratory of Harvard College. *Carnegie Institution of Washington Publication* No. 118 (1909), 39–72.

Riesenfeld, Ernst H. *Svante Arrhenius*. In *Grosse Männer. Studien zur Biologie des Genies*. Vol. 11. Edited by W. Ostwald. Leipzig: Akademische Verlagsgesellschaft, 1931.

Ringer, Fritz. *The Decline of the German Mandarins: The German Academic Community, 1890–1933*. Cambridge, Mass.: Harvard University Press, 1969.

Rocke, Alan J. *Chemical Atomism in the Nineteenth Century : From Dalton to Cannizzaro*. Columbus: Ohio State University Press, 1984.

The Quiet Revolution: Hermann Kolbe and the Science of Organic Chemistry. Berkeley/Los Angeles/London: University of California Press, 1993.

Root-Bernstein, Robert Scott. "The Ionists: Founding Physical Chemistry, 1872–1890." Ph.D. diss. Princeton University, 1980.

Rosenfeld, Léon. "La première phase de l'évolution de la Théorie des Quanta." *Osiris* 2 (1936): 149–96.

Ross, Sidney. "The Story of Volta Potential." In *Selected Topics in the History of Electrochemistry.* Edited by George Dubpernell and J. H. Westbrook. Princeton, N.J. : The Electrochemical Society, 1978, pages 257–71.

Rossini, Frederick D. "Modern Thermochemistry," *Chem. Rev.* 18 (1936): 233–56.

Ruhemann, M., and B. Ruhemann. *Low Temperature Physics.* Cambridge University Press, 1937.

Sackur, Otto. *Ann. d. Physik* 36 (1911): 964–5.

"Die spezifische Wärme der Gase und die Nullpunktsenergie." *Verh. d. D. Phys. Ges.* 16 (1914): 728–34.

Schelar, V.M. "Thermochemistry and the Third Law of Thermodynamics." *Chymia* 11 (1966): 99–124.

Schilpp, P. A., ed. *Albert Einstein: Philosopher-Scientist.* Evanston, Ill.: The Library of Living Philosophers, 1949.

Schimank, W. "Walther Nernst. Neue Grundlagen zur Molekulartheorie und Thermodynamik." *Gestalter unserer Zeit* 3 (1955): 129–38.

Schleunes, Karl A. *Schooling and Society. The Politics of Education in Prussia and Bavaria 1750–1900.* Oxford/New York/Munich: Berg, 1993.

Schmidt-Ott, Friedrich. *Erlebtes und Erstrebtes, 1860–1950.* Wiesbaden: Franz Steiner, 1952.

Seelig, Carl, ed. *Albert Einstein: Eine Dokumentarische Biographie.* Zürich: Europa Verlag, 1952.

Servos, John W. "Physical Chemistry in America, 1890–1933." Ph.D. diss. The Johns Hopkins University, 1979.

"G. N. Lewis: The Disciplinary Setting." *J. Chem. Ed.* 61 (1984): 5–10.

"A Disciplinary Program that Failed: Wilder D. Bancroft and the *Journal of Physical Chemistry,* 1896–1933." *Isis* 73 (1982): 207–32.

Physical Chemisty from Ostwald to Pauling: The Making of a Science in America. Princeton, N.J. : Princeton University Press, 1990.

Shields, Margaret Calderwood. "A Determination of the Ratio of the Specific Heats of Hydrogen at 18C and –190C." *Phys. Rev.* 10 (1917): 525–40.

Simon, F. "Die Bestimmung der freien Energie." *Handbuch der Physik* 10 (1926): 350.

"Vom Prinzip der Unerreichbarkeit des absoluten Nullpunktes." *Z. f. Phys.* 41 (1927): 806.

"Fünfundzwanzig Jahre Nernst'sches Theorem." *Erg. d. exakt. Naturw.* 9 (1930).

"The Third Law of Thermodynamics. An Historical Survey." Reprint of the 40th Guthrie Lecture. *Year Book of the Physical Society,* 1956.

Simon, F. E., N. Kurti, J. F. Allen, and K. Mendelssohn. *Low Temperature Physics: Four Lectures.* Oxford: Pergamon Press, 1961.

Smith, Crosbie, and M. Norton Wise. *Energy and Empire: A Biographical Study of Lord Kelvin.* Cambridge University Press, 1989.

Smith, F. W. F., Earl of Birkenhead. *The Professor and the Prime Minister.* Boston: Riverside Press Houghton Mifflin, 1962.

Sommerfeld, Arnold. *Atombau und Spektrallinien.* Braunschweig: F. Vieweg, 1921.

Stern, Fritz. *Dreams and Delusions: The Drama of German History.* New York: Alfred A. Knopf, 1987.

Stewart, A. W. *Recent Advances in Organic Chemistry.* London: Longman's, Green, 1908.

Taylor, Hugh S. "The Thermodynamic Properties of Silver and Lead Iodide." *JACS* 38 (1916): 2295–310.

"Fifty Years of Chemical Kineticists." *Annual Review of Physical Chemistry* 13 (1962): 1–18.

Taylor, Hugh S., and G. St. John Perrott. "The Thermochemical Data of Cadmium Chloride and Iodide." *JACS* 43 (1921): 484–93.

van der Waals, J. D. *The Continuity of the Liquid and Gaseous States.* Translated by R. Threlfall and F. Adair. A Physical Memoir. London: The Physical Society of London, n.d.

"Theorie thermodynamique de la capillarité dans l'hypothèse d'une variation continue de densité." *Archives Neerlandaises* 28 (1880).

van't Hoff, Jacobus Hendricus. "Vorschlag zur Ausdehnung der gegenwärtig in der Chemie gebrauchten Strukturformeln in den Raum nebst einer damit zusammenhängenden Bemerkung über die Beziehung zwischen dem optischen Drehvermögen und der chemischen Konstitution organischer Verbindungen." Utrecht: J. Greven, 1874.

"Lois de l'equilibre chimique dans l'etat dilue, gazeux ou dissous." *K. Svenska Vetenskaps-Akademiens Handlingar* 21, no. 17 (1886); "Une propriete generale de la matiere dilue." Ibid.; "Conditions electriques de l'equilibre chimique." Ibid.

Physical Chemistry in the Service of the Sciences. Translated by Alexander Smith. Chicago: University of Chicago Press, 1903.

Vorlesungen über theoretische und physikalische Chemie. Brunswick, 1903.

"The Relation of Physical Chemistry to Physics and Chemistry." *J. phys. Chem.* 9 (1905): 81–9.

Varnedoe, Kirk. *A Fine Disregard: What Makes Modern Art Modern.* New York: Harry N. Abrams, 1990.

Vicedo, Marga. "Scientific Styles: Toward Some Common Ground in the History, Philosophy, and Sociology of Science." *Perspectives on Science* 3 (1995): 231–54.

Volterra, Vito, Ernest Rutherford, Robert W. Wood, and Carl Barus. *Clark University Lectures.* Delivered at the Celebration of the Twentieth Anniversary of the Foundation of Clark University. Worcester, Mass.: Clark University, 1912.

Weyer, Jost. *Chemiegeschichtsschreibung von Wiegleb (1790) bis Partington (1970).* Eine Untersuchung über ihre Methoden, Prinzipien und Ziele. Hildesheim: Gerstenberg, 1974. Vol. 3 in *Arbor Scientarum, Beiträge zur Wissenschaftsgeschichte.* Edited by Otto Krätz, Fritz Krafft, Walter Saltzer, Hans-Werner-Schütt, and Cristoph J. Scriba.

Wise, Norton. "On the Relation of Physical Science to History in Late Nineteenth-Century Germany." In Graham Lauren, Wolf Lepenies, and Peter Weingart, eds. *Functions and Uses of Disciplinary Histories*. Volume VII. Dordrecht/ Boston/ Lancaster: D. Reidel, 1983, pages 3–34.

Woodward, Robert S. "The Unity of Physical Science." *Congress of Arts and Science, Universal Exposition of St. Louis, 1904.* Vol. 4. Boston: Houghton Mifflin, 1905.

Zeise, H. "Spektralphysik und Thermodynamik. Die Berechnungen von freien Energien, Entropien, spezifischen Wärmen und Gleichgewichten aus spektroskopischen Daten und die Gültigkeit des dritten Hauptsatzes." *Zeitsch. f. Elektrochemie* 39 (1933): 758, 895; 40 (1934): 662, 885.

Zuckerman, Harriet. *Scientific Elite, Nobel Laureates in the United States.* New York: The Free Press, 1977.

Index

Abegg, Richard, 171
Abel, Emil, 31
absolute zero temperature, 2, 136, 139–40, 143, 166, 174, 178, 212; resistance at, 179; state of matter approaching, 151; unattainability of, 215
academia: and industry, 7
Acheson, Edward, 119
affinity: comparative theory of, 43–4; *see also* chemical affinity
affinity theory, 141–2
Allgemeine Elektrizitätsgesellschaft (AEG), 92, 95, 97f, 98–102, 107
Althoff, Friedrich, 42, 59, 72, 148, 218, 224, 246
Aluminothermie, 118
aluminum, 118–19, 179
ammonia, 3, 150, 176
ammonia synthesis, 5, 23, 129–30, 131, 216
Ampére's laws, 175
Amsterdam Academy of Sciences, 180
analogies, 51–2, 55, 65, 67–8, 70, 176–7; Nernst's model-building by, 178, 179–80, 242
Annalen der Physik und Chemie, 61, 79, 81, 86
Anschaulichkeit, 71, 75, 174, 176, 199
anti-Semitism, 165, 186
argon, 114, 149, 157–8
Armstrong, Henry Edwards, 172
Arrhenius, Svante, x, 1, 9, 10–11, 14, 22, 36, 41, 42, 43, 53, 57, 62–3, 73, 76, 120, 141, 183, 213, 246, 247, 248; dissociation theory, 32–3, 34, 48, 49; independent research, 60–2; as kineticist, 64; and Nernst's Nobel Prize candidacy, 217–28, 232–3, 234–7; Nobel Prize, 208–9, 217; and problem of affinity, 17; relationship with Nernst, 35, 51, 54, 217–28; theory of ionic dissociation, 88, 172

Art Deco, 100
Art Nouveau, 100, 107
Aschkinass, Emil, 219n38
astrophysics, vii, 1, 22, 24, 176, 204, 241, 242
atomic heats, 5, 157, 168, 228, 235, 237; and electrical resistivity, 160; electrons and, 173; low, 210
atomic hypothesis, 29, 77, 78
atomic model, 182, 190, 250
atomic theory, 68; and thermodynamics, 88
atomic weight determinations, 114, 128, 168
atomic weights, 113, 168, 196
atomism, 18, 61, 242; and thermodynamics, 166–7
atoms, 249; quantization of, 180, 182
"Auer mantle," 101
Auerlicht-Gesellschaft, 101, 102
Auskristallisation, 44–5
Austin, Louis, 120
automobiles, 110, 111, 252
Avogadro, Amedeo, 68–9, 71, 113

BAAS (British Association for the Advancement of Science), 181
Badische Anilin-und Soda Fabrik (BASF), 42, 129, 131
Bahr, Eva von, 180
ballast, 94, 95, 104
battery(ies), 91, 110
Bauhaus, 99
Bechstein (piano manufacturer), 243
Becker, C. H., 243
Beckmann, Ernst, 61, 213–14
behavior of gases at high temperatures, 150
behavior of matter, 3; at absolute zero, 2, 212; at extremes of temperature, 151–2; at high temperatures, 111, 129
behavior of solids and gases at high temperatures, 91

Ettingshausen, Albert von, 31, 34, 36–7, 39–9, 40, 52, 228–9
Ettingshausen effect, 37
"Étude gravito-matérialitique" (Solvay), 182
Eucken, Arnold, 157, 167, 180, 201–2, 204, 234
Eulenburg, Philipp, 107
Euler-Chelpin, Hans K. A. S. von, 234–8
"exile from Eden" syndrome, 18
Exner, F., 31
experimentalists, 190
explosion method, 149, 150
expositions, universal, 13
Expresslampe, 101
extremes of temperatures, 162–3, 249, 250; behavior of matter at, 151–2; precise measurement of, 152–3

Falkenhagen, H., 88
Faraday, Michael, 47, 73
Faraday's laws, 29
Favre, A. P., 135, 136, 145
First International Solvay Congress in Physics, x, 2, 161, 164–5, 170, 174, 176, 180, 181–207, 215
First Law of Thermodynamics, 54
Fischer, Emil, 34, 123, 147–8, 187, 224, 225, 226, 235, 245
France, Anatole, 238–9
Frank, Philipp, 24
Franz Schmidt & Haensch, 127
free energy (A), 87, 120, 121, 122, 134, 142–3
free-expansion method for gases, 114
French Physical Society, 182

galvanic cells (batteries), 6, 46, 53–4, 88, 208; Nernst's paper on, 51; theory of, 46, 50
gas equilibria, 111
gas lamp, 101
gas laws, 53, 55, 65, 79, 80; application in gas thermometer, 152; application to solution theory, 11, 81
gas lighting, 93
gas reactions, 3–4, 5, 251
gas thermometer, 152
gas warfare, 23, 238
gases, 62, 173; conductors at high temperatures, 120; dissociation at high temperatures, x, 134; kinetic theory of, 64; temperatures of, at dissociation temperatures, 150–1; vapor densities of, 125
General Electric, 92, 102

German Association of Scientists and Physicians, 191–2; 83rd Meeting of, 177–8
German Chemical Society, 178, 225
German Physical Society, 81, 190
German scientists: excluded from international meetings, 238
Germany, 147, 239, 242–3; demographics, 107; economic crisis, 26–7; educational system, 27–30, 41
Gibbs, J. W., 22, 87, 133, 135, 136
Gibbs-Helmholtz equation, 132
Giese, Wilhelm, 251
Giessen, 59
Goldschmidt, Hans, 118–19
Goldschmidt, Robert, 205–6
Göttingen, 5, 245
Göttingen Academy of Science, 148
Göttingen institute for electrochemistry, 14, 16
Göttingen Science Society, 66
Göttingen University, 15, 58–76, 161; Nernst's physico-chemical institute at, 63; Physics Institute, 58, 59
Graz, 5, 19, 31
Great Depression, 30
Gropius, Walter, 99
Gründerjahre, 30
Grundriss der Allgemeinen Chemie (Ostwald), 51
Grundriss der allgemeinen Thermochemie (Planck), 84, 85
Grüneisen, E., 177, 203
Guldberg-Waage theory, 82
Guthrie Lecture, 3
Guye, Charles-Eugène, 182
Guye, Phillipe-Auguste, 182
Gymnasia, 26, 27, 29–30, 42

Haber, Fritz, 25, 26, 30, 105, 129, 130, 131, 135, 216, 217, 237, 238; and Nernst's Nobel Prize, 231–2; Nobel Prize, 208n1, 229, 232
"Habilitation," 56–7, 73, 152
Habilitation (Nernst), 64
Haeckel, Ernst, 27
Hahn, Otto, 25, 147
Hall, Edwin H., 37, 38
Hall effect, 37–9, 40, 52, 54, 68, 172
Halske, Johann Georg, 98
Hammersten, O., 61, 213
Handwörterbuch der anorganischen Chemie (Dammer), 59
Handwörterbuch der Chemie (Ladenburg), 85
Haniel, Edgar von, 242–3